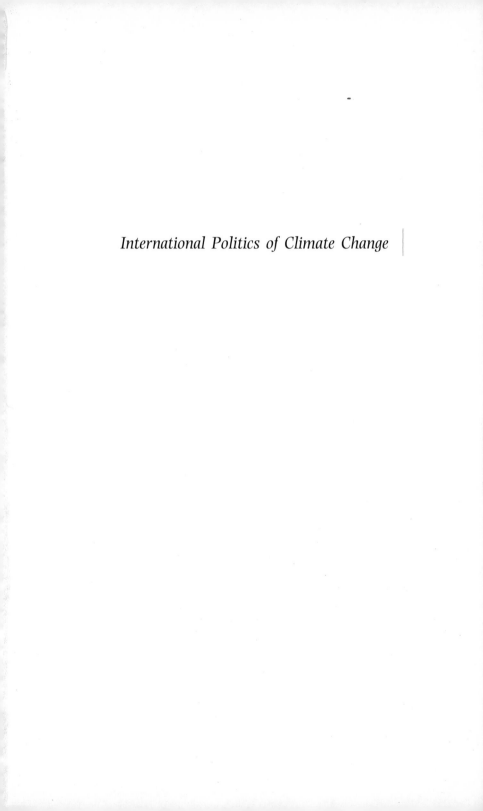

International Politics of Climate Change

International Politics of Climate Change

Key Issues and Critical Actors

Edited by Gunnar Fermann

SCANDINAVIAN UNIVERSITY PRESS

Oslo - Stockholm - Copenhagen - Oxford - Boston

Scandinavian University Press (Universitetsforlaget AS)
P.O. Box 2959 Tøyen, N-0608 Oslo, Norway
Fax +47 22 57 53 53

Stockholm office
SCUP, Scandinavian University Press
P.O. Box 3255, S-103 65 Stockholm, Sweden

Copenhagen office
Scandinavian University Press AS
P.O. Box 54, DK-1002 København K, Denmark

Oxford office
Scandinavian University Press UK
60 St. Aldates, Oxford OX1 1ST, England

Boston office
Scandinavian University Press North America
875 Massachusetts Ave., Ste. 84, Cambridge MA 02139, USA
Fax +1 617 354 6875

Scandinavian University Press (Universitetsforlaget AS), Oslo 1997

ISBN 82-00-22711-1

Published in cooperation with Centre for Environment and Development,
and the Department of Sociology and Political Science, Norwegian University
of Science and Technology (NTNU).

Design: Astrid Elisabeth Jørgensen
Cover illustration: ©NPS/Naoki Okamoto
Typeset in 9 on 12 point Photina by Formatvisual Ltd
Printed on Carat Offset 90gms by Østfold Trykkeri, Norway

Contents

Part II: Critical Actors

Preface

The overriding objective of this book has been to scrutinize a range of political problems inherent in the issue of climate change, and to provide updated analyses of the climate change policies of major industrialized and developing countries. Through thirteen chapters, the contributors to *International Politics of Climate Change* shed light on the development of the international climate change regime; the state of scientific knowledge on climate change and the assessment process; the complex relationship between science and politics; the intricate task of making political decisions under scientific uncertainty; the question of responsibility and burden-sharing, and that of cost-effective abatement efforts. They describe and explain the climate change policies and prospects of Brazil, China, Africa, Germany, Japan, Russia, the European Union, and the United States.

The above issues are essential in the political process developing the 1992 Framework Convention on Climate Change to a more mature international problem-solving regime. Moreover, most of the countries included in the study belong to the rather exclusive, but heterogeneous, group of critical actors capable of contributing substantially to either the solution or the enlargement of the problem of climate change. The actors analysed are collectively responsible for about two-thirds of the total anthropogenic CO_2 emissions and account for an even larger share of the World Gross Product.

The idea for the book was conceived more than two years ago while I was still conducting climate change policy research at the Fridtjof Nansen Institute. Since then, the original concept has developed considerably, not least as a result of numerous probing discussions with individual contributors to this anthology. Needless to say, I have incurred several debts in bringing the project to fruition. I am grateful to the chapter contributors, each of whom in their expert capacity and on a strictly non-profit basis found time for the project in their tight research schedules. I also very much appreciate the patience shown by all the contributors at various stages in the protracted editing and publishing process. Special thanks are extended to my friend Georg Børsting, who volunteered to co-write one of the chapters at a stage when time was running out. He also assisted me on several occasions during the editing process.

Without the financing granted by the Norwegian Ministry of Foreign Affairs and the Ministry of the Environment, as well as the research-friendly facilities and inspiring milieu provided by the Norwegian University of Science and Technology (NTNU), this book project could not have been completed. In particular, my association with the Centre for Environment and Development and the Department of Sociology and Political Science (NTNU) has been critical in the initiation and finalization of the project. I am grateful to Ola Listhaug, Helge Brattebø, and Sigmund Asmervik, who showed their enthusiasm for the project at an early stage and provided the facilities necessary for making the book into what it has become. Thanks are also due to Hans-Einar Lundli, who in the midst of his thesis engaged in numerous discussions relating to the political management of climate change. The proof-reading services provided by Nancy Lea Eik-Nes and Susan Høivik are also appreciated.

In the editing process I have benefited from information from various sources. In this regard, I particularly appreciate the assistance given by the European Science and Environment Forum, Greenpeace International and Greenpeace Norway, the IPCC Secretariat, and the World Energy Council.

Finally, a special thanks to Kristin Wergeland Brekke – loving companion – for always being there with care and advice.

Gunnar Fermann
Department of Sociology and Political Science
Norwegian University of Science and Technology
Trondheim, November 1996

About the Authors

CHRISTIANE BEUERMANN is a researcher at the Climate Policy Division of the Wuppertal Institute for Climate, Environment and Energy, Germany. Her main areas of research are international climate policy, and institutional adjustment and learning processes in response to the issues of climate change and sustainable development.

SONJA BOEHMER-CHRISTIANSEN studied climate change policy formation and the IPCC when a Senior Research Fellow at the Science Policy Research Unit (SPRU), University of Sussex. The project was funded by the British Research Council under its GEC Programme. She was fortunate to be able to analyse climate changes from within several policy communities: atmospheric physics, science policy, and environment, bringing to bear her academic training in geography and international relations. She is a German-born Australian who has studied and worked in the UK since the late 1960s and now teaches at the University of Hull, where she is a member of the Research Institute for Environmental Science and Management. She has published on marine pollution control and environmental policy, including *Acid Politics* (with Skea, 1991/1993) and *The Politics of Vehicle Emission Reduction in Britain and Germany* (with Weidner, 1995).

BERT BOLIN received his PhD at the University of Stockholm in 1956 on a thesis that deals with the development of models of the atmosphere for numerical weather forecasting. His research in the 1960s and 1970s concerned analyses of global bio-geochemical cycles, particularly the global carbon cycle. During the last two decades his research interest has focused on climate issues. Bolin served as chairman of the committee that launched the Global Atmospheric Research Programme (GARP) in 1967. He organized a workshop on global climate issues (1974) that led to the transformation of GARP into the World Climate Research

Programme (WGRP) in 1980. During the 1980s he was instrumental in the creation of the International Geosphere Biosphere Programme (IGBP). He has served as chairman for the IPCC since its establishment in 1988. Bolin has received a number of international awards, among those the IMO Prize (1981), the Thyler Prize (1988) and the Blue Planet Prize (1995).

GEORG BØRSTING works at the Norwegian Ministry of Environment, Section for Climate and Energy, Department for International Cooperation, Air Management and Polar Affairs. When contributing to this book, Børsting was a scholarship holder at the Fridtjof Nansen Institute, where he was engaged as a research assistant from 1993 to 1996. Previously, he conducted research on international fisheries management in the Barents Sea area. Børsting did his main subject in political science at the University of Oslo, specializing in the field of international environmental and climate politics. His cand. polit. thesis is on the North–South politics of the international ozone regime.

EWAH OTU ELERI is Co-ordinator of the Southern African Energy and Environment Programme at the University of Cape Town and Associate Research Fellow at the Fridtjof Nansen Institute in Oslo. He has conducted research on the international political economy of energy and environment in Africa, and has published in such journals as *International Environmental Affairs*, the *Journal of Energy and Development*, and *Energy Policy*.

GUNNAR FERMANN is Assistant Professor at the Department of Sociology and Political Science, Norwegian University of Science and Technology (NTNU), and associated with the Centre for Environment and Development, NTNU, and the International Peace Research Institute, Oslo (PRIO). He received his Magister Artium degree in Political Science at the University of Oslo, and is currently working on a study investigating the political and operational conditions for effective international conflict management in the post-Cold-War era. Fermann has been associated with the Norwegian Institute for International Affairs (1985–1987, 1989–1990) and the Norwegian Institute for Defence Studies (1992). From 1992 to 1994 he was a research fellow at the Fridtjof Nansen Institute. Fermann has published on international conflict management, the United Nations, foreign policy, and the political management of climate change.

SVEN OVE HANSSON is an Associate Professor of Theoretical Philosophy at

Uppsala University. His major research areas are decision theory and belief dynamics, but his publications in international journals also include articles on preference theory, philosophy of science, legal and political philosophy, and the history of philosophy. He has served on the board of the Swedish Natural Science Foundation and several government committees concerned with risk management and environmental issues.

MIKAEL JOHANNESSON is a PhD student at the Natural Resources Management Institute, Department of Systems Ecology, Stockholm University. He has a master of science degree in environmental chemistry and geology. Previously he has been responsible for a study of chemical flow in the technosphere for the Swedish National Chemical Inspectorate. Johannesson has also been the chairman of two environmental organizations in Sweden, the Environmental Federation (now Friends of the Earth, Sweden) and the Swedish NGO Secretariat on Acid Rain. The latter is an umbrella organization dealing with air pollution problems, including climate change.

SJUR KASA is a Research Fellow at the Centre for International Climate and Environmental Research, University of Oslo (CICERO). He received his Magister Artium in sociology from the University of Oslo. Kasa is presently in the process of finishing his PhD dissertation, which is a comparison of the inclusion of environmental concerns into the forest policies of Brazil and Indonesia during the 1985–1993 period. Special fields of research interest are the sociology of development as well as global environmental problems.

FRIEDEMANN MÜLLER, Senior Research Associate at Stiftung Wissenschaft und Politik (SWP), Ebenhausen, a think tank on international affairs, received his PhD in economics in 1973 from Freiburg University. Dr Müller has been a visiting research fellow at Moscow State University (1970/1971), the RAND Corporation, Santa Monica (1980/1981), and the Overseas Development Council, Washington DC (1988/1989), focusing mainly on Soviet/Russian and East European economics. He teaches at Munich's Hochschule für Politik.

ASBJØRN TORVANGER is Research Director of the Centre for International Climate and Environmental Research in Oslo (CICERO). He was previously Research Fellow at Statistics Norway and the Centre for Research in Economics and Business Administration, Oslo. His publications include contributions to the *UNESCO Yearbook on Peace and Conflict*

*Studies 1987, Joint Implementation of Climate Change Commitments –
Opportunities and Apprehensions* (edited by Gosh and Puri, 1994), and *The
New Global Oil Markets* (edited by Shojai, 1994).

JAY WAGNER is a Senior Environmental Consultant at Petroconsultants
(UK) Ltd, where he heads the environmental unit. At Petroconsultants,
Wagner has established and managed the *Environmental Law and Policy
Service (ELS)*, a detailed series of reports on environmental law and pol-
icy pertaining to international oil and gas activities. Prior to joining
Petroconsultants, Jay Wagner worked as a consultant for Environmental
Resources Management (ERM), where he specialized in European envi-
ronmental policy issues. He holds a BA (hon.) in Political Science from
the University of Iowa (1985) and an MSc in International Law and
Marine Policy from the London School of Economics (1989). From 1986
to 1988 he was a research associate at the Norwegian Institute of
International Affairs (NUPI). During his stay at NUPI, Wagner special-
ized on German foreign and security policy. His publications include
"Soviet Naval Strategy in the North: The West German Response", in
J. Skogan and A. Brundtland (eds), *Soviet Strategy in Northern Waters*
(Pinter Publishers 1990); "German Unification: The External
Ramifications", *Internasjonal Politikk* (1991); and various book reviews
on environmental issues for *International Affairs* (London) and articles on
environmental law and the international oil and gas industry.

Abbreviations

ADESG	Association of Diplomats of the Brazilian War College
AGBM	Ad Hoc Group on the Berlin Mandate
AGGG	Advisory Group on Greenhouse Gases
AOSIS	Alliance of Small Island States
AP	Action Programme
BAT	Best Available Technology
BMU	Bundesminister fur Umwelt, Naturschutz und Reaktorsicherheit
CBD	Convention on Biological Diversity
CCAP	Climate Change Action Plan
CEBRES	Brazilian Center for Strategic Studies
CEFIC	European Chemical Industry Federation
CFC	Chlorofluorocarbons
CNCC	Chinese National Climate Committee
CO	Carbon monoxide
CoP	Conference of the Parties
CH_4	Methane
CO_2	Carbon dioxide
DICE	Dynamic Integrated Climate-Economy Model
EA	Environment Agency
ECA	Economic Commission for Africa
EC	European Community
ECE	Economic Commission for Europe
ECEME	School of the Staff of the Army
ECJ	European Court of Justice
EPCSC	Environmental Protection Committee of the State Council
ESG	Brazilian War College
EU	European Union
EUROPIA	European Petroleum Industry Association

FCCC	Framework Convention on Climate Change
FRG	Federal Republic of Germany
FSU	Former Soviet Union
FUNAI	National Indian Foundation
FUNATRA	Pro-Nature Foundation
GAIM	Global Analysis, Interpretation and Modelling
GARP	Global Atmospheric Research Programme
GCC	Gulf Cooperation Council
GCM	Global circulation model
GCOS	Global Climate Observing System
GDP	Gross Domestic Product
GDR	German Democratic Republic
GEF	Global Environmental Facility
GEMS	Global Environment Monitoring System
GNP	Gross National Product
GRID	Global Resources Information Database
GWP	Global warming potential
G-7	Group of 7
G-77	Group of 77
HD	Human Dimensions of Global Change
HFC	Hydrofluorocarbons
IBAMA	Brazilian Institute for the Environment
IBGE	Brazilian Institute for Geography and Statistics
IBDF	Brazilian Institute for Forest Development
ICSU	International Council of Scientific Unions
IEA	International Energy Agency
IGBP	International Geosphere Biosphere Programme
IGCE	Institute of Global Climate and Ecology
IGO	Intergovernmental Organization
IIASA	International Institute for Applied Systems Analysis
INGO	International Non-governmental Organization
INC	Intergovernmental Negotiating Committee
INCRA	National Institute for Colonization and Agrarian Reform
INPE	Brazilian Space Agency
IPCC	Intergovernmental Panel on Climate Change
ISSC	International Social Science Research Council
IUCC	Information Unit on Climate Change
KEPS	Committee for Productive Forces and Natural Resources
LDC	Least Developed Countries
MITI	Ministry of International Trade and Industry
Mtoe	Million tonnes of energy equivalent
NASA	National Aeronautics and Space Administration

NATO	North Atlantic Treaty Organization
NCCCG	National Climate Change Coordination Group
NEPA	Chinese National Environment Protection Agency
NGO	Non-governmental organization
NNSFC	National Natural Science Foundation of China
N_2O	Nitrous oxide
OAU	Organization of African Unity
ODA	Official Development Assistance
OECD	Organization for Economic Cooperation and Development
OPEC	Organization of Petroleum Exporting Countries
PC	Problem-solving capacity
PEAL	Emergency Programme for Legal Amazonia
PP	Problem-solving potential
PPP	Polluter-pays principle
PS	Problem-severity
RFSHE	Russian Federal Service for Hydrometeorology and Environmental Monitoring
SADEN	Secretariat for National Defence
SAE	Secretariat for Strategic Affairs
SAVE	Specific Actions for Vigorous Energy Efficiency
SBI	Subsidiary Body for Implementation
SBSTA	Subsidiary Body for Scientific and Technological Advice
SCOPE	Scientific Committee on Problems of the Environment
SEC	State Education Commission
SEI	Stockholm Environment Institute
SEMA	The Secretariat of the Environment
SEPC	State Environment Protection Commission
SMA	State Meteorological Administration
SOA	State Oceanographic Administration
SPC	State Planning Commission
SUDAM	Superintendency for the Development of Amazonia
TPES	Total primary energy supply
UN	United Nations
UNCED	United Nations Conference on Environment and Development
UNEP	United Nations Environmental Programme
UNCHE	United Nations Conference on the Human Environment
UNGA	United Nations General Assembly
UNI	União das Nacões Indigenas
UNICE	Union of Industrial and Employers Confederation
USCSP	United States Country Studies Programme
WCED	World Commission on Environment and Development

WCIRP	World Climate Impact Assessment and Response Strategies Programme
WCP	World Climate Programme
WCRP	World Climate Research Programme
WEC	World Energy Council
WG I–III	Working Groups I–III (of the IPCC)
WMO	World Meteorological Organization
WRI	World Resources Institute

Chapter 1

Political Context of Climate Change

Gunnar Fermann[1]

1. Introduction

Since the industrial revolution the atmospheric concentrations of carbon dioxide (CO_2), methane (CH_4) and nitrous oxide (N_2O) have increased by some 30, 145 and 15%, respectively. According to the second scientific assessment of the *Intergovernmental Panel on Climate Change* (IPCC) the anthropogenic emissions of greenhouse gases have triggered a process of global warming which is likely to be beyond the range of natural variability (IPCC 1995). In spite of considerable scientific uncertainty – particularly regarding the impacts of human-induced global warming on climate, on ecosystems, and eventually on the lives of individuals and societies at large – the unprecedented scope and potential destructiveness of climate change started to attract the attention of political leaders and decision-makers in the late 1980s. From that time, the issue of climate change has ascended the international political agenda with considerable speed, culminating in the adoption of the United Nations *Framework Convention on Climate Change* (FCCC)[2] in June 1992 at the United Nations Conference on Environment and Development (UNCED) in Rio de Janeiro, Brazil.[3]

Perhaps more than any other environmental problem – including the

---

1 The author gratefully acknowledges the valuable comments and interventions of contributors to this book: Sonja Boehmer-Christiansen, Bert Bolin, Georg Børsting and Jay Wagner. Remarks made by Torunn Laugen, Alf Håkon Hoel and Jon Hovi at the 1996 Norwegian Conference on Political Science are also appreciated. Remaining weaknesses, factual or interpretative, are, of course, solely the responsibility of the author.

2 The full text of the FCCC is presented in Appendix 1 of this volume. For assessments of the Convention, see Bodansky (1993), Grubb (1992), Haas et al. (1992), Thacher (1992), and Underdal (1992a).

3 The FCCC entered into force in March 1994 following the 50th ratification. By October 1996 the FCCC had been ratified by 153 states and the EU.

depletion of the ozone layer and the decline of biological diversity – climate change underlines the extent to which environmental problems have become increasingly more global in scope. Ultimately caused by technological development and diffusion, as well as by accelerating population growth, the process of global environmental change is part of a broader and more general trend towards increasing interdependence among states: The fact that states in the twentieth century have become increasingly sensitive and vulnerable to the consequences of each other's actions in various fields of international affairs has, in turn, provided a strong impetus for international cooperation, as indicated by the tremendous growth in the number of intergovernmental organizations (IGOs) since the Second World War.

Cooperative problem-solving arrangements – whether in form of international regimes, institutions or organizations – vary profoundly in their capacity to solve or manage the particular problems justifying their establishment. For instance, in the field of military security, NATO has been generally successful in deterring attacks on any member of the alliance, while the collective security system envisaged in Chapter VII of the UN Charter has had only minor bearing on the management of international conflict (Fermann 1994a: 297–298). Not surprisingly, the record on environmental problem-solving regimes is mixed as well. As a result of its contribution to the phasing-out of ozone-depleting chlorofluorocarbons (CFC), the 1987 Montreal Protocol (with amendments) is regarded as a success story and a model for environmental cooperation, although its continuing effectiveness should not be taken for granted (Händl 1990: 250–257; Parson 1993: 71–73).

In the case of climate change, however, much remains for the present problem-solving regime (as embodied in the FCCC) to constitute an adequate solution to the threat of global warming. The climate change regime has so far been incapable of curtailing the growth of greenhouse gas emissions. Notable examples of recent CO_2 emission reductions in the former Soviet Union, in the Eastern European countries and in the unified Germany have mainly been due to processes of political disintegration, industrial reconstruction, and economic recession and transformation; they can be expected neither to last nor to offset the general global trend towards increased CO_2 emissions. According to a recent forecast by the International Energy Agency (IEA) (1995: 4–5, 47–53), energy-related world CO_2 emissions will have increased by 30–40% by the year 2010. This would imply that the trend towards global warming is unlikely to be reversed in the next century.

The inadequacy of the FCCC to induce the abatement of greenhouse gas emissions not only indicates a considerable problem-solving deficit in managing climate change, it also illustrates a general feature of international politics, i.e. that international cooperative arrangements are not likely to be more effective than governments allow them to be. As will be discussed later in this Chapter, the signatories of the FCCC are faced with the challenge of circumventing and coming to grips with a host of obstacles if the Convention is to develop into a more mature problem-solving regime capable of coping with and matching the malignant nature of climate change. The first Conference of the Parties in Berlin, April 1995, provided the signatories to the FCCC with the first major opportunity to further develop the climate change regime and give substance to the claim that the Convention should not be considered the end-result of a political process, but rather be conceived of as a framework and a starting point for further elaboration. However, the facts that (i) the 1995 Berlin Conference demonstrated considerable rifts both between and within the groups of industrialized and developing countries; (ii) few, if any, countries have as yet taken decisive action to curtail greenhouse gas emissions; and (iii) climate policy in most countries could still be regarded more of a derivative of other policy goals (most notably, energy security and economic growth) than as a corrective to such policies, imply that the development of the current international climate change problem-solving regime should indeed not be taken for granted.

1.1. Aim and Scope of the Book

The future success of the international climate change negotiations is by no means secure. In particular, the outcome of the negotiations will depend on the capacity and will of certain critical countries and political entities to cooperate on solving various political problems inherent in the issue of climate change. While insight into central political aspects of climate change is interesting for academic reasons, it is also a prerequisite for the constructive development of the present regime. Indeed, the overriding objective of this anthology is to scrutinize a range of political problems inherent in the issue of climate change, and to analyse the climate change policies of some major industrialized and developing countries. To our knowledge, the approach of analysing key political issues of climate change as well as analysing some actors critical to the further development of the international climate change regime, has not previously been undertaken in a single volume.[4]

---

4 For major contributions to the literature on international environmental and climate change politics, see Caldwell (1990), Carroll (1988), Conca et al. (1995), Churchill and Freestone (1991), Grubb (1989), Grubb et al. (1991), Haas et al. (1993),

Through thirteen chapters, the contributors to this book shed light on the current status of scientific knowledge and the process of scientific assessment within the IPCC, the complex and disputed relationship between science and politics, the difficult task of making sound decisions under a state of scientific uncertainty, the intricate and politically complex question of responsibility and burden-sharing, and the equally important challenge of initiating international cost-effective abatement measures. In addition, the climate change policies and prospects of Africa, Brazil, China, Germany, Japan, the European Union, the Russian Federation, and the United States are described and explained.

These issues and actors are essential to the political process of developing the 1992 FCCC into a more effective international problem-solving regime. Most of the countries included in the present study belong to the rather exclusive, but heterogeneous, group of *critical actors* capable of contributing substantially toward either the alleviation or the aggravation of the problem of climate change. The actors analysed are collectively responsible for about two-thirds of the world's total anthropogenic emissions of CO_2 – the major greenhouse gas involved in climate change – and account for more than four fifths of the World's Gross Product (GWP) (Table 1).[5]

Table 1: CO_2 Emissions and GDP of Countries and Regions Analysed in the Book (% of World Total Energy-related CO_2 Emissions and of GWP)[6]

	Share of CO_2 emissions	Share of GWP
United States	21.9	26.3
European Union	12.7	31.9
Japan	4.9	15.4
Russian Federation	9.4	1.8
China	11.9	4.2
Brazil	1.0	1.9
Africa	3.2	1.8
Total share	65.0	83.3

Hayes and Smith (1993), Hurrel and Kingsbury (1992), McCormick (1989), McNeill (1990), Mintzer (1992), Mintzer and Leonard (1994), O'Riordan and Jäger (1996), Orr and Soroos (1979), Porter and Brown (1991), Rowlands (1995), Rowlands and Green (1992), Sjöstedt (1993), Susskind (1992), Thomas (1992), Underdal (1996), and Vellinga and Grubb (1993).
5 GWP is defined as the sum of all states' GDP.
6 Calculated from WRI (1996: Table 14.1.) and UNDP (1995: Tables 20 and 38).

In the last section of this Chapter, the various contributions to the book are outlined in more detail. Before this presentation, the nature of the climate change problem and its rise to the political agenda are briefly accounted for; the gap between the scope of climate change and the problem-solving capacity developed internationally to cope with the threat is identified; and some important barriers to rectifying this problem-solving deficit are outlined. First, however, climate change and efforts to cope with the problem are placed within the wider context of international politics: that is, the increasing interdependence among states that has motivated an incredible growth in international problem-solving institutions during the past 20 years or so, especially in the field of global environmental change.

2. The Nature of International Politics: Coping with Interdependence

One of the most striking features of international politics in the latter part of the twentieth century is the unprecedented growth in interdependence among states. Interdependence is characterized by a plurality of actors, issues and means of influence, and typically expresses itself in an upsurge in the rate and diversity of international interactions. Ultimately driven by technological and demographic developments, interdependence has created a widespread feeling that the world has "grown" both smaller and more complex, resembling a global village. Indeed, in this setting, states are capable of influencing one another to an unprecedented degree and the option of withdrawing from the turmoils of international affairs is a luxury available to an ever-shrinking number of governments. State sovereignty, as traditionally perceived, is also limited by the activities of increasingly powerful transnational corporations involved in production, trade and finance (Keohane and Nye 1977; Strange 1986, 1988).

While states in a situation of high interdependence are more sensitive to the choices and activities of other actors, their manoeuvrability and vulnerability are by no means evenly distributed. Asymmetries of power and vulnerability, as well as diverging interests, prevail among governments. Although the particular distribution of power and vulnerability may vary over time and depend on the issue area in question, some states remain – to paraphrase George Orwell's, *Animal Farm* – more (inter)dependent than others. Hence, international interactions – whether executed within the increasingly overlapping fields of security, trade, finance, resource management or protection of the environment – are still prone to produce winners and losers, arguably more often in relative than in absolute terms (Powell 1991).

Alongside the upsurge in interdependence, there has been a tremendous growth in the number of international regimes and organizations. Political scientists refer to the phenomenon of an international organization as a "formal arrangement transcending national boundaries that provides for the establishment of institutional machinery to facilitate cooperation among members in [various] fields" (Plano and Olton 1988: 308). The criteria that may be used to define international organizations are that their members meet at relatively regular intervals, have specified procedures for making decisions, and have a permanent secretariat or headquarters staff. Using these criteria, one authoritative source estimated that the number of government-controlled international organizations – so-called intergovernmental organizations (IGOs) – rose from less than 100 in 1945 to about 1500 by the early 1990s. Even though four out of five international organizations are now non-governmental (INGOs), the remainder are more important because their members are nation-states.[7]

By describing the evolving multilateral structure in terms of "networks of interdependence", Harold Jacobsen (1984) suggests that the increasing number of international organizations is a defining characteristic of interdependence in itself. The advantage of treating interdependence and international organizations as *separate* phenomena is that it is possible to see the growth of international organizations as a *result of* the increasing interdependence of states (List and Rittberger 1992: 87). In fact, there is much evidence to support the argument that governments establish international organizations for the purpose of coping with the opportunities and dangers of interdependence. From the dawn of civilization, cooperative arrangements, whether in terms of international organizations or in terms of international regimes,[8] have been recognized as a potential solution to common challenges too big, too severe or too

---

7 The number of international non-governmental organizations (INGOs) has reached close to 8000 entities (Yearbook of International Organizations 1993/94: 1698). Other sources applying the criteria of structure and permanence with less flexibility find the number of international organizations to be somewhat lower (Jacobson 1984: 9; Kegley and Wittkopf 1995: 150–152). This does not refute the fact, however, that there has been a tremendous growth in the number of various kinds of international organizations during the past few decades. For a typology of different kinds of international organizations, see Kegley and Wittkopf (1995: 150–152).

8 According to Robert Keohane, international *regimes* are "institutions with explicit rules, agreed upon by governments, that pertain to particular sets of issues in international relations" (1989: 4). Unlike international organizations, international regimes does not require the existence of a organizational structure (e.g. headquarters staff), although many regimes are institutionalized in such a sense.

complex to be adequately managed by a single actor alone. For this reason, the dictum "organize or perish" has established itself as a tacit imperative in many walks of social and political life. With regard to the international state system, the emergence of international organizations and problem-solving regimes may be perceived as no less than a survival strategy in coping with negative aspects of interdependence – whether expressed as instability and chaos, conflict escalation, protectionism, resource depletion, or environmental degradation (Fermann 1995).

Whether international problem-solving regimes are established for the purpose of realizing opportunities or designed to minimize risks and vulnerability for the states involved, there is certainly more to international regimes than their instrumental function. It would be naive to limit the function of regimes to their capacity to realize and preserve collective goods. International cooperative arrangements do indeed in many cases also have long-term (re)distributive implications. Recalling that power asymmetries and diverging interests remain a fact of international politics, and that despite the pattern of interdependence the international political system is still predominantly anarchical, inequality and conflict among governments clearly make a significant imprint on both the quality of international cooperation and the extent to which cooperation is at all possible. To the extent a state is perceived to benefit from an international regime, its government is likely to join and support it. But a government suspecting that the regime may undermine its national goals or political standing internationally is inclined to aim for its reform, leave the regime, or even work for its abolishment. Hence, every problem-solving international regime has to justify itself in terms of its ability to serve the lowest common denominator of interests of its member-states – and, particularly so, the interests of the most powerful ones. International regimes failing in this regard are at risk of being dismantled or falling into political oblivion.

Governments are inherently ambiguous in their support of international regimes: In the effort to realize or protect collective goods, the *efficiency* of regimes is hampered by the fact that member-states may be as eager to use international problem-solving arrangements to serve their own special and immediate interests as they are to subdue such impulses for the benefit of common and long-term interests. States *need* international problem-solving regimes to cope with an increasingly complex political environment. Simultaneously, they also have reason to *fear* that other states might use cooperative arrangements against their interests, or that international regimes might even take on a degree of autonomy, challenging state-sovereignty itself.

3. Cooperative Responses to Global Environmental Change

With varying degrees of enthusiasm, governments are prone to establish more or less effective problem-solving arrangements in order to cope with challenges arising from interdependence. Over the past two or three decades, a new aspect of interdependence has received increasing attention, subsequently becoming a compulsory part of the international political agenda. Under the label of *global environmental change* – or, more appropriately, *degradation* – environmental issues such as nuclear safety, marine and air pollution, hazardous substances, and decreasing marine and terrestrial biodiversity, have been recognized and addressed at a political level, both nationally and internationally (Untawale 1990).

Since the pioneering 1972 *United Nations Conference on the Human Environment* (UNCHE),[9] in Stockholm, Sweden, this process of political recognition has been embedded in an assertive ideology of ecologism, which itself undoubtedly has been fuelled by phenomena such as acid rain and "Waldsterben" in many areas of Europe and North America; the discovery of the "ozone hole" over Antarctica in 1985; the Chernobyl nuclear accident in 1986; the Exxon Valdez oil spill in 1989; and the extraordinary droughts and floodings in the United States and Europe during the 1980s and 1990s. While the institutionalization of international environmental cooperation was still at a fairly primitive stage in the early 1970s, the 1980s and 1990s brought about a virtual take-off in the development of international environmental regimes, at both regional and global level. According to the United Nations Environmental Programme (UNEP), about 150 such treaties are presently in operation.[10] While this estimate obviously includes environmental regimes of varying degrees of institutionalization and effectiveness,[11] the

---

9 For a detailed discussion on the outcome of the 1972 Stockholm Conference, see McCormick (1989).

10 Reviewing the UNEP's 1992 estimate of 124 agreements UNEP 1992, Sand (1992) found that a large majority of these agreements were directed exclusively towards environmental protection; 25 agreements focused on the management and regulation of living resources and 21 covered the interface between environment and development. As to the scope of agreements, Sand (ibid.) found that 50 agreements were open for universal membership, the remaining 74 were classified as regional or subregional.

11 The problem-solving regimes range from the less than satisfactory practising of the 1986 Convention on Early Notification of a Nuclear Accident to the 1985 Vienna Convention for the Protection of the Ozone Layer – including the follow-up 1987 Montreal Protocol, the 1990 London amendment, and the 1992 Copenhagen amendment – which has been described as the "flag-ship" of international environmental regimes.

sheer quantity and scope of structured environmental cooperation as it has developed over the past two decades is a political acknowledgement that the world has reached an unprecedented degree of environmental interdependence and a recognition that the "world community" ought to do something to cope with it.[12]

By the late 1980s, it was increasingly clear that environmental degradation and the corresponding development of environmental problem-solving regimes had made an imprint on international politics. This was epitomized in 1992. Nurtured by what could be interpreted as a *globalization* of environmental degradation and fuelled by the post-Cold War political vocabulary of environmentalism,[13] the gathering of 178 governments, including more than 100 heads of state/government, at UNCED, in Rio de Janeiro, June 1992, demonstrated the extent to which environmental issues pervade the wider agenda of international politics.[14] Just as important as the unprecedented scale of the conference and the media attention it attracted, was the political outcome of UNCED.[15] While much can be said about the weakness of the political commitments made in Rio, the four main documents resulting from UNCED are nevertheless a major diplomatic achievement in that they provide political points of departure, as well as judicial and organizational frameworks for subsequent international negotiations. The most important of these are:

- *The Rio Declaration on Environment and Development*, containing a set of 27 general principles concerning a wide range of aspects related to development and environment.[16]

- *Agenda 21* – a wide-ranging and ambitious blueprint for sustainable development covering more than 100 areas of developmental and environmental concern, from alleviation of poverty to the strengthening of the international community's capacity to protect the atmosphere, oceans, seas and fresh waters.[17]

---

12 For a recent review of the most important international agreements, see *Green Globe Yearbook 1995*.

13 This change was depicted by *South Magazine* (1990), its editor observing that the "Cold War is over, the green war has begun".

14 For a well-informed review of the political process from Stockholm to Rio, see Grubb et al. (1992: 3–12).

15 UNCED had been prepared and negotiated since the submission of the report of the World Commission on Environment and Development – Our Common Future – in 1987. For a comprehensive account of the UNCED process, see Spector et al. (1994).

16 For an assessment of the Rio Declaration, see Grubb et al. (1992: 85–96).

17 Agenda 21 is described and evaluated in Grubb et al. (1992: 97–158).

- *The Convention on Biological Diversity* (CBD), the aim of which is the "conservation of biological diversity, the sustainable use of its components (genes, species and ecosystems) and the fair and equitable sharing of the benefits arising out of the utilization of genetic resources" (Art. 1).[18]

- *The Framework Convention on Climate Change* with the "ultimate objective of [stabilizing] greenhouse gas concentrations in the atmosphere at a level that would prevent dangerous anthropogenic interference with the climate system" (Art. 2).

Having explained the virtual explosion in the growth of international regimes in terms of the increasing interdependence between states within fields as different as military security and environmental change, we are now in a better position to address the issue of climate change and its rise to the international political agenda.

4. Climate Change: From a Scientific to a Political Issue[19]

The Earth's atmosphere consists of a range of gases resembling a "greenhouse", trapping heat from solar radiation with the beneficial effect of securing an energy balance that creates a viable climate for biological organisms, including human life. This *natural* greenhouse effect is a prerequisite for life in that it keeps the Earth's average mean temperature some 33°C higher than it would otherwise be. Since the Industrial Revolution from the mid-eighteenth century, however, this greenhouse mechanism has become a two-edged sword: Whereas natural emissions of greenhouse gases such as CO_2, CH_4 and N_2O help to ensure a climate that enables life to evolve, the *added* anthropogenic emissions from human activities might alter the equilibrium between incoming and outgoing energy in the atmosphere.

The scenario so instrumental in fuelling atmospheric research over the past two decades is that of anthropogenic greenhouse gas emissions from energy production, transport and agriculture trapping so much heat as to provoke a perhaps irreversible process of global warming likely to cause dramatic changes in climatic systems. This, in turn, will probably

---

18 The Convention on Biological Diversity entered into force on 29 December 1993 following the 30 ratification three months earlier. As of October 1994, the Convention has been signed by 167 states and the European Union, and been ratified by 92 parties. For an assessment of the Convention, see Rosendal (1995) and Grubb et al. (1992: 75–84).
19 For a more detailed review, see Chapters 2 and 4.

have serious impacts on ecosystems and consequently on the lives of individuals and on societies at large. Hence, the greenhouse mechanism is both a friend and a foe. In the words of Mintzer and Leonard, the problem of global warming "is a case of getting too much of a good thing" (1994: 6).

The greenhouse theory,[20] as briefly outlined above, was put forward more than a century ago.[21] Conventional scientific wisdom long speculated that most of the increase in anthropogenic CO_2 emissions, the most important greenhouse gas, was absorbed by the oceans. In the late 1950s, however, the validity of this assumption was questioned by atmospheric scientists in the Swedish journal *Tellus*. Most notably, the now famous article of Revelle and Suess (1957), contributed to the triggering of empirical research which by the early 1960s confirmed beyond doubt that atmospheric concentrations of CO_2 were on the rise.[22]

However, it was not until the 1970s that the development of supercomputers and satellite sensing data made it possible to construct sophisticated general circulation models (GCMs) of the atmosphere. The GCMs led US scientists to conclude that a continuing increase in the concentration of CO_2 in the atmosphere could lead to climate changes. This conclusion was supported, if not conclusively confirmed, in the 1980s by studies of the climatological record, which proved to be roughly consistent with the global warming forecasts developed in model research. Based on a detailed comparison of observed and model-deduced changes of climate during the last 50 years or so, a general consensus emerged that planet Earth was warming (Bodansky 1994).

It has long been disputed whether global warming is due mainly to the added greenhouse emissions from human activities, or rather a consequence of natural variability. Most notably, the first scientific assessment report of the IPCC concluded cautiously that "the size of [global] warming is of the same magnitude as natural climate variability" and that "the unequivocal detection of the enhanced greenhouse effect (as predicted by climate models) from observations is not likely for a decade

---

20 As pointed out by Lanchbery and Victor, the analogy of a "greenhouse" is somewhat misleading since greenhouses "work primarily by blocking convection", whereas in the open atmosphere "warm air rises and carries the heat elsewhere in the atmosphere" (1995: 29).

21 For contributions accounting for the history of greenhouse warming science, see Cain (1983), Handel and Risbey (1992), Kellogg (1987), and Revelle (1985).

22 For other key contributions to the scientific debate in *Tellus*, see Altshuller (1958); Arnold and Anderson (1957), Bray (1959), Callendar (1958), and Erikson and Wellander (1956). In the late 1950s, *Tellus* was edited by the present chairman of the IPCC, Bert Bolin.

or more" (Houghton et al. 1990). Earlier than expected, the last IPCC assessment report (1995) would seem to have brought the issue much closer to a settlement, concluding that the recent extent of warming is likely to be beyond the range of natural variability.[23]

Climate change – or "global warming" as the phenomenon initially was termed – became a political issue of serious international concern in the latter part of the 1980s. While the first World Climate Conference in Geneva, 1979, marked the beginning of sustained international high level attention among scientists, the 1988 Toronto World Conference on the Changing Atmosphere represented the breakthrough of climate change as an international political issue. The final statement from the conference included a call for political action and suggestions as to what targets might be adopted for greenhouse gas emissions – most notably a return to 1988 levels of emissions by the year 2000 (Lanchbery and Victor 1995: 31–32).

The institutionalization of the scientific and political process now gained momentum. Influenced by leading scientists and environmentalists who seized the opportunity created by the extraordinary climate conditions (drought and storms) of the 1980s, 1988 saw the establishment of the IPCC by the World Meteorological Organization (WMO) and United Nations Environmental Programme (UNEP).[24] The IPCC was organized in three Working Groups, with the mandates to (i) assess whether and how much the climate might change due to anthropogenic greenhouse emissions; (ii) estimate what the environmental and socio-economic impacts of possible climate change might be; and (iii) formulate response strategies for the mitigation of global warming.[25] The first scientific assessment report of the IPCC was finalized and presented at the 1990 Second World Climate Conference (Houghton et al. 1990).

Prior to this, in December 1988, the United Nations General Assembly (UNGA) adopted a resolution on "the protection of the climate for present and future generations of mankind" (UNGA Res. 43/53 1988), thus giving broad political recognition to the issue. This, along

---

23 For a collection of dissenting voices to the IPCC scientific consensus on climate change, see Emsley (1995). For studies indicating that recent discoveries of soil mineral dust in the atmosphere may have influenced previous assessments of climate change, see two recent articles in *Nature* (Li et al. 1996: 416–419; Tegen et al. 1996: 419–422). 24 For reviews of the international scientific events leading up to the establishment of the IPCC, see Chapters 3 and 4 in this volume, and Boehmer-Christiansen (1994a, b). 25 In 1992, the organization of the IPCC was changed after the completion of the first assessment report and the supplement. In the Second Assessment (1995), Working Group II has been concerned with impacts, adaptation and mitigation options, while Working Group III has focused on socio-economic considerations. See chapters 2–4 for more detailed information.

with the IPCC scientific assessment, prepared the ground for the establishment of an Intergovernmental Negotiating Committee (INC) in December 1990 by the UNGA (UNGA Res. 45/212 1990). The INC provided a negotiating mechanism capable of integrating and channelling the political debate spurred by the first IPCC assessment report. The operational mandate of the INC was to prepare a climate change convention which ultimately was signed by 154 states at UNCED on 5 June 1992.

5. Climate Change: Problem-Solving Regime and Environmental Threat Compared

While some observers have pointed out the evident weaknesses of the FCCC regarding its failure to commit Parties to concrete and binding abatement obligations, others have hailed the convention as a major diplomatic breakthrough in the international capacity-building process towards the mitigation of climate change (Bergesen 1995; Garbo 1995; Grubb 1992; Haas et al. 1992). Both interpretations have considerable merit. Seen from the bright side, the FCCC is a major diplomatic achievement for six reasons.

In the first place, the FCCC was negotiated at comparatively *great speed*. Following the acknowledgement of climate change as a potential global threat at the 1988 Toronto Conference, the Convention was negotiated in 15 months, from February 1991 to June 1992, within the INC (Paterson 1996: 60–61).

Secondly, due not least to the haste of the negotiations, it is the more remarkable that the FCCC received the *broad international recognition* of more than 150 states at UNCED in 1992 – a recognition which, by October 1996, has been confirmed by 153 states and the EU ratifying the Convention; among these are all the key countries included in this book.

Thirdly, the Convention's *long-term goal* as expressed in Art. 2 *is highly ambitious*, aiming at "[the stabilization] of ... greenhouse gas concentrations in the atmosphere at a level that would prevent dangerous anthropogenic interferences with the climate system".

Fourthly, the Convention defines several *principles of basic importance*, establishing that climate change is a serious problem (Preamble); that action should not wait upon the resolution of scientific uncertainties (Art. 3.3); that industrialized countries should take the lead in mitigating climate change (Art. 3.1); and that developing countries should be compensated for additional costs incurred in taking measures under the Convention (Art. 4.3).

Fifthly, all the parties to the Convention are *committed to report* periodically "inventories of anthropogenic emissions" as well as on "measures to mitigate climate change" (Art. 4.1.a–b).

Finally, the convention provides an *institutional framework* – including a Secretariat (Art. 8); the Subsidiary Body for Scientific and Technological Advice (Art. 9); the Subsidiary Body for Implementation (Art. 10); a financial mechanism (Art. 11); and, most importantly, the Conference of the Parties (CoP) (Art. 7) – with the intention of making the agreement on climate change an *ongoing and continuous negotiating process*. Since the adoption of the convention, the INC has met on several occasions. After the first CoP in Berlin, April 1995, the INC was dissolved, the Ad Hoc Group on the Berlin Mandate (AGBM) being established to continue the negotiations.

Although the FCCC includes a promising judicial and organizational framework for further deliberations, the convention nevertheless fails to commit the parties to any time-table or specific abatement target for the mitigation of climate change: The general objective of stabilizing "greenhouse gas concentrations in the atmosphere at a level that would prevent dangerous anthropogenic interference within the climate system" (Art. 2) lacks in specificity, and the recommendation that the industrialized countries should aim to return greenhouse gas emissions to 1990 levels by the year 2000 (Art. 4.2.b) is neither judicially nor politically binding. Hence, in the words of one observer, the FCCC does not constitute a "coherent, quantified, clearly defined and monitored regime which ... [is] ... required if global emissions are to be seriously curtailed" (Grubb et al. 1992: 72).

As a political instrument, the FCCC as yet lacks the teeth necessary to persuade governments to take the action required to mitigate climate change. Therefore, the Convention constitutes no more than a nascent problem-solving regime. This is particularly evident in view of the *nature* of climate change.[26] Climate change represents a new generation of international environmental problems transcending both the local-national and the regional level. The severity of the climate change problem may be described and compared along four dimensions.

Firstly, environmental threats vary in *spatial scope*. Along with the depletion of the ozone layer and the reduction of biodiversity, climate change is the environmental problem widest in scope. Climate change would seem to be a truly *global* problem – although emission responsibility and impacts vary from region to region.

Secondly, environmental problems vary in *durability* (time) and *reversibility*. Some challenges can be solved once and for all, while others

---

26 For a comprehensive review of the climate change challenge, see Chapter 3.

may grow or last for generations, at some point reaching the point of irreversibility, despite measures taken to change the trend. Because of the huge costs and long time-span required to change energy systems (which account for the majority of CO_2 emissions), and the long atmospheric life-time of many greenhouse gases (up to 200 years), climate change is a long-term environmental problem. Even if the FCCC's signatories through concerted action managed to reduce greenhouse emissions immediately by, say, 20%, it would take decades before this translated into a slowing down of global warming.

Furthermore, environmental challenges vary as to their *urgency*. Some problems pose an immediate threat whereas others need to be accounted for only in the long term. While climate change is not as acute a threat as a nuclear melt-down, it is urgent enough for the population of small island states in the Pacific, the Caribbean, and the Indian Ocean, some of whom are evacuating from the rising sea levels and the 100-year waves threatening their existence. Moreover, climate change is also quite real for the sub-Saharan countries currently facing the threat of desertification – a problem possibly linked to the increase in the global mean temperature.

Finally, environmental threats can be distinguished according to the *complexity* of the causes, mechanisms and consequences involved. Climate change is *uniquely* complex: Global warming is caused by multiple kinds of human activity involving several gases. Moreover, it affects most ecosystems of the world through a myriad of direct and indirect linkages and mechanisms (IPCC 1995).

Having briefly considered climate change in terms of the nature (severity) of the threat and the problem-solving capacity developed to cope with it, it is fair to say that climate change, as yet, is characterized by (a) *high problem severity* due to the broad scope (space), the long duration (time), the extreme complexity, and – more arguably – the urgency of the issue; and (b) *low problem-solving capacity* reflected in the failure of the Parties to the FCCC to commit themselves to specific and binding climate change abatement targets. What this implies is that a *wide gap* exists between the probable seriousness of the challenge at hand and the institutional capacity developed internationally to actually cope with it.

6. Problem-Solving Deficit Illustrated: Three Critical Gaps

This gap becomes even more evident when comparing (i) the *emissions reductions required* to actually stabilize the concentration of greenhouse gases in the atmosphere (which, due to time lags, is far more demanding than just stabilizing greenhouse gas emissions), with (ii) the specific *abatement ambitions* advocated in the present climate change regime, as well as with (iii) the *expected emission trends* towards the year 2010 – as summarized in Table 2.

Table 2: CO_2 Emissions and Abatement – Need, Ambition, Trend

Need: *Reduction* of anthropogenic CO_2 emissions required to stabilize the concentration of CO_2 in the atmosphere (IPCC)	÷ 60–70%
Ambition: Recommended CO_2 abatement target for industrialized countries according to the FCCC	Stabilization
Trend/projection: Expected world CO_2 emission *increase* from 1990 to the year 2010 (IEA)	+ 30–40%

The gaps resulting from the evident discrepancy between abatement need, abatement ambition and actual CO_2 emission trends clearly indicate the inadequacy of the climate change problem-solving regime and shall be granted further scrutiny.

The aim–need gap (comparing current abatement ambitions under the FCCC with abatements required to stabilize CO_2 concentrations in the atmosphere): Although the relatively affluent industrialized countries are not judicially committed to any specific abatement target, the FCCC does include a recommendation that the industrialized countries stabilize their greenhouse gas emissions at 1990 levels by the year 2000.[27] While the professional negotiators of the FCCC may have reason to be satisfied by the fact that it was at all possible to include such a recommendation in the agreement, atmospheric scientists pointed out prior to the negotiations that, because of the long life of major greenhouse gases (including CO_2), radical reductions in the anthropogenic emissions of CO_2 in the order of 60–70% would be necessary to stabilize the atmospheric concentration of CO_2 and arrest the trend towards global warming (Houghton et al./IPCC 1990; Isaksen 1995: 56–58).[28] In the not-too-distant future, *some* amount of emission reduction will be required if the

---

27 Confer Art. 4.2.b.

28 In the forseeable future, a 60–70% reduction of CO_2 emissions is probably neither economically nor politically feasible because of the huge abatement costs involved. In the short run, decision-makers would prefer to continue to increase greenhouse emissions and bear the costs of adapting to the impacts of climate change than pay for substantial abatements – not least due to the fact that abatement costs are much easier to quantify than the impact costs of global warming. The relationship between adaptation and mitigation of climate change is elaborated on in Chapters 3 and 7.

main objective of the FCCC, as set out in Art. 2, is to be fulfilled. However, the FCCC does not as yet compel signatories to take the necessary action, thus leaving a substantial gap between the scientifically calculated emission reductions required to stabilize CO_2 emissions in the atmosphere and the ambitious long-term objective of the convention on the one hand, and the particular, and much weaker recommendation of stabilizing the CO_2 emissions of the industrialized countries, on the other. Admittedly, at the time of the adoption of the Convention, the CO_2 emissions stabilization target was considered only a first step. The insufficiency of stabilizing emissions in the industrialized countries at the 1990 level was, moreover, recognized at the first Conference of the Parties in Berlin, April 1995, and negotiations on the question of more stringent targets initiated. As yet, however, the specific abatement target of the FCCC clearly does not meet the abatement needs required to stabilize the level of CO_2 emissions in the atmosphere and to arrest global warming.

The aim-trend gap (comparing the operational abatement ambition of the FCCC with actual and projected CO_2 emission trends): Since 1990, Eastern Europe and the former Soviet Union have experienced a decrease in CO_2 emissions due to industrial and economic decline. The same pattern is evident in Germany and Great Britain, although the causes for the decreasing CO_2 emissions vary: In Germany, the decrease is a direct result of unification of East and West Germany, which sparked off industrial transformation as well as the dismantling of coal-fuelled power plants in the former GDR; in Britain, the decrease in CO_2 emissions is explained by the fuel switch from coal to natural gas.[29] These reductions, however, have been more than offset by rising CO_2 emissions from other parts of the world, particularly in Asia and North America. Overall, world CO_2 emissions increased by 0.8% from 1990 to 1994.[30] It should be noted that this moderate increase is already in conflict with the recommendation of the FCCC that the industrialized countries should stabilize their CO_2 emissions. The discrepancy between aim and emission developments is likely to grow radically larger in the future. According to the International Energy Agency (IEA), world CO_2 emissions will grow from 21.6 billion tons in 1990 to 25.1 billion tons in the year 2000 (IEA 1994). This amounts to an 18% increase in CO_2 emissions. According to the IEA, this negative trend will continue beyond the year 2000: In its last annual report, the organization predicts that the energy-related

---

29 Despite the decrease in German and British CO_2 emissions, EU emissions are expected to increase by nearly 10% between 1990 and the year 2000, and 13% if the former GDR is kept out of the calculation. Estimated from Grubb (1995: 45, table 1).

30 In the same period, the net increase of the CO_2 emissions of OECD is estimated at 0.5%. Calculated from Jefferson (1995: 92–94).

world CO_2 emissions will increase by some 30–40% from 1990 to 2010 – depending on what assumptions are made (IEA 1995: 4–5, 47–53).[31] This forecast could be understood to imply that the IEA does not expect the FCCC to have any impact on the structure of energy supply and demand in the foreseeable future; in turn, it will have no impact on the amount of energy-related CO_2 emissions either.[32]

The need–trend gap (comparing abatement requirements with expected emissions trends): Even wider than the gap between abatement need and abatement aim, is, of course, the gap between abatement need and expected emission trends. As explained above, world emissions of CO_2 are at present increasing at a rate which is likely to bring the CO_2 emissions to a level some 30–40% higher in 2010 than the 1990 base year. The gap between the CO_2 emission reductions required if global warming is to be averted and the projected emission trends is growing ever wider.

In conclusion, the story of the FCCC is not just the tale of a world community incapable of setting the abatement targets required to arrest, or even slow down the process of global warming. It is also the story of signatories unwilling or unable to implement the modest recommendations already agreed upon. Indeed, in most countries, climate change policies are still more of a *derivative* of, rather than a *corrective* to, other policies of greater priority (e.g. concerns about energy security and economic growth). Lacking both the support of a strong abatement commitment and the mandatory instruments required to enforce or persuade member-states to comply with its intentions and recommendations, the present regime clearly suffers a huge *problem-solving deficit* in the management of climate change. Stated differently, the climate change regime cannot be deemed *effective* since the parties to the FCCC (i) remain to agree upon emission regulations that are strong enough to *solve or substantially reduce* the problem of climate change *over time*, and, moreover, (ii) have *failed to comply* with the modest recommendation of the FCCC that the industrialized countries shall *stabilize* their greenhouse gas emissions.[33] This is indeed unfortunate, especially since a consensus within the scientific community is emerging that climate change is a real and serious environmental problem (IPCC 1995).

---

31 Energy-related CO_2 emissions account for 80% of total anthropogenic CO_2 emissions, which again are responsible for about 60% of the human-induced greenhouse effect.
32 For other scenarios, see World Energy Council (1995).
33 For other specifications of regime effectiveness, see Chayes and Chayes (1993), Kay and Jacobson (1983: 14–18), Mitchell (1994) and Underdal (1992b: 231).

7. Causes of Climate Change Problem-Solving Deficit

What are the fundamental causes of the grave mismatch between the current rather weak climate change problem-solving regime and the potentially severe threat of climate change? And what are the obstacles to improving the present international regime so that this gap can be reduced? These questions should be granted serious consideration, if for no other reason than to avoid spoiling intellectual energy and political capital on wishful-thinking and unrealistic "solutions" to the problem of climate change. For instance, it would be naive to expect that improved understanding and knowledge of the seriousness of the climate change problem in line with the recent IPCC assessment would *in itself* be sufficient to get politicians and negotiators rushing into progressive collective action. The knowledge that climate change is probably both a real and a serious long-term threat is but one condition for concerted international action. Unfortunately, a host of other realities work in the opposite direction, and the overly optimistic political activist should be aware of what he/she is up against. In this section, several factors related to the structure of international politics, the motivation of states, as well as to particular features of the climate change problem itself, are briefly considered in terms of their capacity to impede attempts at reducing the problem-solving deficit currently characterizing the international climate change regime.

7.1. The Anarchical Nature of the International Political System

In the study of international politics it is a truism that the international political system lacks a supranational structure capable of enforcing recalcitrant states to adapt or comply with international regulations they perceive to conflict with their national interest. The development of international regimes accounted for in Section 2 has done little to modify the basically anarchical nature of the international political system. This state of affairs puts a clear limit on the extent to which concerted international action is possible. In theory, the potential for international cooperation within a given issue-area is limited to the smallest common denominator of what states can agree upon. This means that in many fields of international politics, including the politics of climate change, the scant potential for cooperation granted within the anarchical structure of the international political system *may not be enough* to solve the collective problem in question.

This does not necessarily imply that the potential for cooperation in the field of climate change is now exhausted. What it means, however, is that somewhere on the road to improving the present climate change problem-solving regime – whether it be in terms of stronger and more binding abatement commitments, or in terms of stricter control on compliance – those favouring such improvements are likely to stumble upon the fact that there are distinct, although not *a priori* defined, limits as to how far state-leaders are willing to go in sacrificing pressing national goals (e.g. energy security, economic growth) for long-term and much vaguer collective interests such as the mitigation of climate change. In order to convince sceptical leaders who are responsible primarily to their national constituencies, proponents of a more effective climate change regime are forced by the logic of the anarchical international political system to develop approaches for mitigating climate change that minimize the conflict between immediate and long-term interests, between economic and environmental concerns, and between the particular interests of nations and the collective interest of all nations.

The extent to which such a reconciliation is politically and practically possible, remains to be seen, however. On the one hand, the optimist may take the various "no-regret" measures[34] suggested as an indication that a reconciliation is already underway. A more cautious and pessimistic observer may, on the other hand, suspect that the world community is still locked in the dilemma of choosing between a weak climate change "problem-solving" regime that does not really solve the problem, and a regime with regulations so demanding that states, in the end, are unwilling to bear the costs of complying with them.

7.2. The Free-Rider Incentive: Get Something for Nothing

With the lack of a supranational authority capable of imposing solutions in the best long-term interest of the collective of states, international cooperation on the management of climate change is limited by the least common denominator of what states are able to agree upon. It was argued above that in the case of climate change such a minimal consensus is probably not sufficient to solve the problem if one by "solution" implies a substantial *reduction* in the global emissions of greenhouse gases.

---

34 "No-regret" refers to measures which lead to emission abatement while *simultaneously* contributing to the attainment of other policy-goals (e.g. increasing energy security).

Just as alarming as the anarchical structure of international politics is for the prospects of improving the present climate change problem-solving regime, however, is the incentive of states to unduly exploit and thereby erode treaties they have already agreed upon: Even in situations where states perceive it in their best interests to restrain themselves for the purpose of solving a common problem, they might nevertheless try to cheat in the anticipation that other states will still pay the costs of implementing regulations. Such free-rider behaviour invites other states to act likewise. If a critical mass of states decides to engage in free-riding, the international regime will become ineffective and the problem will not be solved – at least not in any meaningful sense of the word.

The free-rider incentive explains the abortion of several cooperative ventures in international politics and also why many problem-solving regimes, owing to *expectations* of free-riding, never reach take-off. Moreover, free-riding in the implementation phase explains why many problem-solving regimes do not become very effective. Considered in terms of collective rationality, such outcomes are obviously suboptimal, and the lesson to be learned is that what are individually rational actions for each state in a specific situation may lead to outcomes harmful for all in the long run. Why then, are decision-makers so short-sighted?

Aside from the likelihood that certain aspects of free-riding can be ascribed to irrationality, the temptation to reap the benefits of international agreements while not sharing the costs of implementing the regulations can be explained by the fact that short-sightedness in decision-making, even at the top national level, is *institutionalized*. As representatives of states, state-leaders are responsible primarily towards their own nation, not towards the suffering of people in other states. In democratic societies, politicians are, moreover, elected by present generations to defend their interests, not by generations yet to be born. A third factor explaining why state-leaders are willing free-riders, thus sacrificing the long-term collective interest for immediate national returns, is the competitive perspective dominant in the foreign policy decision-making clusters of many important capitals of the world; decision-makers are often strongly focused on the distribution of *relative* gains from an agreement. Such an emphasis certainly contradicts the notion that states enter into international agreements primarily to solve common problems. In the realist perspective, states value international regulatory regimes rather as a means of changing the behaviour of *other* states and take part in the benefit of other states' action, while not contributing fully themselves.[35] In the case of climate change, such mechanisms provide the particular

---

35 I am grateful to Jon Hovi for clarifying this point.

state-leader with additional incentives to be strong in speech, e.g. in making a general call for action, while at the same time attempting to delay or minimize implementation of costly abatement measures at home.

7.3. Some Critical Characteristics of the Climate Change Problem

The anarchical nature of the international political system and the self-serving and often short-sighted motivation of states outlined above are features that complicate international cooperation in general. One should therefore not expect efforts to improve upon the present climate change problem-solving regime to escape the cooperative limitations imposed by these characteristics of international politics. *Indeed, the climate change problem contains some additional and probably unique features which are likely to make it more difficult to manage than any other environmental problem.* These features, accounted for below, pose a tremendous challenge to those working for a reduction of world-wide greenhouse gas emissions.

7.3.1. Comparative Mitigation Costs

In Section 5 it was concluded that climate change is an extremely severe problem due to the broad scope, the long duration, and the extreme complexity of the issue. Climate change, ozone depletion and the reduction of biological diversity are all truly global problems. Mitigating climate change will probably involve much higher costs, however, than those involved in effectively managing the threat to biodiversity and the ozone layer. This is due to the fact that in contrast to for example ozone depletion, which is caused by emissions of CFCs and other ozone-depleting substances produced in a fairly small number of factories around the world, and which can be quite cheaply substituted, climate change is caused by the emissions of several gases originating from a wide range of human activities linked, *inter alia*, to agriculture, transportation, power-generation and industry. While the ozone layer can be saved through relatively minor adjustments with limited costs, climate change will require huge investments in new energy and transportation systems, as well as change in lifestyles in a much more profound sense. In this process, capital may be destroyed and economic growth may suffer.

Several comparative studies on the ozone and climate change regimes indicate a strong correlation between the expected costs involved in mitigating the problem in question and the extent to which effective regimes are likely to be established (Beukel 1993; Rowlands 1995; Lundli 1996). The effectiveness of the ozone regime is in part attributed to the limited costs involved in substituting CFC substances. The unwillingness of signatories to the FCCC to commit themselves to specific and binding abatement targets can to a considerable degree be explained by the member-states' expectations that the implementation of measures required to reach such targets could be extremely costly.

The point being made here is that the economic consequences involved in effectively combating climate change are unprecedented; the expectations of high abatement costs are as likely to curb member-states' future inclination to adapt ambitious abatement targets as they have curbed it in the past. As argued below, however, the greatest obstacle to an effective international climate change regime may not be the assumption of high overall abatement costs, but rather the political implications of cost impact asymmetries (vulnerability) between various regions of the world.

7.3.2. Asymmetrical Distribution of Cost and Gain in Space and Time

The effective management of climate change is likely to inflict huge costs upon the world community, thus requiring great social and political flexibility. In the previous section, the willingness of member-states of the FCCC to bear the economic burdens of committing themselves to specific and demanding abatement targets, and to actually complying with the required regulations on national emissions was questioned. In the present section, two more attributes of climate change which complicate the development of a more effective climate change problem-solving regime even more are pointed out. Again, it will prove useful to draw upon the comparison of climate change and ozone depletion.

7.3.2.1. Spatial Asymmetries

Climate change as a true global problem epitomizing the extent to which the world has developed into a state of environmental interdependence does *not* imply that *vulnerability* to the negative impacts of climate change is evenly distributed among states and peoples. On the contrary, there are indications of a North–South divide also in the field of climate change: Owing to the greater sensitivity of ecosystems in the tropics and sub-tropics to climate change, the developing South is probably more vulnerable to global warming than the industrialized North – even though the mean temperature is expected to increase more in the temperate zone than in the tropical and sub-tropical zones (IPCC 1995). This

unfortunate spatial (geographical) distribution of vulnerability is aggravated by the fact that the countries most vulnerable to climate change
(developing South) are those that bear the least responsibility for creating the problem in the first place.[36] Whether measured in terms of share
of current or accumulated CO_2 emissions, combined emissions of CO_2 and
remaining greenhouse gases, or CO_2 emissions per capita, every indicator points to the industrialized North as having the prime responsibility
for anthropogenic emissions causing climate change (Subak 1993:
55–60).[37] Even more serious than the evident negative correlation
between vulnerability and responsibility, however, is the fact that the
industrialized North, which obviously has the superior capacity[38] to combat climate change, lacks the crucial incentive of great vulnerability to
actually mobilize their resources in a struggle against global warming
(see Table 3). This is further amplified by the fact that some representatives have expressed the hope that their countries might harvest a net
gain from climate change.

Table 3: The Responsibility–Capacity–Vulnerability Triangle and the Climate
Change North–South Divide (in % of world total)

	Industrialized countries	Developing countries
Responsibility:[39]		
- Share of current energy-related CO_2 emissions	72	28
- Share of cumulative energy-related CO_2 emissions	86	14
Capacity:[40]		
- Share of GWP[41]	83.5	16.5
Vulnerability:[42]		
- Impact on ecosystems	Low-moderate	High

36 See Chapter 6 for an elaboration.
37 It is nevertheless clear that the South has the biggest growth potential in future
greenhouse gas emissions.
38 Whether measured in terms of share of World Gross Product or average GDP/
capita.
39 Based on Subak (1993: 59).
40 Calculated from UNDP (1995: Tables 20 and 38).
41 GWP (Gross World Product) is defined as the sum of all states' GDP.
42 Based on the last IPCC assessment (1995).

In my view, the particular relationship between (i) the distribution of vulnerability to the impacts of a problem, and (ii) the distribution of the capacity to do something about it, goes a long way in explaining why some environmental regimes become effective while others remain dead letters of intent (see Lundli 1996). This is most clearly illustrated by comparing the international ozone regime and the climate change regime.

As noted previously, several observers consider the ozone regime a success story. The production and consumption of CFCs and other ozone-depleting substances has been significantly reduced, and the present regime is likely to reverse the depletion of the ozone layer in the long run if fully complied with. In explaining the effectiveness of the ozone regime one cannot avoid observing that the constellation of vulnerability and capacity is indeed favourable. In contrast to climate change, ozone depletion is as threatening, if not more, to the industrialized countries in the temperate regions of the world as it is to developing countries in the tropical and sub-tropical regions of the world.[43] This implies that the industrialized countries, which are producing and consuming most of the ozone-depleting substances, have also had a strong incentive of vulnerability in their attempt to solve the problem. Applying their superior scientific resources, Western manufacturers soon developed less aggressive substitutes. Thus, in the case of ozone depletion, the industrialized countries felt threatened and therefore had the incentive to mobilize their huge resources to solve the problem – which they are actually in the process of doing (Børsting 1996).

In the case of climate change, however, there is little correspondence between vulnerability and capacity. Those countries possessing the resources to build an international problem-solving capacity do not as yet feel sufficiently threatened by the prospects of climate change to take serious action. Along with the huge costs involved in combating climate change compared to for example ozone depletion, this explains to a considerable degree why greenhouse gas emissions continue to increase. The negative correlation between vulnerability and capacity in the case of climate change certainly is a serious obstacle to current attempts at rectifying this state of affairs.

---

43   It has to be acknowledged, however, that the industrialized countries are much better equipped with resources to adapt to the negative impacts of global atmospheric change, so the vulnerability of industrialized countries is reduced. This, in turn, would tend to change the distribution of vulnerability between industrialized and developing countries to be the benefit of the former in the case of ozone-depletion too.

7.3.2.2. Temporal Asymmetry

State leaders, believing that they can externalize themselves and their countries from the negative impacts of climate change, are unlikely to play an active part in improving upon the climate change problem-solving regime. This is particularly unfortunate in the case of climate change, since those countries possessing the capacity to seriously invest in abatement measures are the same countries as yet seeming to lack the incentive of vulnerability.

This, however, is not the only *mechanism of externalization* at work. The situation is further aggravated by the fact that, owing to the long life-span of many greenhouse gases (up to 200 years), there is a considerable time-lag between cause and effect in climate change; between greenhouse gas emissions and climate change impacts; and between abatement measures implemented and the actual curbing of the global temperature increase. This implies that returns on investments made today in mitigating climate change will be harvested by generations yet to be born.

For reasons already mentioned (Section 7.2), one can only to a limited extent count on present decision-makers to look after the interests of future generations. The logic of current political systems induces politicians to act in the best interests of their own countries. Moreover, they are representatives of present-day voters and under siege from current interest groups. Since today's generation of decision-makers have good reason to believe that they can externalize themselves from the negative effects of global warming, they are much more concerned about the *present costs* of taking steps to combat climate change than they are about the probable *future* devastating effects of global warming if preventive action is delayed.[44]

7.3.3. Scientific Uncertainty – An Excuse for Not Taking Substantive Action?

The remaining uncertainties as to how *serious* the problem of climate change is tend to aggravate the problem of mobilizing support for improving the international capacity to combat climate change. The negative incentives provided by (i) the anarchical structure of the international political system, (ii) the self-serving and short-sighted motivation of states, (iii) the lack of correspondence between the distribution of

---

44 For two normative analyses of present generations' responsibility for the well-being of future generations, see Malnes (1995: 94–114) and Gower (1995: 49–58).

vulnerability and the distribution of abatement capacity, as well as (iv) the significant time-lag existing, causing a delay between abatement investments and actual returns, are amplified by remaining scientific uncertainties; these uncertainties serve as an additional incentive for decision-makers to *delay substantive abatement efforts*. While further research on climate change is clearly required, and probably should be intensified, especially on the spatial distribution of impacts, there are indications that research funding has become a substitute for adopting substantial abatement measures. Luring behind such priorities may be hopes that the problem of climate change might not be quite so real as indicated by the IPCC, and that one might gain from climate change or at least escape the worst impacts. As pointed out by Hansson and Johanneson in Chapter 5, it is hard to determine what new knowledge might result from future scientific research. Research may increase or decrease uncertainty; it may confirm or radically change our present knowledge of the distribution of impact costs. At present, however, the remaining uncertainties seem to function as an additional reason for not spending a substantial amount of money on mitigating climate change. Or, put differently: Although policy-makers would seem to be risk-aversive while supporting the precautionary principle as a basis for a future protocol on climate change, they are considerably more risk-prone on the part of future generations when it comes to practical climate policies and actual implementation of abatement measures in their own countries.

For those favouring political decision-making based on scientific advice, the Second Conference of the Parties in Geneva, June 1996 (CoP-2), would seem to represent a step in the right direction: Stating that "science calls upon us to take urgent action" and dismissing opponents to the recent IPCC assessment, the US Undersecretary of State Tim Wirth surprised most participants at the CoP-2 by calling for a legally binding protocol to reduce greenhouse gas emissions from industrialized countries. While favouring a flexible protocol to be signed at the CoP-3 scheduled for December 1997 in Kyoto, Japan, one that allows for international cooperation on abatement efforts, US Undersecretary of the Environment Eileen Carlson emphasized that the industrialized countries should commit themselves to equal percentage reductions of greenhouse gas emission. Failing to specify the level and time-table for emission reductions, the new American position nevertheless changed the dynamics of the CoP-2 and made its imprint on the political outcome of the conference. Supported by most Parties, the Geneva Declaration goes beyond the 1995 Berlin Mandate in three respects: Firstly, it strongly endorses the IPCC scientific assessment, rejecting complaints by some industrial lobbyists and OPEC countries. Secondly, it specifically confirms the findings of the IPCC that the continued rise in greenhouse gas concentrations in the atmosphere "will lead to dangerous interference with the climate system", which, according to the FCCC, should be avoided. Finally, the Geneva Declaration calls for "legally binding" objectives for emission limitations and "significant" reductions.

It would be misleading to interpret the recent change in the American policy towards international regulation of greenhouse gas emissions and the novel elements of the Geneva Declaration as indications that the "world community" is in the process of overcoming the problem-solving deficit so clearly suffered in the field of climate change. What the new American position *does* illustrate, however, is that the US government has started to take the IPCC assessment and the majority view of their own scientific community seriously, fearing that the impacts of climate change may inflict huge costs upon the American economy in the long run.

But fear of impact costs is probably not sufficient to explain the change in United States climate policy, the sincerity of which remains to be confirmed. More important is the fact that the United States has a host of "no-regret" abatement options available which can reduce American greenhouse gas emissions without inflicting additional net costs on the American economy. There has always been a gain potential for emission reductions in the United States because of the country's heavy reliance on fossil fuels and its huge potential for energy conservation. This means that the US may gain a comparative advantage over other industrialized countries if binding emission reduction targets were adopted, which in turn is an indication that the American position is based on realpolitical grounds and thus should be taken the more seriously (for an elaboration on the US climate change policy, see Chapter 13).

For anyone eager to see the present climate change problem-solving regime improved, the American policy change is the best thing that has happened since the FCCC entered into force in March 1994. The change in US climate policy will bring pressure to bear upon the EU, Japan and Canada to follow suit, and isolate obstructionist countries like Australia and New Zealand, and OPEC. It remains to be seen, however, whether the political platform agreed upon in Geneva will lead to the adoption of binding and substantial emission reduction targets at the CoP-3 in December 1997, in Kyoto, Japan, and, moreover, whether the targets decided upon actually will be implemented. The structural, motivational, and issue-specific mechanisms explained above function as the tide through which any improvement of the present climate change problem-solving regime must force its way.

8. Contributions to the Book

A crucial question in years to come will be how states can circumvent the restrictions to improving the current climate change problem-solving regime: in particular, obstacles related to the basically anarchic logic of the international political system. Furthermore, how can we compensate for the self-serving and often short-sighted motivation of national decision-makers when a wider and long-term sense of interest and rationality is required?

Needless to say, this book is not a blueprint for solving these problems. However, by shedding some light on the priorities and capabilities of critical political actors, on key issues relating to the state of climate change science, on the role of science in political decision-making, and on the necessity of legitimate and cost-effective abatement efforts, the contributors to this anthology attempt to clarify some of the *epistemological and political conditions* for developing a more efficient climate change, problem-solving regime – which indeed is required if the recent predictions of the IPCC on human-induced climate change is correct.

Climate Change Turning Political: Conference-Diplomacy and Institution-Building to Rio and Beyond

In Chapter 2, Georg Børsting and Gunnar Fermann summarize the response of the international community to climate change up to the second conference of the parties in Geneva in July 1996. The emphasis is on the period from 1988 onwards, when it became increasingly understood that the problem of global warming required a serious political response.

In four sections, the chapter traces the climate change issue from (i) being recognized as a potentially serious problem by the international scientific community in the late 1970s and early 1980s; (ii) via the 1988–1991 period when climate change increasingly was acknowledged as an international political issue, thus contributing to the establishment of the IPCC and earning the attention of several conferences; to (iii) the 15-month period of negotiating a climate change convention which was signed by 153 governments and the EU at UNCED in June 1992; and, finally, (iv) the period since the signing of the FCCC which has been characterized by difficult post-agreement negotiations where the parties have battled over the necessary completion and specification of the FCCC, including the need to strengthen the commitments in the agreement.

Scientific Assessment of Climate Change

In Chapter 3, Bert Bolin gives an account of the most important scientific findings that have led to the conclusion that a human-induced climate change is under way because of increasing emissions of greenhouse gases into the atmosphere. It is still difficult to assess how rapidly this change may come about and how serious it may prove to be. However, with the aid of models for the global carbon cycle it is possible to assess the restrictions imposed on future emissions in order to prevent atmospheric carbon dioxide concentration from surpassing prescribed thresholds, for example the doubling or tripling of pre-industrial concentrations, and thereby stabilize the climate. It is noteworthy that most greenhouse gases have long life-spans in the atmosphere, so the change of climate that has occurred will therefore disappear only slowly, even if drastic reductions of greenhouse gas emissions were to be achieved. The chapter concludes with an account of the role that the Intergovernmental Panel on Climate Change (IPCC) has played in synthesizing available scientific-technological knowledge on the issue of climate change.

Uncertainty in the Service of Science: Between Science Policy and the Politics of Power

Sonja Boehmer-Christiansen (Chapter 4) views scientific institutions as political actors and discusses their role in global climate politics. The IPCC is seen as representing the global research enterprise, which is largely concentrated in a handful of Northern countries. Boehmer-Christiansen argues that tacit alliances, not scientific advice alone, drive climate change politics – leaving the IPCC with the difficult task of negotiating policy-relevant advice. In this battle over concepts, truth and priorities, science – as synthesized by the IPCC – becomes a tool of all contending parties as long as it remains sufficiently uncertain. Moreover, it is argued that the climate change issue, as integrated in the "sustainability" debate, has become part of the political process through which post-communist forces compete for global influence by offering to transform the global energy supply systems. If so, what is the political function of the global research enterprise?

The responsibility for using science fairly and appropriately cannot lie with the large, publicly funded institutions of science and their managers acting as IPCC spokesmen, but must belong to all political forces. Hence, Boehmer-Christiansen concludes that debates about the accountability of research and the nature of the interaction between policy-making and knowledge-creation should be nurtured.

Decision-Theoretical Approaches to Climate Change

In Chapter 5, Sven Ove Hansson and Mikael Johannesson analyse three aspects of climate change that make this particular environmental challenge an extremely difficult and intricate problem for policy-makers to come to grips with: One is the fact that climate change is still surrounded by scientific uncertainty regarding the causes, mechanisms and impact of the problem. The second problem is coordination, arising from the fact that climate change can only be effectively acted upon through concerted international measures. Finally, there are problems connected with the time factor.

As a preamble to their elaboration on factors complicating climate change in terms of a decision-making problem, the authors distinguish between three practical approaches to managing the problem of global warming resulting from the anthropogenic emissions of greenhouse gases: The first option is to prevent or reduce the build-up of such gases in the atmosphere through mitigation; that is to say, by reducing greenhouse gas emissions from human activities and/or by increasing carbon sinks through measures of, for example, reforestation. The second, and probably more impracticable and risk-taking, strategy is that of compensation: Through various efforts at "climatic engineering", such as the application of large mirrors placed in space to reflect sunlight, the warming effect of the continuous emissions of greenhouse gases could, at least in theory, be offset. A final approach is for human civilization to adapt to the impacts of climate change. This can be done by building protection against rising sea level, constructing irrigation systems to cope with the changing patterns of precipitation, and by developing new crops that are more suited to a changed climate. Adaptation is an inevitable approach to the management of climate change as long as anthropogenic greenhouse gas emissions are still on the rise.

Practical decision-making will be severely complicated by the veil of uncertainty surrounding climate change science, the political problems inherent in initiating and coordinating an effective international response, as well as the inter-generational character of climate change. Concerning the question of scientific uncertainty, Hansson and Johannesson conclude that there are good reasons for policy-makers choosing decision-making principles that are less risk-taking. Concerning the problem of justice, the authors suggest that every single country be assigned a permissible emission level of greenhouse gases that is proportionate to its part of the world population and that the sum of all national quotas be compatible with the principle of sustainable development. Finally, as to the long-term nature of the climate change problem, the authors find that the principle of sustainable development, the scientific uncertainty, the risk of irreversible damages and the long time-lag from negotiations to the point in time when most of the reversible damages will be required, all imply that anthropogenic emissions of greenhouse gases should be reduced at the earliest possible time. "No-regret" measures, i.e. measures that are strongly motivated also for reasons other than the concern for climate change should be implemented immediately.

The Requirement of Political Legitimacy: Climate Change Burden-Sharing Criteria and Competing Conceptions of Responsibility

In Chapter 6, Gunnar Fermann proposes that an effective climate change problem-solving regime is unlikely to develop unless its regulations and burden-sharing schemes are perceived of as fair and equitable. Because of the potentially huge costs involved in mitigating climate change, however, the question of what a fair and equitable burden-sharing scheme looks like has already spurred violent debate. Depending on their particular abatement conditions and emission responsibilities, divergent countries are likely to favour and argue along more or less incompatible lines. This is made possible not least by the fact that "equity" and "burden-sharing" have no unequivocal meaning.

The author attempts to structure this interpretive "chaos" by first considering six divergent criteria for burden-sharing: (i) "polluter pays", (ii) "efficient energy-use", (iii) "willingness-to-pay", (iv) "ability-to-pay", (v) "distributional implications", and (vi) "reasonable emissions". He finds the first-mentioned criterion, linking the distribution of abatement costs to emission responsibility, to have considerable merit and moral appeal. The task of actually assigning responsibility among countries for greenhouse gas emissions, however, is a complex venture. From a "polluter pays" perspective, at least four factors need to be considered: the time frame applied, the greenhouse gases included, the emission sources accounted for, and the precise meaning of national responsibility. It is concluded that these four dimensions give rise to multiple conceptions of responsibility and that the specific time-frame, combination of greenhouse gases, emission sources, and the approach to national responsibility applied, will heavily influence the distribution of responsibility among various categories of countries, thus becoming a highly political question.

The Requirement of Cost-Effectiveness: Climate Change and the Notion of an Effective Abatement Policy

Asbjørn Torvanger (Chapter 7) observes that the FCCC focuses on aims and principles for climate policy cooperation between countries, and, moreover, that climate policy measures should be cost-effective to ensure global benefits at the lowest possible cost. Defining cost-effectiveness in relation to climate policy, the author argues that the global cost of reaching a global emissions abatement target is minimized if policy options are implemented according to an increasing cost-effect quotient independent of national borders.

Studies of climate change impact indicate cost estimates of 1 to 2% of GDP by the middle of the next century, but the estimates are prone to substantial uncertainty. The impacts will be unevenly distributed between countries. Studies estimate that the cost involved in stabilizing carbon dioxide emissions at the 1990 level could amount to 0.5–1% of GDP. The main options available for climate policy decision-makers are taxes, tradable quotas, joint implementation under the FCCC, and efficiency standards; these are in addition to emission constraints. From a cost-effectiveness perspective, taxes and tradable quotas are most attractive. The distribution of costs and benefits associated with climate policies between countries is likely to be the most demanding issue to be handled on the international political scene. In the long term, this is an issue of distribution between generations.

China and Climate Change

China is the first country of prevalent importance to the climate change issue to be dealt with in this book. Christiane Beuermann (Chapter 8) analyses China's climate policy, starting with an account of the Chinese responsibility for and vulnerability to climate change. From a *per capita* perspective, Chinese emissions are significantly lower than the world average. In current total terms, however, they are among the highest in the world. According to the IPCC, China is among the most vulnerable countries in the world to climate change, although the impact of climate change will vary considerably between climate zones.

As to China's willingness and ability to implement abatement measures, the country has so far denied international demands for the adoption of targets and time-tables. This position has been justified in terms of the minor responsibility of developing countries as well as on the grounds that China suffers financial and economic constraints. Despite the rejection of specific commitments for abatement measures, international negotiations have had an impact on Chinese policy-making: Legal and organizational infrastructures have been adjusted, and climate change response measures have been carried out with financial support from bi- and multilateral funds. China was also the first country to adopt a national Agenda 21. However, the present development in the Chinese energy sector and the challenge of population growth are likely to increase the tensions between economic and environmental goals.

Brazil and Climate Change

Sjur Kasa describes the development of Brazilian positions and policies related to climate change during the 1987–1995 period (Chapter 9). The author's main focus is on the political conflicts connected with the most important Brazilian source of greenhouse gas emissions, namely deforestation in Brazilian Amazonia.

Kasa argues that the election of Collor as new Brazilian President in 1989 spurred the transition to a more cooperative position on the issue of climate change. This happened because Collor perceived the improvement of Brazil's international environmental record as an essential precondition for the fulfilment of his economic liberalization programme, and because his position as the first democratically elected president since 1961 facilitated an increased political distance to the military and to the regional business groups sceptical to such reforms. However, partly as a result of resistance from these interests and the powerful Ministry of External Affairs (Itamaraty), and partly because of the weakness of the sections of the Brazilian environmental movement with an interest in the Amazon region, Collor's attempt to change Brazil's international role in the UNCED negotiations and to launch a forest policy for the Amazon region inspired by considerations for climate change were partially crippled.

Africa and Climate Change

Ewah Otu Eleri observes that Sub-Saharan Africa's past and present contribution to the build-up of greenhouse gases in the atmosphere is comparably insignificant (Chapter 10). At present, the region does not pose a major threat to the global climate. Nevertheless, climate change may pose a formidable challenge to already deteriorating conditions for human development in the region.

Despite the potential destructiveness that global warming has on African development, climate change has not as yet been granted high political priority. The author argues that this is not least due to the fact that African policy-makers are struggling with more immediate problems of development and environment.

Following the first Conference of the Parties in Berlin (April 1995), however, there are indications that change may occur. Of special importance to African countries is the initiation of the pilot phase of joint implementation and the adoption of the Global Environmental Facility as the key financial transfer mechanism to the South. Very much depending on the volume of the transfer of capital and technology industrialized countries make available, African states are likely to step up investments in the energy and land-use sectors that could have a positive effect not only on the amount of greenhouse emissions, but on economic and agricultural performance as well. As crucial to the inclusion of the climate change issue in African development planning, however, is the integration of climate concerns into development assistance. While the emergence of climate conditionalities on aid is a highly controversial issue in the ongoing debate between donor and recipient countries, such conditionalities might lead to the incorporation of the issue of climate change into mainstream African economic and developmental policies. In particular, such a development might stimulate the willingness of African states to set emission targets, which, according to the author, in turn, will force countries like South Africa and Nigeria to engage more actively in international climate change diplomacy.

Russia and Climate Change

In Chapter 11, Friedemann Müller seeks to identify determinants of Russian climate change policy during the current transition period, which in many ways defies the search for such constants. This is especially true for the issue of climate change, which plays a less than subordinate role in the perception of the recession-stricken population and in the priorities of decision-makers. Even ecologists put little emphasis on climate change because of other more acute and pressing environmental problems, such as air and water pollution, and nuclear contamination.

Nevertheless, Russia is the only major country experiencing substantially reduced emissions of CO_2 during the 1990s. The author emphasizes, however, that this reduction is not due to any policy of climate change, but relates solely to the deep economic recession and industrial transformation Russia has undergone since the dismantling of the Soviet Union. If Russia successfully completes its transition towards a market economy, it will be in a position to utilize its vast energy conservation potential, thus increasing energy efficiency and, in turn, reducing CO_2 emissions even further. However, such a development is threatened by the fact that only a minority of Russian politicians are at present in favour of transforming Russia into a Western market economy, and also by the Russian and old Soviet perception that economic growth is unthinkable without a corresponding increase in the input of energy.

The Climate Change Policy of the European Union

Jay Wagner provides an overview of the evolution of the EUs climate policy and assesses its prospects for the future (Chapter 12). To this end, climate change is placed within the wider context of EU environmental policy. Emphasis is put on discussing the instruments underlying EU policy, assessing their legal status and in presenting the obstacles facing EU climate change policy. Overall, an attempt is made to assess whether the EU has been a catalyst or a hindrance on the road to implementing the FCCC.

The author describes the process leading up to the declaration of the Community's target to stabilize greenhouse gases at their 1990 levels by the year 2000. Wagner argues that despite repeated reaffirmation by the Commission that the EU will meet this target, it is increasingly doubtful that the Community will meet its emission objectives. Emphasizing the importance of the Community's stance on climate change at the international level, the author concludes that serious economic and political stumbling blocks have emerged that are obstructing Community climate change policy. These include differences over burden-sharing, subsidiarity, waning public pressure, fierce industry opposition to carbon taxes; in addition are such institutional obstacles as EU voting procedures, deregulation, and limits to policy integration.

The chapter concludes that despite these obstacles there remains a substantial impetus towards concerted environmental action within the EU, particularly in the field of climate change. However, whether or not the EU achieves its targets, and manages to provide the necessary leadership to ensure that the forthcoming negotiations on the protocol to the FCCC are successful, depends on whether agreement on the scope and pace of European integration can be achieved. At a minimum, subsidiarity and scientific uncertainty surrounding climate change will delay the initiation, agreement, implementation and enforcement of EU climate change measures. Indeed, the author stresses, since efforts to curb greenhouse gas emissions have major implications for economic policy and social habits, it is likely that the scope and nature of Community action on climate change will be hotly contested within the EU.

Political Leadership and Climate Change: The Prospects of Germany, Japan and the United States

In Chapter 13, Gunnar Fermann provides an analysis of the leadership potential of Germany, Japan and the United States within the field of climate change. From this, the most promising avenue for each of the countries to take a leadership role in the years to come is inferred. Based on the empirical investigation of the various dimensions constituting leadership potential (responsibility, energy conservation and fuel-switching potential, current climate change policies including abatement target, commitment, measures), the author makes the following predictions.

Japan is unlikely to fulfil its quite modest abatement target (*per capita* stabilization of CO_2 by the year 2000) due to worsening abatement conditions and high marginal costs of abatement efforts. Since Japan is reluctant to remedy this state of affairs by adopting significant energy and CO_2 taxes, its prospects for leadership are not found within the national abatement target approach. Rather, Japan's most promising avenue for taking a leadership role, thus contributing to the development of the international problem-solving capacity, seems to be in international cooperation elaborating on the mechanism of joint implementation. Joint implementation is the idea that one country, facing high marginal costs on abatement efforts at home, provides assistance (technological, administrative, financial) to less energy-efficient and less developed countries, thus securing both a more efficient use of their resources and gaining some credit for their foreign abatement efforts. What makes this approach promising is not just the fact that Japan has a strong incentive to deflect the foreign criticism arising from its inability to fulfil its national abatement commitment, but also the fact that it is one of the world leaders in energy conservation technology and also possesses the financial resources required to transform the abatement needs of developing countries into actual purchasing power by means of subsidies and financial aid.

Germany's most promising avenue for taking a leadership role is likely to be found within the national abatement target approach: Even without the adoption of a CO_2 tax, Germany is likely to reduce its emissions of CO_2 by 10% by the year 2005 compared to the 1987 level. While this achievement is considerably less than the very ambitious German abatement target of 25% reductions, the German abatement effort might nevertheless prove to be a "world record", and satisfies the requirement of the FCCC (stabilization) with a good margin. In what way might this achievement contribute to making Germany a leader of the future? And how might the specific quality of German leadership reduce the current mismatch between the seriousness of the climate change problem and the problem-solving capacity of the climate change international regime? Firstly, it should be acknowledged that a 10% German reduction of CO_2 emissions would translate into some 0.5% reduction in total world CO_2 emissions compared to the business-as-usual scenario. While such an achievement would not be enough to solve the collective problem of climate change, the substantive impact of the German CO_2 reduction could be multiplied to the extent that this achievement triggered more extensive abatement efforts from other countries. But Germany need not content itself with the example set by its substantial reductions in national CO_2 emissions: The impact and effectiveness of Germany's climate change policy would increase considerably to the extent that a CO_2 tax would be adopted. This would not only increase the substantive impact and persuasive power of the German abatement effort, it would, moreover, single out Germany for entrepreneurial leadership along with the five small European states already having adopted such a tax. This kind of leadership would put pressure on both the United States and Japan to act likewise. Equally important, however, the adoption of a German CO_2 tax would bring tremendous pressure to bear on reluctant EU member-states to accept an EU-wide CO_2 tax. The unilateral adoption of a CO_2 tax should, of course, be followed up by a German diplomatic offensive within the EU, demonstrating that Germany was in earnest. Germany's potentially greatest achievement is not what it can manage by itself. German leadership rests even more on the extent to which it is able to promote its own achievements so as to trigger collective action.

While Germany's and Japan's most promising avenues for taking a leadership role (national abatement target approach and the abatement approach of joint implementation) may contribute toward increasing the problem-solving capacity of the present climate change regime, the American leadership potential is mainly to be found in its superior capacity to enhance knowledge of the problem itself. The United States' scientific community currently accounts for a substantial share of the total research effort in the field of climate change. Much of this research is basic science directed towards reducing the prevailing uncertainties regarding how severe and real the problem of global warming is. Proper assessment of the challenge facing us remains a necessary condition for the effective design and calibration of response strategies capable of mitigating the problem of climate change. The United States is likely to continue executing intellectual leadership by producing generative systems of thought. Moreover, American top scientists are likely to play a vital role within the IPCC process assessing and popularizing basic research on climate change. Moreover, American intellectual and entrepreneurial leadership is likely to continue in their research on impacts of global warming (advanced modelling) and effective means of mitigation. However, because of the existence of strong institutional and political restrictions, the United States is unlikely to extend its intellectual and entrepreneurial leadership to costly political action in the foreseeable future. The only factor possibly capable of altering this conclusion is a new sense of vulnerability created by scientific research showing that the impacts of climate change on the US are significantly more costly than previously perceived. Such a development would imply that the one-dimensional concern about abatement costs still dominating US climate change policy-making be replaced by a more balanced conception making impact costs part of the cost-benefit equation. In its policy shift at the CoP-2 in Geneva, July 1996, the United States government linked the equal percentage reduction of greenhouse gas emissions to the international application of tradeable emission permits. This indicates that the United States is in the process of taking a leading role in the practical elaboration of the joint implementation strategy to mitigation.

Part I
Key Issues

Chapter 2

Climate Change Turning Political: Conference-Diplomacy and Institution-Building to Rio and Beyond

Georg Børsting and Gunnar Fermann

1. Introduction

On 5 June 1992, the United Nations Framework Convention on Climate Change (FCCC) was opened for signature at the United Nations Conference on Environment and Development (UNCED) in Rio de Janeiro. After ratification by the necessary 50 nations, the FCCC became international law on 21 March 1994. Although important milestones, these events were only two of several steps being taken in the international process of managing climate change. While elaborations on the status of the scientific knowledge as well as analysis of the political nature of scientific advice is offered in subsequent Chapters (3–4), the present chapter summarizes the response of the international community to climate change up to the second conference of the Parties (CoP) in Geneva in July 1996. Emphasis is on the period 1988 onwards, when it was increasingly understood that the problem of global warming requires a serious political response. It is practical to divide such a political chronology into four distinct phases.

Firstly, 1972, the time of the United Nations Conference on the Human Environment (UNCHE), to about 1988 was a period of increasing scientific recognition and concern that anthropogenic emissions of greenhouse gases were probably a major cause of climate change, leading to increased international scientific cooperation to determine and develop the state of knowledge.

Secondly, from 1988 until 1991 global warming was also becoming an issue of international *political* concern. In this period, a series of meetings and international conferences on climate change were initiated, addressing the issues of both science and policy. The level of scientific assessment was stepped up during the same period, both quantitatively and qualitatively, with the establishment of the Intergovernmental Panel on Climate Change (IPCC) in 1988 generating the first comprehensive assessment report, and so establishing a degree of international scientific consensus on climate change.

In the third phase, from late 1990 up to UNCED in June 1992, the international political effort was channelled through the Intergovernmental Negotiating Committee (INC) and aimed at producing a convention on climate change.

The period since the signing of the FCCC in Rio has been characterized by difficult post-agreement negotiations during which the Parties to the FCCC have battled over the necessary completion and specification of the Convention, including the need to sharpen and strengthen the commitments in the agreement. The question of how to implement the commitments already agreed upon is also central in these negotiations. Parallel to the political conferences in Berlin in March 1995 (CoP-1) and Geneva in July 1996 (CoP-2), scientific study has continued. Despite the ensuing scientific dissent, the IPCC has strengthened its position as the authoritative provider of scientific knowledge on climate change.[1]

2. From Scientific Identification of Problem to "Call for Action"

Although the greenhouse effect was identified and described a century ago (1896) by the Swede Svante Arrhenius, little systematic research on climate change was done until the late 1950s, when measurement of carbon dioxide (CO_2) concentrations in the atmosphere were begun as part of the International Geophysical Year (1957–58). Other main greenhouse gases, such as methane (CH_4), chlorofluorocarbons (CFCs), and nitrous oxide (N_2O) have been measured continuously only since the late 1970s. Following the International Geophysical Year in 1957–58, the first theoretical models of the circulation of atmospheric air currents were developed in the 1960s. These were forerunners to the sophisticated General Circulation Models (GCMs) now used for scenarios of possible climate change.

The 1972 UNCHE was both an expression and the catalyst for increased scientific and political attention to local and transboundary environmental threats. After this pathbreaking conference, environmental protection became a compulsory part of international political discourse, although most countries were slow to integrate environmental assessment with economic activities. For the environmental sciences in general and atmospheric science in particular, the UNCHE created a

---

1 For a selection of articles questioning the IPCC scientific assessment and thus the scientific consensus on climate change, see Emsley (1995).

fertile political climate for instigating new research projects on global warming. By the end of the 1970s, the scientific community had increasingly begun to see climate change as a real and potentially serious problem (Lanchberry and Victor 1995: 30–31).

As an expression of the increased scientific concern about the atmospheric commons, the World Meteorological Organization (WMO) called together the first World Climate Conference in Geneva in February 1979. Essentially a scientific conference, this was the first time climate change was recognized as a serious problem by a major intergovernmental meeting. In retrospect, it might be argued that this conference marked the beginning of sustained high-level attention among scientists. The main issue debated at the conference was how climate change might influence human activities. The declaration from the conference recognized that there was a "clear possibility that ... anthropogenic increases in carbon dioxide may result in significant and possibly major long-term changes of the global-scale climate" (quoted from Lanchberry and Victor 1995: 31). Moreover, the conference called upon the world's governments "to foresee and prevent potential man-made changes in climate that might be adverse to the well-being of humanity" (*Environmental Policy and Law*, Vol. 6 (1980), No. 2, p. 103). The conference led to some concrete institution-building as well: The World Climate Programme (WCP) was established to strengthen the research and coordination of research on climate change (Cain 1983: 75).

Following a series of scientific gatherings between 1980 and 1983, the next big step towards scientific consensus on climate change came with the October 1985 Villach Conference, co-sponsored by the WMO, the United Nations Environment Programme (UNEP), and the International Council of Scientific Unions (ICSU). The conference gathered top scientists from 29 industrialized and developing countries and focused on the impact of greenhouse gas emissions on the world's climate. The conference statement presented a consensus on the scientific findings on the issue of climate change, and comprised a detailed description of recommended priorities for further scientific research, an urge for intensified research into the causes and impacts of climate change, and a call for increased support from governments and international funding agencies for climate change research (WMO 1986). While the Villach Conference was a milestone in building scientific consensus, it also had political significance. The conference statement offered politicians some general proposals for action and encouraged them to consider an international convention on climate change. According to Paterson, the scientific conference in Villach marked a "shift of emphasis away from solely the need for more research, towards including assertions of the need for political action" (1992: 176).

In September–November 1987, follow-up workshops were held in Villach and Bellagio – the former science-oriented, the latter for policy-makers (WMO/UNEP 1988). Like the Villach Conference in 1985, these workshops urged the need for political action. In particular, the joint report from the workshops recommended that immediate steps be taken to (i) limit greenhouse gas emissions by promoting implementation of the Montreal Protocol on Substances that Deplete the Ozone Layer and adopting new energy and deforestation policies; (ii) limit the impact of sea-level rise by elaborating river and coastal zone policies based on studies of the vulnerability of particular areas; and to (iii) promote further understanding of the greenhouse effect. The main political significance of the Villach–Bellagio workshops can be attributed, however, to the fact that this was the first time a gathering of both scientists and policy-makers called for "the examination ... of the need for an agreement on a law of the atmosphere as a global commons or the need to move towards a convention along the lines of that developed for ozone" (quoted in Paterson 1992: 177). At this point it was clear to most observers that the issue of climate change was in the process of being politicized.

The breakthrough of climate change as a political issue came at the June 1988 Toronto Conference on the Changing Atmosphere, which brought together more than 300 scientists and policy-makers from 48 countries, several United Nations organizations and other international bodies, as well as non-governmental organizations. The conference addressed climate change together with the related problems of ozone layer depletion and the long-range transport of toxic and acidifying substances. This comprehensive approach to global environmental problems was a novelty in international conference diplomacy – an approach which would be taken to the extreme four years later at UNCED. Convened by the Canadian government, the Toronto Conference can be deemed a watershed, as it placed the climate change issue on the political agenda of governments and policy-makers. The Conference enjoyed broad international and high-level political representation and for the first time specified the international response required to mitigate climate change. The Conference issued an ambitious "call for action" to governments, industry and international organizations to (i) reduce CO_2 emissions by 20% of the 1988 levels by the year 2005,[2] and eventually by 50%; (ii) improve energy efficiency by as much as 10% by the year 2005; (iii) initiate the necessary technological changes to reach these

---

2 This abatement target went beyond targets recommended by later international conferences as well as tentative targets set by the 1992 UN Framework Convention on Climate Change.

goals; (iv) prepare principles and components of a framework treaty for the protection of the atmosphere in time for the 1992 UNCED; and (v) promote the establishment of an international scientific assessment panel on climate change (Toronto Conference Statement 1988; *American Journal of International Law and Policy*, Vol. 5 (1988), No. 5, p. 515). The Conference also called upon governments to establish a World Atmosphere Fund financed in part by a levy on the fossil fuel consumption of industrialized countries (Paterson 1992: 177). While the 1979 World Climate Conference in Geneva marked the beginning of sustained international high level attention among scientists, it is safe to say that the 1988 Toronto Conference, because of its specific and demanding recommendations, put climate change on the political agenda of most governments in the industrialized world.

3. Deepening the Political Commitment Through Institution-Building

A large number of international conferences on climate change were convened in the period after the Toronto Conference. They addressed issues of both science and policy and were attended by government representatives, scientists and environmental groups. Some of the meetings took place under the auspices of the United Nations and its specialized agencies. Others have been held within regional and global fora such as the European Community (EC), the Commonwealth and the South Pacific Forum, or have been convened by individual governments. A number of meetings have been dedicated to the particular concerns of small island states and of developing countries.

3.1. Proliferation of Climate-Change Conference Diplomacy

In June 1989, 80 legal and policy experts met in a personal capacity in Ottawa, Canada, to develop the idea of a "law of the atmosphere treaty" and a narrower convention on climate change. Thus the Ottawa International Meeting of Legal and Policy Experts on the Protection of the Atmosphere was an expert response to the 1988 Toronto Conference "call for action". The main output of the conference was drafts modelled on the approach taken by the 1982 Law of the Sea Convention and by the 1985 Vienna Convention on the Protection of the Ozone Layer and its 1987 Montreal Protocol – including a provision that "states should consider the possibility of establishing a World Atmosphere Trust Fund" (Ottawa Meeting Statement 1989: 7). The wide approach adopted at the

Ottawa International Meeting was later criticized for being too ambitious. While discussed at various international meetings, it was not taken up by the Intergovernmental Negotiating Committee (INC) as a model for a framework convention on climate change.

As part of an effort to make developing countries more active participants in the political discourse on climate change, the Tata International Conference on Global Warming and Climate Change was held in New Delhi, India, in February 1989. The conference was sponsored by UNEP and the World Resources Institute (WRI). The 1989 Tata Conference was the first international conference dedicated to the particular concerns of developing countries in relation to climate change. In addition to providing a detailed analysis of climate change and its impact on developing countries, the statement from the meeting (i) made it clear that the industrialized countries had to bear most of the responsibility for human-induced climate change; (ii) acknowledged that developing countries should participate in an international response to climate change, but not at the expense of economic development; and (iii) emphasized the importance of research and training in developing countries (Tata Conference Statement 1989).

On the initiative of the Dutch, French and Norwegian governments, the Hague Ministerial Conference on Atmospheric Pollution and Climatic Change was convened in March 1989 and attended by representatives of 24 countries – most of whom were heads of state. The resulting Hague Declaration emphasized the need for new institutional mechanisms that could enable international decision-making on climate change to be more effective and the regulations more enforceable (*Environmental Policy and Law*, Vol. 19 (1989), No. 2, p. 45). In particular, discussions centered around the need for an institution within the United Nations system to take the main responsibility for combating global warming. Whether such an institution should be built upon existing UN organizations such as the UNEP, or whether a new body was required, was left open after some debate. Nevertheless, the Hague Declaration spelt out certain specific features of such an institution (Hague Declaration 1989). The significance of this conference should not be overrated, since major countries such as the United States, the USSR and China were not invited because of their unenthusiastic attitude towards climate change policy commitments. Great Britain was invited, but declined to attend for the same reason (*International Environment Reporter* 1989, No. 12, p. 215).

The threat of human-induced climate change was addressed by several international conferences and meetings of a more exclusive regional, political or socio-economic nature – including the Meeting of Non-Aligned Countries in Belgrade, September 1989; the Tokyo Conference on Global Environment and Human Response Towards Sustainable Development, September 1989; and the Commonwealth Heads of Governments Meeting, October 1989 (Paterson 1992: 178). Most notably, the July 1989 G-7 Annual Summit in Paris issued a statement that called for "common efforts to limit emissions of carbon dioxide", and stated furthermore that a "framework or umbrella convention ... [was] ...urgently required" (G-7 Economic Declaration 1989).

Of great significance for the further development of climate change as a political issue was the November 1989 Nordwijk Ministerial Conference on Atmospheric Pollution and Climate Change. Attended by representatives of 67 countries, 11 international organizations, and the Commission of the European Community (*Environmental Policy and Law*, Vol. 19 (1989), No. 4, p. 229), the Nordwijk Conference became a critical milestone on the road to international CO_2 emission targets. The resulting declaration from this broadly based conference was to become a reference point for subsequent debate and conferences on climate change, e.g. the 1989 Cairo Conference and the 1990 Bergen Conference. The declaration includes, *inter alia*, a statement on the stabilization of CO_2 emissions which has made a lasting impact on the international climate change negotiating process, including the 1992 FCCC itself:

> The Conference recognizes the need to stabilize, while ensuring stable development of the world economy, CO_2 emissions and emissions of other greenhouse gases not controlled by the Montreal Protocol. Industrialized nations agree that such stabilization should be achieved by them as soon as possible, at levels to be considered by the IPCC and the Second World Climate Conference of November 1990. In the view of many industrialized nations such stabilization should be achieved as a first step at the latest by the year 2000 (Nordwijk Declaration 1989).

Concerned about the destructive impacts a rise in sea level would have on Egyptian coastland and agriculture, Egypt convened the December 1989 World Conference on Preparing for Climate Change in cooperation with UNEP. Attended by more than 400 participants from all regions of the world, the conference gave its political support to the recently established IPCC (see below) and to the idea of a framework treaty on climate change. The conference statement (i) urged governments to take action on the national and regional level; (ii) emphasized that wealthy nations should provide developing countries with the means to cope with climate change; (iii) stated that measures should focus on efforts to reduce the "historically unprecedented population growth" which is "a driving force behind the increase in greenhouse gas emissions"; and, finally, as an early concession to realism, (iv) acknowledged that as a certain degree of climate change is inevitable, necessary adaption measures had to be taken (Cairo Compact 1989).

The May 1990 Bergen Conference was one in a series of regional meetings held in advance of UNCED. Hosted by the Norwegian government and co-sponsored by the United Nations Economic Commission for Europe (ECE), the conference was attended by the environment ministers of 34 countries and the EC Commissioner for the Environment. Although a variety of environmental issues were discussed, climate change was granted primary attention. The ministers, recognizing that the EC countries consumed a large share of the world's energy, declared their willingness to "assume a major responsibility to limit or reduce greenhouse gases" (*Environmental Policy and Law*, Vol. 20 (1990), No. 6). Moreover, the Bergen Declaration (i) pointed at specific measures necessary for combating the causes of climate change; (ii) expressed strong support for the work of the IPCC; (iii) lent support to the "precautionary principle" and the "principle of common but differentiated responsibilities" of states in responding to climate change; and (iv) advocated new ways and means for providing financial resources to developing countries in combating and adapting to climate change (Bergen Ministerial Declaration 1990). The Bergen Declaration was an early indication that several EC members were willing to take concrete steps to fight climate change. Most notably, many EC members made statements to the effect that they were willing to stabilize CO_2 and other greenhouse gas emissions by the year 2000. Owing to opposition from the United States and a few other countries, the stabilization target was not made part of the official statement. The ministerial meeting was also unable to agree on a specific financial pledge for the developing countries.

Sponsored by the WMO, UNEP, and other international organizations, the Second World Climate Conference in Geneva in October–November 1990 gathered some 750 scientists from around the world in a series of non-governmental scientific sessions in 18 Expert Panels and Task Groups, and a Consultation Group on the Special Needs of Developing Countries. The main task of the scientific conference, which was parallelled by an intergovernmental conference, was to review the WCP established by the first World Climate Conference in 1979 (*Environmental Policy and Law*, Vol. 20 (1990), No. 6). Expectations for the conference were high, and subsequent international climate conferences have since stressed its importance. The 1990 World Climate Conference was held at a crucial time: the IPCC had just completed its First Assessment Report (1990), which, in turn, was to provide critical input for the INC attempting to negotiate a convention on climate change.[3] In its statement, the

Scientific and Technical Sessions noted that since the First World Climate Conference in 1979 a clear scientific consensus had emerged on the estimated range of global warming that was to be expected through the twenty-first century, and, accordingly, agreed that it was time for the world community to take strong measures to reduce sources and to increase sinks of greenhouse gases, despite the remaining scientific uncertainties (Statement of the Scientific and Technical Sessions 1990).

The political advice from the scientific part of the conference was taken up by the political sessions of the conference, which was attended by heads of government and ministers from 137 states and the EC. After hard bargaining on a number of difficult issues, the resulting Ministerial Declaration managed to establish some basic principles that later became part of the FCCC; among these were the conception of climate change as a "common concern of humankind"; the "principle of equity" and the "common but differentiated responsibility" of countries at different levels of development, the concept of "sustainable development", and the "precautionary principle". Furthermore, the declaration (i) stressed the need for further scientific research; (ii) recommended response measures to be adopted without delay, despite remaining scientific uncertainty; (iii) urged the industrialized countries to establish abatement targets; (iv) recognized that emissions from developing countries would continue to grow to accommodate development needs, but urged action with support from the industrialized nations and international organizations; and finally (v) called for the elaboration of a framework treaty on climate change and necessary protocols in time for adoption by the 1992 UNCED (Ministerial Declaration 1990). Although some participants and observers were disappointed that the political output of the Second World Climate Conference did not offer a higher level of commitment, most notably the failure to specify any emission abatement target, the high political level and broad approval granted within the Ministerial Declaration nevertheless pushed the international political process a crucial step towards a convention on climate change.

3.2. The Role of UN Bodies

Having reviewed the international conference diplomacy on climate change up to the fall of 1990, it may at this point be justified to depart temporarily from the chronological logic that has so far been the basis of our presentation. On the insistence of several quarters and

---

3 The INC was established by the United Nations General Assembly, 21 December 1990, obviously in order to provide the climate change negotiating process with the broadest possible political basis (UNGA Res. 45/212, 1990).

conferences, various UN bodies have, in different ways, facilitated and influenced the process of transforming climate change from being merely a scientific issue to becoming a political one as well. Before continuing the historical presentation towards the adoption of the FCCC at the 1992 UNCED, the roles of various UN and UN-linked bodies are granted particular consideration.

3.2.1. The UN General Assembly: Securing Political Legitimacy

The first United Nations General Assembly (UNGA) initiative relevant to climate change was the establishment in 1983 of the World Commission on Environment and Development which led to the 1987 report "Our Common Future", also known as the "Brundtland Report". The report analyzed the state of the global environment and identified areas of priority for an institutional and legal response. It also called for the negotiation of a climate treaty, further research and scientific monitoring of climate change, as well as international efforts to reduce emissions of greenhouse gases. In 1988, the UNGA adopted a resolution that recognized climate change as a "common concern of mankind". It (i) urged the world to treat climate change as a priority issue; (ii) reaffirmed a 1987 resolution approving an initiative by the UNEP and the WMO to cooperate on this issue; and (iii) endorsed the establishment of the IPCC (UNGA Res. 43/53, 1988).

In 1989, the UNGA decided to convene the UNCED in June 1992 and granted the issue of climate change top priority within this framework. The UNGA also welcomed the various international conferences that had been held to discuss climate change and endorsed the decision by UNEP and the WMO to work on a climate treaty. The UNGA's 1989 resolution on climate change emphasized the concerns of developing countries and of low-lying coastal and island states, urging the world community to help the states most affected by climate change impacts to adapt to the situation (UNGA Res. 44/207, 1989).

Most importantly, however, in late 1990 the UNGA adopted its "climate resolution" establishing the INC as responsible for the negotiation of a convention on climate change to be signed at UNCED (UNGA Res. 45/212, 1990). The UNGA reaffirmed its endorsement of the negotiating process in December 1991 and called upon member-states to continue contributing to the funds established for financing the participation of developing countries in the work of the IPCC and the INC. When the INC succeeded in preparing and adopting the FCCC in June 1992, the UNGA welcomed this achievement in a resolution that also set out some guidelines for future action (*Environmental Policy and Law*, Vol. 23 (1993), No. 1, p. 44).

3.2.2. UNEP and WMO: Scientific Catalysts and Institution-Builders

Since the early 1980s, UNEP has become increasingly active in the issues of stratospheric ozone depletion and climate change. The policies and activities of UNEP are decided by its Governing Council, and in the last decade climate change has been high on the Council's agenda. Together with the WMO, UNEP plays a leading role in United Nations work on climate change. A number of units and programme activities within the UNEP Secretariat deal extensively with climate change: The Global Environment Monitoring System (GEMS) links some 25 major global monitoring networks, thus providing the scientific basis for UNEP's political activities, as well as for environmental programmes of governments, international agencies and institutions. Founded by UNEP in 1985, the Global Resources Information Database (GRID) was established to make scientific findings of GEMS available to decision-makers. A similar initiative was taken in 1992 when UNEP set up the Information Unit on Climate Change (IUCC) in Geneva to provide information on climate change for both decision-makers and the public at large.

The primary responsibility of the WMO is the world-wide coordination of meteorological activities. The WMO also plays a leading role in ensuring the provision of authoritative scientific information on the state of and behaviour of the global atmosphere, promotes research on climate change, and supports a number of predominantly science-oriented climate programmes and institutions. WMO's policies and activities are determined by the World Meteorological Congress, while WMO's Executive Council monitors the implementation of the Congress's decisions by the WMO Secretariat.

UNEP and WMO joined forces on the issue of climate change for the first time in 1979, when they participated in convening the First World Climate Conference; this was followed by their co-sponsorship of the Villach Conference in 1985 (see above). In 1988, UNEP and WMO jointly established the IPCC (see below), and both organizations subsequently followed and gave guidance to the IPCC's work. In 1990, they sponsored the Second World Climate Conference and set up an intergovernmental working group whose activities were later taken over by the IPCC.

The UNEP–WMO partnership continues to be a leading force in the international response to the threat of climate change. In 1991, the WMO Congress and the UNEP Governing Council reaffirmed the importance of the ongoing work of the IPCC and of the results of the Second World Climate Conference. The WMO Congress asked WMO's Secretary-General and UNEP's Executive Director to consider convening a third world climate conference at a later stage, while confirming their intention to commit more of their financial and organizational resources to the climate change issue.

3.2.3. The IPCC: Generating a Scientific Consensus

The IPCC was established by UNEP and WMO in 1988 as a response to a call by the Toronto Conference the same year.[4] Scientific uncertainty plays a major role in the international politics of global warming (see Chapters 3–5), and the establishment of the IPCC has meant a great deal in the alleviation of this state of affairs.[5] In particular, the IPCC has played a vital role in providing assessments of climate change research to policy-makers. The IPCC was initially provided with a threefold mandate: (i) to assess the state of scientific knowledge on climate change, (ii) to examine the environmental economic and social impacts of climate change, and (iii) to formulate response strategies for the management and mitigation of climate change.

Meeting for the first time in November 1988 in the WMO offices in Geneva, the IPCC set up three working groups reflecting the above-mentioned mandate. The IPCC continued meeting throughout 1989 and 1990. In October 1990, it issued the highly influential First Assessment Report for the Second World Climate Conference (see Section 3.2), reflecting what was widely accepted to be the scientific consensus on what the problem of global warming entailed. Together with a 1992 Supplement Report, the First Assessment Report served as an important basis for the climate treaty negotiations up to Rio. Of particular importance to the political momentum of the negotiating process was the conclusion of IPCC Working Group I (WG I) that the atmospheric concentrations of greenhouse gases were increasing, a trend which would probably lead to a warming of the Earth's surface. Projections of the sort of warming one could expect were given, considering both a "business-as-usual" scenario and various possible responses. The magnitude of reductions in the emission of each greenhouse gas necessary to stabilize atmospheric concentrations at current levels was also estimated (Houghton et al. 1990: 8).

---

4 The IPCC succeeded the World Climate Programme (WCP), which was established at the first World Climate Conference in 1979 (see Section 2).

5 For a general discussion of the early period of the IPCC, see Nitze (1989) and Lanchberry and Victor (1995).

The Impacts Working Group (WG II) concluded that climate change could have highly destabilizing effects on human society. The WG II carried out analyses of possible effects on agriculture, natural ecosystems, water resources, human settlements, oceans, and coastal zones (Tegart et al. 1990). The Response Strategies Working Group (WG III) proposed specific measures for limiting and adapting to global warming, emphasizing the common but differentiated responsibilities of developed and developing countries (IPCC Response Strategies 1990). The IPCC also established a Special Committee to explore how the full participation of developing countries could be promoted. In its conclusions, the committee emphasized the developing countries' need for financial and technical assistance, also pointing out that the industrialized countries should take the lead in adopting policy measures, since about 75% of greenhouse gas emissions originate in these countries.

In November 1992, the IPCC established a new structure and a plan for its continuing work: The WG I was to continue to concentrate on science. The WG II, however, now became responsible for both impacts and response options, while WG III was to focus on economics. This restructuring was intended to improve efficiency and enable the three WGs to help governments meet the challenges of Agenda 21 and the requirements of the FCCC. Since then, the IPCC has finalized its Second Assessment Report (1995), further strengthening some of the conclusions of the 1990 First Assessment Report and the 1992 Supplement Report. There is evidence, however, that the processes of producing scientific knowledge (the IPCC) and political negotiation (INC/CoP) have been drifting apart since the finalizing of the FCCC. It has been claimed that the IPCC has become more distant from the negotiations, reflecting decreasing relevance of formal science to the political process (Victor and Salt 1995: 37).

4. Negotiating the Framework Convention on Climate Change

4.1. The Intergovernmental Negotiating Committee

During the period 1988 to 1989, two lines of development were important in forcing formal negotiations on climate change. Not only did numerous international conferences devoted to climate change issue urgent calls for a binding global treaty addressing the problem of climate change,[6]

---

6 In addition to those mentioned above, some of the most important were the Global Forum on Environment and Development in Moscow, January 15–19, 1990, at which

also many industrialized countries made unilateral commitments in relation to their future CO_2 emissions.[7] Responding to these political signals,
the UNEP and the WMO in 1990 established an intergovernmental
working group to prepare for treaty negotiations (UNEP Dec. SSII/3.C,
1990; WMO Res. 8, 1990). This group proposed that all negotiations
under UN auspices should take place within a single forum. This
proposal was confirmed by the UNGA in December 1990, setting up the
INC to draft a framework convention (UNGA Res. 45/212, 1990).
Unexpectedly, with this resolution the UNGA took direct control over
the negotiation process. The INC's task was to have an "effective"
convention ready in time for the Rio "Earth Summit" (UNCED) in June
1992. Becoming the central forum for the international effort to develop
a climate treaty, some 150 states and numerous IGOs and NGOs
participated in the negotiations within the INC.

Not surprisingly, progress was slow at the beginning, since much time
was spent on formal aspects and procedural points. There was much criticism from environmental NGOs that time was being wasted.[8] There
were worries about the fact that the UNGA had taken direct control over
the negotiations rather than delegating this to UNEP. Pointing to the failure of the UN Conference on the Law of the Sea, some felt that UN multilateral diplomacy would encumber the negotiation of a climate change
treaty unduly (Sebenius 1991; Ramakrishna 1990).

Compared to other global environmental negotiations, the INC must,
however, be said to have made relatively rapid progress: the FCCC was
negotiated during the five sessions (INC-1 to INC-5) between February
1991 and May 1992.[9] During these sessions, the INC discussed the
belligerent issues of (i) binding commitments, targets and timetables
for the reduction of CO_2 emissions, (ii) financial mechanisms, (iii) technology transfers, and (iv) "common but differentiated" responsibilities of

---

President Gorbachev suggested the UNCED Conference to be held at "summit level"
(Keesing's Record of World Events, Vol. 36 (1990), No. 1, p. 37202); The Conference
for Small Island States in the Maldives, January 1990; a conference at the White House
in Washington DC, April 17–18, 1990; and a conference on African Perspectives on
Global Warming in Nairobi, 2–4 May, 1990 (Paterson 1992: 179–180).

7 For a listing of these commitments, see Table 1.3 in O'Riordan and Jäger (1996:
22–23).

8 See Hayes and Smith (1993: 10-12) for a review of the specific problems that were
encountered during the INC sessions. For a contribution explaining the barriers to an
effective climate change regime in more general terms, see Chapter 1 in the volume.

9 Sessions where held at Chantilly/Washington DC, 4–14 February 1991 (INC-1); in
Geneva, 19–28 June 1991 (INC-2); in Nairobi, 9–20 September 1991 (INC-3); in
Geneva, 9–20 December 1991 (INC-4); and in New York, 18–28 February and 30
April to 8 May 1992 (INC-5).

industrialized and developing countries. Building upon the lessons from past negotiation processes, the INC sought to achieve a consensus that could be supported by a broad majority. This was considered more important than going for a more specific and demanding treaty that could limit the participation of crucial actors such as the United States. An agreement in such a short time was greatly facilitated by the work of the IPCC and by international meetings such as the Second World Climate Conference. The way the negotiation process was integrated into the larger UNCED process was probably also important in making the Convention possible in this short time: the FCCC was seen as the treaty flagship of the UNCED, thus creating tremendous pressure on the last INC meeting in May 1992 to reach an agreement.

4.2. The Framework Convention on Climate Change[10]

After only 15 months of intensive negotiations within the INC, the FCCC was adopted in New York on 9 May 1992, and opened for signature at UNCED on 3–14 June. Here it was signed by the heads of state of 153 nations. The Convention comprises 26 articles. Being a framework agreement, its wording concerning commitments is relatively vague, but despite this many commentators described the FCCC as one of the major achievements of the UNCED process. This Section presents some of the most important parts of the Convention, and at the same time some of its limitations and weaknesses (see also Chapter 1).

The Parties agreed that the *objective* of the Convention was to achieve "stabilization of greenhouse gas concentration in the atmosphere at a level that would prevent dangerous anthropogenic interference with the climate system", and that this should be done quickly enough to "allow ecosystems to adapt naturally to climate change, to ensure that food production is not threatened and to enable development in a sustainable way" (Art. 2). As guiding principles, the Convention emphasized in particular the application of the "precautionary principle", and the principles of "cost-effectiveness" and "equity" – between industrialized and developing countries as well as among these two groups of countries (Art. 3). Acknowledging industrialized countries' predominant responsibility for human-induced global warming, and their superior capability to take action, the Convention states that these countries should take the lead in combating climate change and its impacts.

The *obligations* in the FCCC can be divided into two groups: the first concerns the actual abatement of greenhouse gases; the second, the need for

---

10 The full text of the FCCC is reproduced in Appendix 1.

transfer of financial resources and technology to the developing countries. Targets and timetables for the reduction of CO_2 emissions have been at the hub of the climate debate since the 1988 Toronto Conference. Regarding such abatement commitments, industrialized countries (Annex I countries)[11] committed themselves to only "adopt national policies and take corresponding measures on the mitigation of climate change, by limiting the anthropogenic emissions of greenhouse gases and protecting and enhancing [their] greenhouse gas sinks and reservoirs", with the ambiguous recommendation of "returning individually or jointly to their 1990 levels [of] these anthropogenic emissions of carbon dioxide and other greenhouse gases not controlled by the Montreal Protocol" (Art. 4.2). This means that the Convention contains no substantial commitments to reduce CO_2 emissions or other greenhouse gases at a specific percentage by a specific date. Nor was there any agreement on the appropriate "burden-sharing" between the Annex I countries, except that some flexibility be provided to economies in transition in the former Soviet Union and Eastern Europe.

Developing countries undertook no commitments to reduce their greenhouse gases. In fact, throughout the whole negotiation process they claimed their right to *increase* their emissions. The INC process was favoured by an emerging general awareness of the importance of the link between environment and development, and the industrialized countries clearly recognized that developing countries must be allowed economic development within a new international climate change regulatory regime. All of the signatories, however, including the developing countries, did commit themselves to provide an inventory of greenhouse gas emissions and sinks, as well as a detailed description of policies and measures adopted to implement recommended emission controls.

Furthermore, all Parties committed themselves to work out a national climate change response strategy (Art. 4.1). There are many methodological problems involved in the reporting requirements, and for the reports to be comparable and valuable it is crucial that the Parties use common methodologies. The gathering of the required data is in itself challenging for any country. For the developing countries, the implementation of these reporting commitments was made contingent upon the industrialized countries demonstrating their intent to take the necessary first steps, and upon "the effective implementation by developed country Parties of their commitments under the Convention related to financial resources and transfers of technology" (Art. 4.7).

---

11 The Annex I countries consist of all 24 OECD countries and the EU, as well as 11 countries undergoing the process of transition to a market economy ("economies in transition"). For a complete list, see Appendix 1.

The transfer of resources and technology is therefore a necessary condition for the above-mentioned commitments of the developing countries
to be implemented. Thus, the commitment of the industrialized countries
to "provide new and additional financial resources to meet the agreed
full costs incurred by the developing country Parties" (Art. 4.3) for their
national reports is of high importance, as is their pledge to "provide such
financial resources, including for the transfer of technology, needed by
the developing country Parties to meet the agreed full incremental costs
of implementing measures" necessary to comply with commitments
under Art. 4.1 (Art. 4.3). Furthermore, the industrialized countries have
undertaken to "promote, facilitate and finance, as appropriate, the transfer of, or access to, environmentally sound technologies and know-how"
to developing countries, but also to "support the development and
enhancement of endogenous capacities and technologies of developing
country Parties" (Art. 4.5). Finally, the industrialized countries must
"assist" developing countries in their adaption to some climate changes.
However, the financial commitments do not extend to emission controls
by developing countries that might later be agreed upon. It is likely that
this question will become central in the debate at a later stage, because
emission forecasts predict that developing countries' emissions in the
future will be greater than those from industrialized countries (Hayes
1993: 144–168).

The FCCC established five *permanent institutions*: the Conference of the
Parties; the Secretariat; two subsidiary bodies to assist in questions on
implementation and advice on science and technology; and a "financial
mechanism". The Conference of the Parties (CoP) is the supreme decision-making body of the Convention. It is responsible for the regular
review of the implementation of the Convention and for taking the decisions necessary to promote such effective implementation (Art. 7). The
Convention also established an interim Secretariat, to be made permanent by the CoP at its first session (Art. 8). In addition to these central
organs, the FCCC established two subsidiary bodies to assist the CoP in
its work. The Subsidiary Body for Scientific and Technological Advice
(SBSTA) was made responsible for the provision of timely information
and advice on scientific and technological matters relating to the
Convention (Art. 9). Although its functions and terms of reference were
to be further elaborated, it was clear from the start that the SBSTA was
to have some kind of communicative buffer function between the CoP
and the IPCC, and other parts of the scientific community. Also, a
Subsidiary Body for Implementation (SBI) was established to assist the
CoP in its assessment and review of the implementation of the
Convention (Art. 10).

The FCCC also defines a mechanism for the provision of financial resources based on grants or concessionary financing (Art. 11). It was decided that this *financial mechanism* was to be accountable to the CoP and its operation entrusted to one or more existing international entities. This responsibility was given to the Global Environmental Facility (GEF),[12] but because developing countries were not satisfied with their influence in the GEF, this was only on an interim basis. The major disagreement goes back to the developing countries' wish for the GEF to be linked more closely to the UN system, against the donor countries' wish to sustain GEF's independence. The FCCC stipulates that "the financial mechanism shall have an equitable and balanced representation of all Parties within a transparent system of governance", and it was therefore emphasized that the GEF "should be appropriately restructured and its membership made universal" (Art. 21.3). As this was decided, a restructuring and replenishment of GEF was already underway.

Although many observers expressed disappointment over a lack of substantial commitments in the FCCC, the signing of the Convention has to be seen as only the first step on the long path to the creation of an effective climate regime.[13] As a framework convention the FCCC contains important principles and sets out general obligations: for example, the commitment to work toward the realization of certain goals (such as stabilization of greenhouse gas concentration in the atmosphere at a "safe" level) and to let their policies be guided by certain principles (e.g. the "precautionary principle"). In line with other international environmental agreements,[14] additional commitments may be agreed upon later in one or more protocols. Two other ways of developing the FCCC further are also possible, however: amendments of the Convention itself, or a development by decisions and declarations of the Parties to the Convention (Oberthür and Ott 1995). In a parallel process, Parties to the Convention will have to meet the commitments contained or implicit in the Convention, and at the same time continue the negotiations to further improve and strengthen the regime, e.g. by binding protocols. It might also be argued that international negotiations often carry with them important effects that in sum may be more important than the substantial decisions that come out of the process (Underdal 1992: 2). It is therefore significant that the FCCC contains several review clauses to advance such a dynamic evolution of the regime.

---

12 For an assessment of the GEF, see for instance Fairman (1996) and Sjöberg (1995).
13 For various early assessments of the FCCC, see Bodansky (1993); Grubb (1992); Haas et al. (1992); Thacher (1992); and Underdal (1992).
14 For instance the 1979 Geneva Convention on Long-Range Transboundary Air Pollution and the 1985 Vienna Convention for the Protection of the Ozone Layer.

Important elements of the FCCC are also the *mechanisms for information gathering and reporting* on sources and sinks of greenhouse gas emissions, as well as on policies and measures taken by Parties for the implementation of their commitments (Art. 12). Such "performance reviews" are considered vital for the effectiveness of international environmental agreements.[15]

5. The post-FCCC Negotiations: From "Rapid Progress" to Indecision and Backsliding?

5.1. *The Intergovernmental Negotiating Committee: Preparing for Berlin*

Numerous quarrelsome issues had to be side-stepped for the negotiating Parties to be able to reach an agreement before the meeting in Rio. To carry on the process of resolving these issues, the INC continued to meet after the adoption of the Convention. In an interim phase it prepared for the first session of the Conference of the Parties (CoP-1) to the Convention in Berlin, 28 March to 7 April 1995, holding six meetings (INC-6 to INC-11).[16] The INC worked on recommendations to the CoP and other decisions and conclusions requiring action by the CoP.

Much of the INC's time was devoted to matters relating to *abatement commitments*: (i) a review of information communicated by Annex I Parties to the Convention; (ii) methodological issues; (iii) review of the adequacy of commitments (Art. 4.2), including proposals related to a protocol and decisions on follow-up; (iv) criteria for joint implementation; (v) the roles of the subsidiary bodies established by the Convention, including their programmes of work and calendars of meetings; (vi) report on implementation; and (vii) first communication from non-Annex I Parties to the Convention.

Since the national reports (or "national communications") will have to be developed over time, much attention has been given to the establishment of a systematic review and assessment process. Nevertheless,

---

15 For an elaboration on the concept of environmental performance review, see Lykke (1992). On the problems of compliance and verification in international environmental agreements, see for instance Chayes and Chayes (1991) and Ausabel and Victor (1992). Wettestad (1990), DiPrimio et al. (1992) and Katscher et al. (1994) treat verification of commitments in atmospheric conventions specifically.

16 7–10 December 1992 in Geneva (INC-6); 15–20 March 1993 in New York (INC-7); 16–27 August 1993 in Geneva (INC-8); 7–18 February 1994 in Geneva (INC-9); 22 August to 2 September 1994 in Geneva (INC-10); and 6–17 February 1995 in New York (INC-11).

Annex I countries tried to ensure that the first round of reports (from Annex I Parties only) was promptly reviewed before the CoP-1 in Berlin, March–April 1995. Because of the time pressure, however, this initial review had to be limited to a compilation and synthesis of the reports with a limited consideration. The precedents for effective reporting and review processes are not encouraging (Børsting 1993), but, nevertheless, 15 countries did manage to deliver their national communications in time for the first review.

No doubt one of the most difficult questions facing the Parties after finalizing the FCCC has been the debate on how to express future international commitments. Some of the Parties have a strong belief in the potential role of emission targets and timetables. Subsequent to the INC-11 (February 1995), Trinidad and Tobago, on behalf of the Alliance of Small Island States (AOSIS), submitted a draft protocol to the Interim Secretariat. Building on the "Toronto targets" (see above), this protocol calls for a reduction in emissions of greenhouse gases by "at least 20% by the year 2005" (*Earth Negotiations Bulletin* 1995a).

The INC also worked on matters relating to arrangements for the *financial commitments*, that is, implementation of Article 11, paragraphs 1–4 of the Convention: consideration of the maintenance of the interim arrangements (Art. 21.3); modalities for the functioning of operational linkages between the CoP and the operating entity or entities of the financial mechanism; and guidance on programme priorities, eligibility criteria and policies, and on the determination of "agreed full incremental costs" (Art. 14.3).

Among the other issues the INC handled, the most important were provision to the developing country Parties of technical and financial support, the designation of a permanent secretariat and arrangements for its functioning, the consideration of the establishment of a multilateral consultative process for the resolution of questions regarding implementation (Art. 13), and the review of lists of countries included in the annexes to the FCCC.

What was accomplished in these post-agreement negotiations? Of the above-mentioned issues, the most difficult for the INC was achieving progress on the questions of adequacy of commitments, criteria for joint implementation, voting rules, and location of the permanent secretariat. These were all important questions that had to be submitted to the CoP-1 with no consensus among the Parties. Still, agreement on many of the procedural matters concerning for example the financial mechanism, the subsidiary bodies, reporting and methodological issues, was important in preparing the ground for the process of operationalizing and strengthening the FCCC.

5.2. The CoP-1 and the "Berlin Mandate"

Berlin was the host of the first session of the Conference of the Parties (CoP-1) from 28 March to 7 April 1995. By the end of the session, 117 countries and the EU were Parties to the FCCC – all but one of these were represented at the "Berlin Summit". More than 50 states participated as observers together with almost 200 observer organizations (UN specialized agencies, IGOs and NGOs) and more than 2000 journalists.

An evaluation of the achievements of the CoP-1 very much depends on the group of countries or observers asked, as well as the aspect of the Conference considered. Rather than rank the meeting's degree of success, we therefore point at the degree of progress on some of the issues that were most debated. Ranking highest on the agenda in Berlin, and to be considered more thoroughly below, were (i) the adequacy of commitments; (ii) criteria for joint implementation; (iii) the financial mechanism; (iv) the rules of procedure; (v) the designation of the permanent Secretariat of the Convention, and (vi) other matters relating to the mechanisms for reporting and monitoring, and the elaboration of the functions and rules of the Subsidiary Bodies.

(i) *The Adequacy of Commitments*: According to Article 4.2. (d) of the FCCC, the CoP-1 had to consider whether the commitments under Article 4.2 (a) and (b) were adequate for reaching the Convention's objective of stabilizing the "greenhouse gas concentration in the atmosphere at a level that would prevent dangerous anthropogenic interference with the climate system" (Art. 2). At the end of 1994, the IPCC released an interim report stating that even if current global emissions of CO_2 were stabilized, atmospheric concentrations of this gas would continue to rise for at least two centuries (Oberthür and Ott 1995: 355). At the same time, a first review of "national communications" from 15 industrialized countries clearly indicated that most of the countries were not even on the road toward meeting the existing commitments for stabilization.[17]

Even though the *inadequacy* of current commitments has been recognized by the majority of the Parties, there has been widespread opposition to further commitments. Some of the countries, especially Russia

---

17 This was documented in a first synthesis report concentrating on the adequacy of commitments (UN Doc. A/AC.237/81). As for CO_2 emissions, nine Parties projected an increase from 1990 levels in 2000 if no additional measures were taken, while five Parties projected either stabilization or reductions from that level. One Party did not provide 2000 figures but projected a 5% decrease from 1990 to 2005 (*UN Climate Change Bulletin* 1995).

and the OPEC countries, questioned the scientific basis for the statement that the commitments included in the FCCC were inadequate. Since signing the FCCC, the United States has insisted on the participation of developing countries in any next steps in the process. Developing countries have also opposed strengthened commitments, fearing that this will lead to new obligations on their part too. They have been stressing the requirement for industrialized countries to fulfil their current commitments before discussing new ones.

In need of a mandate that could take the process forward, the CoP-1 finally reached a compromise in the decision later designated the "Berlin Mandate" (FCCC/CP/1995/L.14), where it is explicitly stated that the current commitments under the Convention are not adequate for its objective to be reached. The Berlin Mandate was not decisive on future actions, however; it merely states that the Parties agree to begin a process that *may* lead to the adoption of a "protocol or another legal instrument" (*Earth Negotiations Bulletin* 1995b). The Berlin Mandate is also ambiguous as to whether future negotiations would be on specific targets and timetables for the reduction of greenhouse gas emissions. However, the document is explicit in stating that the process will not introduce any new commitments for developing countries, thereby confirming the primary responsibility of the industrialized countries. The AOSIS draft protocol is explicitly mentioned to be included for consideration in the process "along with other proposals and pertinent documents" (ibid.).

One of the most important decisions for the follow-up of the Berlin Mandate was the establishment of the open-ended "Ad Hoc Group on the Berlin Mandate" (AGBM). This group was given a deadline to complete the negotiations as early as possible in 1997 with a view to adopting the results at the CoP-3. The establishment of such a body should improve the prospects of reaching tangible results within the deadline.

(ii) *Criteria for Joint Implementation:* Joint implementation[18] of abatement measures was not high on the agenda in Berlin, but was much discussed during INC-7 to INC-11. In these discussions comments from the Parties were invited on such vital issues as objectives, criteria and operational

---

18 The concept of "joint implementation" refers to the idea that one country, facing high marginal costs on abatement efforts at home, provides assistance – technological, financial, administrative – to less energy efficient and less developed countries securing both a more efficient use of its resources and gaining some credit for its foreign abatement efforts.While this may be the essence of joint implementation, the practical content is by no means agreed upon. See Chapter 7 for an elaboration.

guidelines, functions and institutional arrangements, as well as on review and early experiences. While the FCCC clearly allows joint implementation to take place between Annex I Parties (Art. 4.2 (a)), discussions on broadening the concept to include developing countries was initiated at INC-8 in August 1993 (*Earth Negotiations Bulletin* 1995a). This raised the concern of developing countries, as they have tended to view joint implementation as a possible means for industrialized countries to avoid national action to meet existing abatement commitments, thereby also shifting the responsibility away from the industrialized world. There was an additional concern that joint implementation should become a substitute and not a supplement for funding and the financial mechanisms established under the FCCC.

Although jointly hesitant under the INC negotiations, there was a split within the G-77 as to whether developing countries should participate in joint implementation activities. Brazil, China and India remained critical, while some of the newly industrialized countries in South-East Asia and some Latin American countries were generally more favourable. At CoP-1, however, there was agreement to test the concept of joint implementation under a *pilot phase* that can include developing countries on a "voluntary basis", under the provision that "no credits shall accrue to any Part as a result of greenhouse gas emissions reduced or sequestered during the pilot phase" (UN Doc. FCCC/CP/1995/L.13; *Earth Negotiations Bulletin* 1995b). Simultaneously, the delegates agreed to change the name and refer to "activities implemented jointly" instead of "joint implementation", implying that there was still no acceptance on the criteria for joint implementation. Brazil remained the only major G-77 country strongly opposing participation of developing countries, whereas the United States continued to insist on emission credits until finally accepting the above-cited formulation. Hence, while the CoP-1 did not establish criteria for joint implementation projects in much detail, and while there was no agreement on implementation mechanisms, the issue of joint implementation of abatement measures was brought a step further.

(iii) *The Financial Mechanism*: The FCCC designates the GEF as the unit temporarily operating the financial mechanism for transfers from industrialized to developing countries. The GEF has been restructured and replenished for that purpose, including the creation of a powerful governing council and new voting rules (Sjöberg 1995). These reforms have brought the GEF several steps towards meeting the FCCC's criteria for the operating unit on questions of fair and equitable representation, transparency and universal participation. While the industrialized countries have wanted to make this arrangement permanent, developing countries still feel that representation in the GEF is not adequately fair and equitable – the main reason being that a double majority applied within the GEF gives a factual veto to the major donor countries. In the end, both sides agreed on a compromise to let the restructured GEF continue to function on an *interim* basis, thereby formally reserving the right to change the financial mechanism at a later stage.

It has been more difficult for the Parties to agree on questions concerning *guidance* on the operating entity, including programme priorities, eligibility criteria and policies, as well as the definition of the concept of "agreed full incremental costs". The status of the issue of adaption to climate change has been especially contentious, the AOSIS countries requesting action on the matter while, in contrast, the OECD countries have been averse because of the conceivably incalculable costs of such an obligation. Once more, a compromise was reached as the CoP-1 adopted a three-stage approach proposed by the INC.[19] As for the development of an operational strategy and initial activities in the field of climate change, the Conference decided to accept a proposal from the GEF Council on a two-track approach: First, the GEF Secretariat should attempt to develop a long-term comprehensive operational strategy; second, some project activities would be initiated to allow a smooth transition between the pilot phase and the restructured GEF (FCCC/CP/1995/L.1). The deadlock on the very important issue of defining the concept of "agreed full incremental costs" was not broken, however. Awaiting further development of guidelines, it was decided that the concept should be applied in a "flexible and pragmatic manner and on a case-by-case basis" (Oberthür and Ott 1995: 9).

(iv) *The Rules of Procedure*: Both the INC and the CoP-1 were unable to arrive at a consensus on the rules of procedure, the two major outstanding issues being the composition of the Bureau of the Conference and the voting procedures. The major obstacle was the demand of several oil-producing developing countries (especially Kuwait and Saudi Arabia) for a permanent seat in the Bureau. Combined with their insistence on consensus as voting procedure for the adoption of protocols and other substantive decisions, this would give the countries in question an effective veto over a more effective international climate change regime.

---

19 In stage one, particularly vulnerable areas and suitable policy options would be identified. Measures to prepare for adaption would be taken in stage two, whereas measures to facilitate adaption would be carried out in stage three (UN Doc. A/AC.237/91/Add.1.Rec.9).

At the same time, many OECD countries (especially the United States and France) sought a consensus rule on financial matters, without which they could be left without any formal possibility of blocking unwanted decisions. As the time for the CoP-1 deliberations was running out, it was decided to send the rules of procedure to the CoP-2 for further consideration. At the time, many observers were concerned about the effects that this stalemate would have on the future work of the CoP and its subsidiary bodies, especially concerning the negotiations on strengthened commitments.

(v) *Decisions on the Permanent Secretariat*: The Parties to the FCCC also had a hard time deciding on the location of the permanent secretariat, which had previously been agreed would start operating on 1 January 1996. Canada (Toronto), Germany (Bonn), Switzerland (Geneva) and Uruguay (Montevideo) had offered to host the Secretariat. Effectiveness criteria such as convenience of access by delegations to the Permanent Secretariat and meetings, possible budgetary savings by having the Secretariat located near other UN offices or secretariats, as well as contributions offered by the potential host government, were considered important when assessing offers to host the Secretariat. Since many of the Parties decided to make the decision an important prestige issue, it became very difficult to reach a consensus decision based on such "rational" criteria. Apparently, Germany took some advantage of its position as the host of the CoP-1 in its lobbying for support for Bonn, and some of the other candidates felt that it would be fairer to take the decision later at a neutral place. However, the Chairman insisted that the original agreement was for a decision to be taken at CoP-1. The Chairman proposed settling the issue by an "informal confidential survey" to help the Conference find a consensus solution. Carried out in three rounds, this "survey" made it possible for the Conference to adopt a decision accepting Germany's offer to host the Secretariat in Bonn (FCCC/CP/1995/L.12). Once this decision was taken, it was possible to make a final revision of the FCCC budget and to establish and determine the institutional linkages between the Permanent Secretariat and the United Nations.

(vi) *Other Matters*: Art. 12 of the FCCC establishes the mechanisms for *information gathering and reporting* on sources and sinks of greenhouse gas emissions, as well as on policies and measures taken by Parties for the implementation of their commitments. These mechanisms are considered very important for the effectiveness of the FCCC, in relation to both the choice of control measures and compliance control. According to the reporting requirements, the developing countries will first have to make their initial communication in 1997 (within three years after the entry into force of the FCCC for the country), as compared to the requirement that industrialized countries submit their first reports six months after the FCCC entered into force for them (Art. 12.5). As a follow-up to this, the CoP-1 adopted a decision stating that the guidelines for these reports should be developed by the Subsidiary Bodies. As mentioned, 15 communications had been submitted prior to the Berlin Summit.[20] In its first review of these communications, the Secretariat highlighted the insufficiency of the measures that had been applied by industrialized countries so far (Doc.A/AC.237/81). The debate at INC-11 made it clear that there were differing views among the Parties as to the frequency of reporting and whether Parties should be named or not. Methodological problems concerning measurements and comparability were also unveiled.

The two *subsidiary bodies* established by the FCCC also needed further elaboration as to their roles and their functioning. At the INC-11 it was agreed that the SBSTA should be responsible for providing and updating the required scientific knowledge for the CoP: that is to say, summarize scientific, technical and other information provided by the IPCC and other competent bodies; consider scientific, technical and socioeconomic aspects of the in-depth reviews of national communications; and carry out various tasks related to technology transfer. It was also decided that the SBSTA should deal with methodological questions concerning inventories, projections, effects of measures, adaption and so forth. As for the SBI, its task is to consider policy aspects of the in-depth reviews, effects on emission trends of steps taken by Parties, and any further commitments. Furthermore, the SBI has an advisory role on the adequacy of commitments, the financial mechanism and technology transfer. There was no agreement, however, on how the SBSTA and the SBI should relate to the IPCC. The solution to the question was delayed to the CoP-2 in Geneva, July 1996, when the subsidiary bodies' bureaus were asked to make a proposal.

The CoP-1 also adopted a number of decisions of an administrative and organizational character, among these a request for the Secretariat to produce an annual "Report on Implementation" (UN Doc.A/AC.237/91/Add. 1, Rec. 1), the adoption of the financial arrangements under the Convention (UN Doc. FCCC/CP/1995/L.4/Rev. 1 and L.2/Rev. 1), and a decision to review the situation with regard to the transfer of technology

---

20 France, Italy and the EU had still not submitted the required communications (Doc.A/AC.237/INF.16/Rev. 2).

at each session of the CoP. Reviewing the situation of technology transfer could probably have a positive effect, since operative commitments in this field seem distant.[21]

5.3. The CoP-2: Paving the Way for Specific and Binding Abatement Commitments?

The second session of the Conference of the Parties (CoP-2) met in Geneva on 8–19 July 1996. Gathering 1500 delegates and observers from 147 ratifying countries, 14 non-Parties to the FCCC, 9 IGOs, and 107 NGOs, the CoP-2 marked the mid-point in the schedule towards fulfilling the core objective of the Berlin Mandate: namely, to strengthen the abatement commitments of Annex I Parties (FCCC/CP/1995/L.14). Facilitated by the Second Scientific Assessment Report of the IPCC, and the work and input provided by the AGBM on likely elements of a protocol, the CoP-2 adopted a ministerial declaration – the Geneva Declaration – that not only endorsed the IPCC report as "the most comprehensive and authoritative assessment of the science of climate change", but specifically confirmed the findings of the IPCC that the continued rise in greenhouse gas concentrations in the atmosphere would "lead to dangerous interference with the climate system" (FCCC/CP/ 1996/15/Add.1). Equally important, the Geneva Declaration explicitly "instruct[s] representatives [within the AGBM] to accelerate negotiations on the text of a legally-binding protocol ... to be completed in due time for adoption at the third session of the CoP", including "quantified legally-binding objectives for emission limitations and significant overall reductions within specified time-frames" (ibid.). This diplomatic breakthrough was made possible through a significant shift in the position of the United States, which for the first time supported a legally binding agreement to fulfil the Berlin Mandate, on the condition, however, that "activities implemented jointly ... and international emissions trading must be part of any future regime" (FCCC/CP/ 1996/15).

While the Geneva Declaration can be interpreted as a signal that sufficient consensus has formed to accelerate the AGBM negotiating process towards enhanced and more specified abatement commitments, the

---

21 Commitments to technology transfer have generally been more difficult to put into operation than for example commitments to financial transfers, one of the reasons being that no consensus exists on the meaning of "technology transfers". The main concern of the major industrialized countries such as the United States, Germany, France and Great Britain is to fend off any attempt at weakening the "patent-rights" institution on which huge economic interests depend.

actual *outcome* of this process is indeed open. Sixteen countries, including Australia, New Zealand, Saudi-Arabia, Russia, and Venezuela, reserved themselves against the majority view of accepting the IPCC's scientific assessment as a basis for political action, and, furthermore, against legally binding emission targets (FCCC/CP/1996/15/Annex IV). These reservations to the Geneva Declaration more than indicated that the rift and differences in opinion had intensified. This was to be expected as the negotiating process increasingly moves from discussing topics of a diagnostic and general character to debating issues – procedural and substantial – of a much more specific nature, thus making the cost- and burden-sharing implications for each country much more evident. Approaching the point in time when the question of binding and more specific abatement commitments are to be decided, the political constraints within which the Parties to the FCCC operate will surface in various forms of conflict of interest (see Chapter 1, Section 7):

Besides the question of strengthened commitments, another issue concerns the rules to be applied by the GEF in determining the financial requirements for the implementation of the FCCC. This question was left unresolved by the CoP-2, clearly because the nature of these rules would directly influence the transfer of economic resources from the industrialized to the developing countries. Delegates from the developing countries expressed dissatisfaction with what they saw as an attempt by the industrialized countries to shift the burden of implementation away from themselves. Also unsettled was the question of the extent and manner in which activities of joint implementation should be part of a future strengthened agreement. Approaching the CoP-3 in Kyoto, December 1997, the negotiators face the challenge of reconciling the United States' position that tradeable emission permits be part of the climate change regime, and the developing countries' concern that such a mechanism would leave poverty-stricken developing countries without the emission permits required to facilitate economic development.

Another unresolved issue concerns the *rules of procedure* for adopting a protocol or another legal instrument that would strengthen the commitment to limit greenhouse gas emissions. There is no agreed-upon procedure for the adoption of such an agreement within the FCCC, and the AGBM has so far failed to reach an agreement. The main disagreement is between the majority view that decisions should be taken by a two-thirds majority (the practice within the UN system), and the minority view that substantive decisions should be made only when full consensus can be achieved. If the latter position wins it would imply that a small minority of states, including Australia, New Zealand and OPEC countries, would be able effectively to block the adoption of a strengthened agreement. This, in turn, would confirm what every diplomat has learned from experience: Rules of procedure may heavily influence the substantial outcome of political processes. It is fair to say that the failure to agree upon rules of procedure continues to dangle like the sword of Damocles over the negotiations.

6. Conclusion

Like most environmental problems, climate change was recognized by scientists long before it attracted the attention of decision-makers. While the first World Climate Conference in Geneva, 1979, marked the beginning of sustained international high level attention among scientists, the 1988 Toronto Conference was a breakthrough for climate change as an international political issue. From then on, the double-track processes of scientific problem assessment and attempts at political problem-solving gained speed. Established by the UNEP and the WMO in 1988, the IPCC produced its first scientific assessment report in time for the second World Climate Conference in October–November 1990. The IPCC report also provided important input to the negotiations starting up within the INC, a political body established by the UNGA in late 1990 to prepare a treaty on climate change in time for signing at UNCED in 1992. The FCCC was signed by 153 governments and the EU on 5 June 1992.

Hence, in less than five years, climate change went from merely attracting the attention of the international political community to becoming recognized as a serious problem deserving the adoption of a framework convention. The speed with which the FCCC was negotiated within the INC (15 months) was unprecedented and created expectations as to what could be achieved in the years to come. Although a rather weak and ambiguous convention, at the signing ceremony it was nevertheless clear that the FCCC ought not to be judged for what it was at the time, but for what it could contribute to future negotiations. In our view, the most important dimension of the FCCC is not the rather weak and ambiguous abatement recommendations made, but the institutions established in order to institutionalize a continuous negotiating process – including the establishment of the Secretariat, the SBST, the SBI and, most importantly, the CoP.

As per October 1996, the FCCC has been ratified by 154 states and the EU. While the unanimity of approval granted the FCCC is impressive, the 1995 Berlin Conference (CoP-1) and the 1996 Geneva Conference (CoP-2) demonstrated a considerable rift regarding the further development of the climate change problem-solving regime, which indeed is required if the FCCC is to become instrumental in curbing the present trend of increasing global greenhouse gas emissions. Although a large majority of states stood behind the 1996 Geneva Declaration, thus committing the Parties of the FCCC to adopt binding emission targets at the CoP-3 in Kyoto, December 1997, the CoP-2 also revealed that the Parties have encountered serious difficulties in agreeing upon (i) the specificity and level of emission targets; (ii) the mechanism for financial transfers to the developing countries; and (iii) the rules of procedure for adopting a protocol (or another legal instrument) under the FCCC. The increasing political tension is not suprising, given the fact that the Parties to the FCCC through the CoP-1 and CoP-2 have moved the international political process from the initial stage of demonstrating good will and establishing general principles for action to the much more committing and, by implication, potentially costly stage of adopting binding abatement targets. Reaching this critical phase in regime development, governments are bound to reconsider their preferences in view of the relative costs of action/inaction for *their own* countries. For the developing countries, the cost-benefit analysis may turn out to be even grimmer: If not compensated by the industrialized countries for the additional costs of abatement measures, governments in developing countries may be forced to give priority to more immediate issues of economic and social development over their long-term interest of mitigating climate change. This despite the probability that developing countries are more vulnerable to climate change than the industrialized world is.

The Parties to the FCCC will probably manage to adopt an agreement strengthening the commitment to curb greenhouse gas emissions. However, it remains a wide open question whether the abatement commitments actually agreed upon will be enough to substantially reduce the problem-solving deficit so evident in the climate change issue (see Chapter 1), and, moreover, whether the agreed commitments will be supported by instruments ensuring the actual implementation of abatement measures at the national level. The lack of supranational bodies in international politics, the ever-present incentive to harvest the gain while escaping the costs, and the uneven distribution of vulnerability to climate change in space and time are realities negotiators somehow must come to grips with if an effective international climate change regime is to evolve.

Chapter 3

Scientific Assessment of Climate Change

Bert Bolin

1. Introduction

Are we changing the climate on earth? This question is being asked by an increasing number of people today. Although we lack conclusive proof, there is increasing evidence that a human-induced climate change is on the way. How serious then is such a prospect? Scientists are unable to give a firm or clear answer, even though it is likely that the climate on earth may become significantly warmer than at any time during the past 100,000 years and with major impacts in many parts of the world. Some people may feel like responding by asking: Why should we really worry about it now? Others feel deeply disturbed about the possibility that we may actually be changing the global environment in a significant way.

The obvious conclusion is that we need to analyse the issue carefully, to try to determine how serious this threat might be and find the best ways and means of dealing with it. And we have also to remember that the possibility of surprises, unforeseen events, can never be excluded. As the scientific community has a special responsibility in this context, several questions arise: How can science best serve in clarifying these issues? How can scientists and politicians cooperate and still keep their different roles distinct? In other words: What questions should be considered by the scientists, and what are the value judgements that would better be left to the political process?

The following account of the assessment of current knowledge is based largely on the work of the Intergovernmental Panel on Climate Change (IPCC) (1990; 1992; 1994 and 1996). There have been some objections to this work. Different views in science should be respected and welcomed, if they are scientifically well founded. One reason why views become controversial, however, is that it is difficult for scientists as individuals to remain detached and not to become engaged in the issues,

socially and politically. In the latter part of this article I outline how the
IPCC has conducted its work, and address the issue of the role of science
in politics more specifically.

2. The IPCC Assessment of Current Knowledge

2.1. What Is Known?

We know that global mean temperature has increased by 0.3–0.6°C during the twentieth century (IPCC 1990, 1992, 1996). The seven warmest
years ever directly observed have occurred since the early 1980s (see
Figure 1). As far as we know, the last fifteen years on average would
seem to have been warmer than any period of that length during the
past 600–800 years.

We know that the average global temperature of the earth is dependent on the balance between incoming solar radiation and the return
flow of infrared radiation to space. This balance is highly dependent on
the presence of so-called greenhouse gases in the atmosphere, primarily
water vapour and carbon dioxide (CO_2). Indeed, our earth would be
uninhabitable if these gases were not there to improve the energy

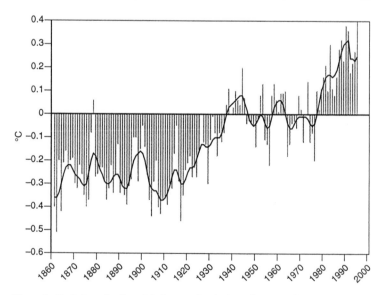

Figure 1: Variations in the global mean land air and sea surface temperatures
relative to 1951–80 averages. The solid curve has been produced by filtering to
highlight longer-term trends (Jones 1994).

balance significantly, yielding a climate about 30°C warmer than would otherwise prevail.

Furthermore, we know that water vapour is by far the most important greenhouse gas, accounting for more than 90% of the improved heat balance that is due to greenhouse gases. Humans cannot directly influence the amount of water vapour in the atmosphere, which is determined primarily by the atmospheric temperature. However, there is a positive feedback mechanism in that warming – for example due to increasing concentrations of other greenhouse gases – is likely to increase the amount of water vapour in the atmosphere and thereby enhance the direct warming due to human emissions of other greenhouse gases.

Moreover, we know that the amounts of greenhouse gases are being enhanced owing to human emissions. The concentration of carbon dioxide (CO_2) has increased by about 30% over the past 200 years (see Figure 2). Also methane (CH_4), nitrous oxide (N_2O) and CFC gases (chlorofluoro-carbons) have increased markedly; the total combined effect of all these greenhouse gases is at present equivalent to an increase of the concentration of CO_2 by 50% over the industrial level (see Figure 3).

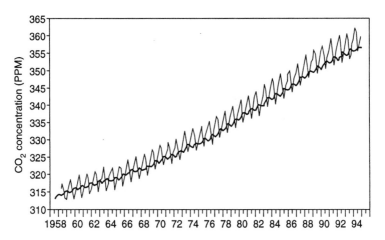

Figure 2: Changes of atmospheric CO_2 on Mauna Loa, Hawaii, 1957–1994 (Keeling et al. 1995). Preindustrial concentrations were about 280 ppmv. In addition to the regular seasonal variations due to photosynthesis and respiration in the terrestrial biosphere, we may note a slow increase with minor variations on the time-scales of 3–5 years. The most significant of the latter occurred in 1991–1994, probably related to the volcanic eruption of Pinatubo in the Philippines in June 1991, which initiated major anomalies of weather and terrestrial photosynthesis.

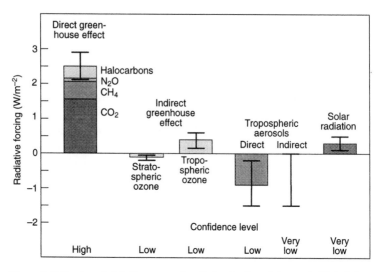

Figure 3: Estimates of globally averaged radiative forcing of the earth (W/m²) due to anthropogenic changes of greenhouse gases and aerosols from preindustrial times to the present. The heights of the bars indicate a mid-range estimate of the forcing, while horizontal bars show the possible range of values. A qualitative indication of relative confidence in the estimates is also given (IPCC 1994).

We can compute with the aid of theoretical models of the atmosphere that such an increase would in an equilibrium state cause a rise of 0.8–2.2°C in the global mean temperature, but also that the inertia of the climate system, particularly due to the great heat capacity of the oceans, has probably delayed such a change by 30–50%. Still, the observed change, 0.4–0.6°C, is in the lower range of what would theoretically be expected (0.4–1.4°C) if accounting for such a delay. These model computations also include the probable enhancement of the greenhouse effect because of the likely increase of water vapour in the atmosphere. This, however, has not yet been possible to confirm by direct observations because of the great spatial and temporal variability of atmospheric water vapour.

We know, also, however, that humans emit sulphur dioxide into the atmosphere when fossil fuels are burned, as well as other gases and particles due to biomass burning, particularly in the tropics. This enhances the atmosphere's content of particulates, i.e. aerosols.[1] They increase the

---

1 Sulphur dioxide is oxidized by solar radiation to sulphate, which forms small acid particles.

reflection of solar radiation and tend to cool the earth. Best estimates indicate that the global warming due to increased greenhouse gas concentrations may at present be diminished by 20–40% in this way. Also, as much as 90% of human emissions of sulphur occur in the Northern Hemisphere, which presumably is more affected than the Southern. The lifetime of aerosols in the atmosphere is short, merely a few weeks, and their spatial distribution therefore quite non-homogeneous. This means that their radiative effects do not offset the warming due to greenhouse gases in a simple way. It is important also to recall that the industrialized countries have decreased their sulphur emissions during the last decade quite significantly, because the sulphuric acid that is formed is a major cause of acidification of fresh water and soils. Developing countries, by contrast, are now increasing their emissions rapidly. As yet, they have taken only limited precautionary measures against pollution and acidification. In conclusion, we may say that if the role of aerosols is also taken into account, the mean temperature of the earth as a whole can be expected so far to have risen by 0.3–0.9°C. Further, this warming should exhibit marked spatial variations.

We know that the key greenhouse gases emitted into the atmosphere stay there an average of 50–100 years or more – except CH_4 which has a lifetime of 14 +/- 3 years. Aerosols, on the other hand, disappear from the atmosphere quickly. This implies that they will be present only as long as emissions continue. Aerosol concentrations would therefore decrease rapidly if measures were taken to curb the use of fossil fuels. CO_2, however, would remain in excess for a long time. In such a case we might therefore see a gradual unmasking of the warming due to greenhouse gases that may now be hidden because of the presence of aerosols.

We should also note that the decline of stratospheric ozone (O_3) due to emissions of CFC gases partly compensates for the warming caused by these emissions, as ozone is also a greenhouse gas. A change of climate might well enhance the rate of decline in biodiversity. Atmospheric pollution enhances the ozone in the low troposphere, which reinforces the warming. This brings us to an important point: the major global environmental issues are all interrelated. That must be borne in mind in developing a strategy for combating them.

The climate system is very complex. Even the most advanced models describe key processes in a necessarily simplified way. Surprises cannot be ruled out; they may enhance or temper the changes as we imagine them now.

Past emissions of greenhouse gases and aerosols are reasonably well known. A number of research groups have used climate models to

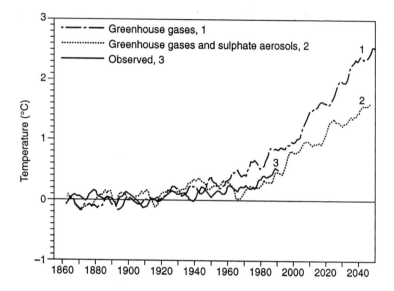

Figure 4: Model projections of globally averaged changes in surface air temperature, 1860–2050, and observed changes, 1860 to the present day (Mitchell et al. 1995).

simulate the expected changes in the climate as a result of their gradual increase until today (see Figure 4).[2] Today we may be on the point of circumventing the "noise" in the temperature record due to the natural and largely unpredictable variations that characterize the climate system. Different models have different sensitivities to changes in the radiative characteristics of the atmosphere. Calibration of the models against past observed changes will soon be possible. Despite these uncertainties the IPCC has concluded it is likely that at least some of the observed rise in global mean temperature during the twentieth century is due to human emissions of greenhouse gases.

2.2. *Complexity and Chaotic Behaviour*

In trying to understand the climate change issue we are concerned with the interaction of two very complex systems:

---

2 Key research groups engaged in climate modelling experiments include: Hadley Centre for Climate Prediction and Research, Meteorological Office, Bracknell, United Kingdom; Max Planck Insitute for Meteorology, Hamburg, Germany; and the Geophysical Fluid Dynamics Laboratory, Princeton, NJ, USA.

- The global climate system, including the atmosphere, the oceans with their marine ecosystems, land with its topsoils and terrestrial ecosystems, and the crysosphere – continental ice sheets, glaciers and sea ice.

- The global socio-economic system: the people of the world, grouped in a number of basically very different countries, and their infrastructures as they have evolved over centuries.

Such systems are characterized by inherent instabilities, the occurrence of non-linear interactions and oscillations, as well as the possible existence of several equilibria, around and between which they may oscillate. For these reasons they are commonly termed chaotic. The predictability of their behaviour is limited, although the range of variations of key variables is usually reasonably well constrained by the overall characteristics of the systems. Important in this context are external conditions, such as the forcing of the climate system by solar radiation, the distribution and topographic characteristics of land, the bottom configuration of the ocean basins, etc. None of these has changed or will change on the time-scales with which we are concerned.

Instabilities may occur on all time and space scales in a complex and chaotic system. Energy transfer from one scale to another implies, for example, the possible activation of large-scale instabilities by small-scale processes. We can take the hurricane as a typical example: Small-scale convection over the tropical ocean becomes organized into a large-scale structure that permits a formidable release of kinetic energy. Similarly, the development of worldwide economic depressions shows how more limited economic crises may become of global importance through an interplay within the socio-economic system. Old societal structures collapse and make room for new developments. Our ability to predict such changes will always remain limited.

Is it then at all possible to develop concepts and models that can deal with such complex systems in a meaningful way? Will we be able to consider their interactions, and expect useful results? The answers are not simple. Some aggregate features may be predictable, whereas detailed changes probably often can only be derived statistically. We can explore and possibly define the most likely ranges of such changes. Likewise, to distinguish between the natural variabilty of such systems, and possible responses to external forcing such as the continuously increasing emissions of greenhouse gases, is another task that might be possible to address. As will be accounted for in the next section, the climate system may now be on the verge of drifting out of the range of natural variability observed in the past. This is the time when validation of climate models against observations is becoming possible and their reliablity can be assessed. Still, however, natural stochastic variations of the system mean that full predictability can never be achieved.

These features of the climate change issue have not been adequately recognized and properly considered in the public debate. Uncertainty is a reality, and when scientists refer to it, this must not be interpreted as attempts to exaggerate uncertainty unduly in order to secure continued funding, as has at times been indicated (Boehmer-Christiansen 1994). Nor should it be used as an excuse to reject the general conclusions so far reached by the scientific community. We simply must try to understand the basic features of the climate system better.

2.3. What Can Be Told About the Future?

Scenarios of future emissions have been deduced on the basis of assumptions about a reasonable economic development in the world during the next 50–100 years. While it is not possible to foresee changes in the world far into the future with much accuracy, it is meaningful to try to establish a likely range of future changes that will occur if no preventive measures are taken. On the basis of a set of such scenarios, climate model computations have been performed to project likely future changes of climate, to enable us to assess possible impacts.

Changes due to a central projection of future emissions and the associated global distribution of temperature change by 2050 have been derived by the Hadley Climate Centre in Bracknell, UK (Mitchell et al. 1995). It is interesting to note the differences obtained between experiments that include or do not include aerosol emissions. The inclusion of aerosols yields a better fit between computed and observed changes in the past, which in turn gives better credence to projections into the future. Similar experiments and analyses have been carried out at the Max Planck Institute for Meteorology in Hamburg, Germany (cf. Hegerl et al. 1996).

It is important first of all to emphasize that the results obtained so far are but examples of possible future developments: they are not predictions. The uncertainty of a climate projection based upon a given scenario of future concentrations of greenhouse gases and aerosols in the atmosphere is about 50%. In addition, there have in the past been temporal variations of the order of 0.2°C on time-scales of a few decades to a century, which we cannot explain and which may well be random and unpredictable. Such natural variations are likely to occur in the future as well. This natural-variability will always partially obscure a possible signal of a human-induced change, making assessments of the impacts of climate change difficult.

On the basis of quite a large number of climate change experiments, however, we can state with some assurance that warming is likely to be quicker over land than over sea. Maximum warming can be expected in high northern latitudes in winter, whereas initially little warming is foreseen over the Arctic in summer. The hydrological cycle would probably be intensified: this would mean increased likeliness for heavy precipitation events, but the simultaneous enhancement of evaporation might still imply a greater risk for droughts elsewhere, and such events might also last longer.

Because of the difficulty of projecting the likely geographical distribution of a climate change, it is not yet possible to foresee the specific impacts on countries and people. But we note that (i) sea level is expected to rise by 20–100 cm during the next century and will continue to rise, even if the climate stabilizes; (ii) climatic zones will shift and influence agriculture and forestry, which may hit subtropical developing countries particularly hard; (iii) water scarcity may well become a more urgent problem in some parts of the subtropics; (iv) diseases, like malaria, may spread to regions not now vulnerable.

Because people and human societies have adapted reasonably well to the prevailing climate, any change would primarily have negative effects – although not exclusively so. There is still considerable uncertainty about how quickly impacts will occur and how severe they may become. We cannot escape the conclusion, however, that *uncertainty does not reduce the risk of significant impacts due to a climate change. Uncertainty only makes it more difficult to assess how serious this issue might be.*

In complex analyses such as those that serve as a basis for the IPCCs' assessments, we can never exclude the possibility that new insights will be gained that justify changes of earlier conclusions. The possibility of surprises can never be excluded, and we cannot know *a priori* whether such events will make the issue more or less urgent than now envisaged.

2.4. Stabilization of Greenhouse Gas Concentrations and Climate

The Climate Convention has set as its prime objective:

> Stabilization of greenhouse gas concentrations in the atmosphere at a level that would prevent dangerous interference with the climate system (Art. 2).

The interpretation of the word "dangerous" is of course not a scientific task, but a political one. However, scientists can assemble information that might be of importance in this context. Before turning to the issue of how best to handle the climate-change issue, we need to understand better the difficulties and implications of trying to stabilize atmospheric greenhouse-gas concentrations.

Let us first consider the stabilization of CO_2. Quantitative models are available that describe the interactions of processes that regulate the distribution of carbon between the main natural reservoirs, the atmosphere, the oceans and the terrestrial ecosystems on land. We may then ask: *Which emission scenarios would lead to stabilization of atmospheric carbon dioxide in the future at some prescribed concentration levels?*

The outcome of one such analysis is available (IPCC 1994; 1996) (see Figure 5). It should be noted that the preindustrial CO_2 concentration was about 280 parts per million (ppmv); at present it is about 360 ppmv. Emissions due to fossil-fuel burning are currently about 6.1 Gt C/year and those due to deforestation and changing land use about 1.1 Gt C/year: in total about 7.2 Gt C/year (1 Gt C = 109 tons of carbon). Doubling of atmospheric concentration implies (according to model computations) an increase of the mean global equilibrium temperature by

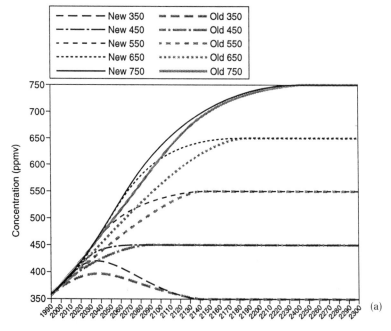

Figure 5: (a) Assigned profiles of the atmospheric concentration of carbon dioxide leading to stabilization at 350, 450, 550, 650 and 750 ppmv. Doubled pre-industrial concentration is 560 ppmv.

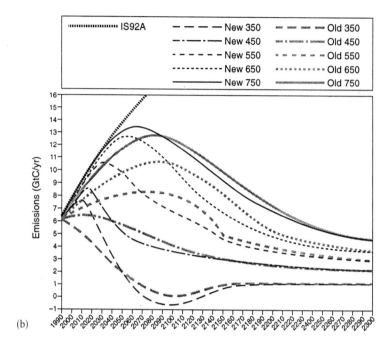

(b)

Figure 5: (b) Anthropogenic emissions leading to stabilization of concentration as given in (a) using a mid-range global carbon cycle model (IPCC 1994). IS92a and 92e are two IPCC scenarios that represent a central and a rapid growth scenario, assuming that no preventive measures are implemented (IPCC 1992). The estimated uncertainty range due to model approximations is represented as a shaded area for the 450 ppmv scenario.

1.5–4.5°C. To provide a broad perspective, it has been assumed that stabilization at the alternative levels of 350, 450, 550 (which is close to doubling), 650 and 750 ppmv might be aimed at. The associated emission scenarios can then be computed and compared with the central IPCC scenario (IS92a), assuming no interventions to reduce emissions.

Different stabilization paths might, of course, be chosen. Permitting a more rapid increase of the concentrations for another decade or two would mean that emissions would continue to increase as in scenario IS92a, but the ultimate decrease of emissions in order to stabilize concentrations would have to begin earlier and be faster (cf. Raper et al. 1995).

The key factor in the stabilization scenarios is the level of stabilization and the associated accumulated emissions over time (see Figure 6). The uncertainty of the figures is about 20%, increasing with time and level of stabilization. The general features are, however, robust. It is striking how slowly the curves for the different scenarios diverge with time – a result of both the slow response characteristics of the carbon cycle, and the necessarily gradual stabilization and ultimate decrease of emissions that is the only realistic assumption. The accumulated emissions for scenario IS92a are above all stabilization scenarios within a few decades, the more so the lower the level chosen for stabilization.

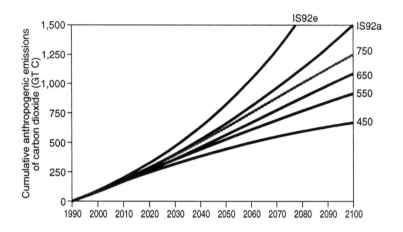

Figure 6: Accumulated emissions of carbon dioxide corresponding to the different cases of stabilization as given in Figure 5.

For comparison, the accumulated emissions due to fossil-fuel use from the middle of last century until 1994 have been about 240 Gt C, and those due to deforestation/changing land use 80–120 Gt C. The known and estimated stocks of oil and natural gas contain 150–250 and 120-250 Gt C respectively. These figures might be larger, but further extraction is likely to be considerably more expensive. The amount of carbon in coal stocks is probably at least ten times larger. There are plenty of fossil fuels available, permitting even a trebling of atmospheric concentrations.

We should also note that world population is increasing and will continue to do so. According to UN analyses it is expected to reach about 9,500 millions in 2050 and about 11,300 millions in 2100, with most of the increase in the developing countries. On the basis of available projections we can estimate approximately the globally averaged *per capita* emissions that are permissible in order to achieve the alternative stabilizations as assumed (see Figure 7). The uncertainty of the curves is of course considerable also because of the uncertainty about population scenarios, but the diagram still illustrates the stringent limitations of the emissions imposed by nature, particularly for stabilization below an approximate doubling of carbon dioxide.

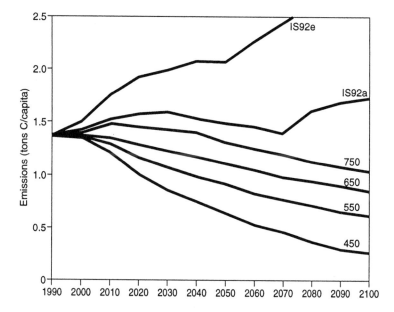

Figure 7: Projections of per capita emissions of carbon dioxide based on the emission profiles leading to stabilization as given in Figure 5 and an assumption of world population changes according to UN estimates.

It should be recalled that at present about one quarter of the world population is living in the industrialized countries (including countries currently in economic transition). Their *per capita* emissions are on the average about 2.8 ton C per year, with emissions from individual countries varying between 1.5 and 5.5 ton C per year. Developing countries, on the other hand, emit on the average merely about 0.5 ton C *per capita* and year, varying for individual countries between 0.1 and above 2.0 ton C (see Figure 8). If we were to aim for stabilization at or below doubling of pre-industrial atmospheric concentrations, emissions might temporarily somewhat exceed present global average *per capita* emissions,

but only for a few decades, and would have to be at or considerably
below that level in 2050 and beyond. It is simple to deduce, from the
curves in Figure 8 and the expected increase of the population of developing countries, that the developing countries on average would not be
able to more than about double their present *per capita* emissions (to
about 1.0 ton *per capita* annually), before emissions would have to be
reduced, if the target were to stabilize at twice pre-industrial concentrations. This maximum is only about one-third of the current average *per
capita* emissions in the industrialized world.

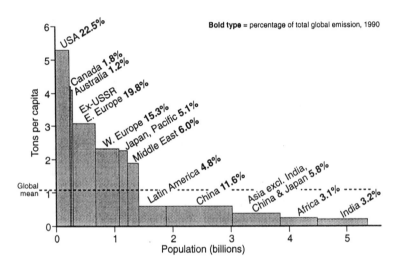

Figure 8: Per capita emissions of carbon dioxide due to fossil-fuel emissions in
1990 for various regions (vertical co-ordinate), plotted against accumulated population from the region with the largest per capita emission to that with smallest
emissions (horizontal co-ordinate). Areas shown for different regions are proportional to total emissions in these. Percentages of total global emissions for the different regions are also given.

Even if an agreement were reached on stabilization levels of 650 or
750 ppmv, the limitation of future emissions would have to be stringent;
but such higher levels would of course also provide more flexibility. Thus
energy for development would soon have to be supplied from sources
other than fossil fuels, if the developing countries are to be able to aim
for an energy supply of the magnitude that the industrialized countries
have enjoyed for the past four to five decades. The implications is that it
would not be possible to use most of the coal resources. The obvious
question arises: When will the countries with major coal reserves – e.g.

China, and also the USA and Australia – refrain from using them, and what primary energy will be used instead?

Finally, we should note that CO_2 is but one of several greenhouse gases whose concentrations are increasing due to human activities. Even if the atmospheric concentrations of these other gases (primarily CH_4, N_2O and O_3) were to be stabilized at present levels, CO_2 must not exceed about 500 ppmv if the total greenhouse effect from exceeding that equivalent of doubling of CO_2 is to be prevented. If the total greenhouse effect were to be permitted to reach a level equivalent to a CO_2 concentration of 750 ppmv, then CO_2 concentrations would have to be stabilized at about 650 ppmv.

2.5. Is It Urgent to Act?

First, we will have to rephrase one important aspect of the key findings presented above: The climate system has characteristic response times to external forcing, such as increase of greenhouse gas concentrations and the induced global warming, of several decades to a century. This conclusion is drawn from climate model experiments involving both natural and human-induced, stable and radioactive, tracers, in the atmosphere and in the oceans. It is well founded and broadly accepted, and implies that human interference with the system will be noticeable only gradually. Moreover, it will necessarily take a long time to stabilize greenhouse gas concentrations in the atmosphere. In other words, *from a practical and political point of view, any major change of the climate system is virtually irreversible.* Admittedly, we do not know well the impacts of global climate change, but there seems to be a considerable risk that at least some regional changes may be serious.

One political option would be to take action only to the extent that there would be other reasons for doing so, and to delay further steps until it can be shown that the impacts indeed appear to become serious. This approach is often called the *"no-regret policy"*, since there would be no direct costs during the initial phase to be ascribed to mitigation of the increase of greenhouse gas. We would have nothing to regret if the seriousness of climate were not confirmed. However, more rapid and presumably also more costly responses might well be required if damages due to climate change should prove to be serious. On the other hand, perhaps more advantage could then be taken of technological innovations developed in the meantime.

Another alternative would be to begin implementing measures that would be least costly soon and assume that society would be ready to pay for an insurance policy. This is the so-called *precautionary principle.*

As a matter of fact, many initiatives can be taken now that are not very costly, and some might even be cost-effective.

Economic considerations should certainly enter into mitigation decisions. It is extremely difficult, however, to weigh the costs of the damage that may be avoided decades to a century into the future by taking early measures, against the costs incurred to society by taking preventive measures now (IPCC 1996, III). To avoid the costs caused by discarding past investments prematurely, renewal of the infrastructure should be gradual – another argument for getting started early.

However, many impacts due to climate change cannot be evaluated in terms of money. Strict cost-benefit analyses are therefore difficult and would probably be of limited value (see Chapter 5). Decisions may instead have to be based on broad judgements of the issues at stake and taken stepwise. The almost irreversible character of climate change then also speaks in favour of early, preventive measures.

That international agreements are desirable is obvious. It will be necessary to harmonize measures and agree on how to share the burdens, which includes issues of financial assistance and technological transfer from industrialized countries to developing countries. It is, however, also obvious that most countries are considering the climate issue on the basis of their own assessments of what might be best for *their* national safety and well-being. The perspective of a long-term truly global risk has not yet featured prominently in political negotiations.

It is generally agreed that energy is not provided and used today in an optimal and efficient manner. Even though efficiency is constantly being increased, much more can be achieved. Because energy use is still very limited in most developing countries, it is, however, not likely that in the long term improved efficiency as such would suffice to limit the future use of fossil fuels and to stabilize greenhouse concentrations in the atmosphere. The *rate of increase* might be reduced, however, and that could give more time for the development of long-term substitutes for fossil fuels. The different attitudes to the climate issue in different countries imply that it will take time to reach international agreements about how to share responsibility for future actions (see Chapters 1 and 6).

Let us recall the scope of the issue. Supply and use of energy in the world today amounts to about 5% of the world economy. The capital stock of past investments is also huge and its characteristic renewal time long, perhaps on the average about 30 years. Clearly, change to other energy supply systems will necessarily take time if capital destruction is to be avoided. Furthermore, fossil fuels have become the prime source of energy (75–80% of the total) because they are comparatively cheap. The costs of oil and natural gas might, however, increase 20–30 years into the next century because of increasing costs for their extraction. On the other hand, new technological inventions might permit more complete exploitation of known resources at competitive costs than is now possible.

To judge how urgent it is to initiate more stringent restrictions on emissions of greenhouse gases into the atmosphere is of course a political issue. The remarks made above, as well as the more complete assessment of socioeconomic factors that the IPCC has attempted (IPCC 1996: III) and that need consideration in this context, might be helpful for national delegates negotiating within the Climate Change Framework Convention (FCCC). These began in August 1995 and are to be completed by the end of 1997, according to a decision taken at the first Conference of the Parties to the FCCC in Berlin, March–April 1995.

3. The Scientific Community in the Politics of Climate Change

3.1. The Climate Change Issue Emerges Politically

The possibility of human-induced change of climate first caught the attention of politicians in the late 1970s. In 1979 the United States National Academy of Sciences (NAS) took the initiative to assess available scientific findings – partly as an outcome of the activities of the task force on issues of "quality of life" established under President Jimmy Carter. I myself served on the small NAS committee that worked for one week under the chairmanship of late Professor Jule Charney. It issued a short report that in essence endorsed the general conclusions that so far had been reached by the scientific community, particularly those based on computations with global climate models developed at the Geophysical Fluid Dynamics Laboratory in Princeton, New Jersey. (Manabe and Welherald 1975). The likely delay due to the inertia of the climate system, primarily caused by the oceans, had at that time received scant consideration; and this inadequacy of existing studies was emphasized in the assessment.

Soon thereafter I was entasked to carry out a similar brief *international* assessment, supported by the International Council of Scientific Unions (ICSU), the United Nations Environmental Programme (UNEP) and the World Meteorological Organization (WMO). In 1982, I initiated a much more thorough study, supported by the same three organizations. The work was carried out at the International Meteorological Institute in Stockholm during the years 1983–1985, with broad participation of key scientists worldwide (Bolin et al. 1986). An international conference in

Villach, Austria, in 1985 concluded the work. Scientists from some 30 countries attended, and the outcome was an appeal to the world to address seriously the issue of human-induced climate change.

In the course of this study, some material was also made available to the UN Commission on Sustainable Development, which was simultaneously at work under the chairmanship of Gro Harlem-Brundtland. The climate issue was considered by the Commission and was brought up in the debate of the report at the UN General Assembly in 1987 (World Commission 1987).

These developments set the stage for more concerted efforts at the political level, as well as for the creation of the Intergovernmental Panel on Climate Change (IPPC). The Villach report had shown convincingly that concentrations of a number of greenhouse gases in addition to CO_2 were increasing due to anthropogenic emissions. Their enhanced concentrations were deemed almost as important in forcing the climatic system as the increase of CO_2. The likely time for a possible change of climate was no longer the latter part of the twenty-first century, but perhaps only a few decades into the next century.

Both the 1986 report and the first IPCC assessment 1989/90 (IPCC 1990), however, also stressed the uncertainties of projections into the future. Continued research would be required in order to permit more firm conclusions. It is hardly surprising that a surge of new research projects came to demand attention from funding agencies. This was particularly so in the USA, where there was a more flexible and responsive organization for research funding than in many other industrialized countries.

3.2. The Intergovernmental Panel on Climate Change

3.2.1. The Formation of the IPCC

The IPCC was created on the initative of the Executive Director of UNEP, Dr Mustafa Tolba and the Secretary General of the WMO, Professor Patrik Obasi, to assess available scientific knowledge relevant to the issue of a possible human-induced climate change. This was in part a response to national initiatives, particularly from the United States (cf. Boehmer-Christiansen 1994).

The terms of reference agreed to by the two organizations were quite general, but from the very beginning there was recognition of the need to include socio-economic analyses of the implications of climate change.

The IPCC was created as an *intergovernmental* body, to be guided by its plenary sessions of government representatives. As chairman of the IPCC, I soon recognized the value of working with representatives of governments, but also sensed the need to retain an independent status for the panel in order to get the very best scientists engaged in the assessment work regardless of their affiliation. However, top-rate scientists might not be willing to become involved in cumbersome and time-consuming bureaucratic procedures. It therefore seemed most important to create a working level within the IPCC structure that could be characterized by an open, lively and genuinely scientific debate.

A step in this direction was the formation of three working groups, chaired by respected scientists. The wide range of topics to be analysed had to be recognized. The fields of responsibility were divided as follows: Working Group I – Science (Chairman: Professor, Sir John Houghton, UK); Working Group II – Impacts (Chairman: Dr Yuri Izrael, Soviet Union); Working Group III – Response Strategies (Chairman: Dr Fredrick Bernthal, USA).

It gradually turned out, however, that the difficulties encountered were of a far more complex and fundamental nature than was initially perceived. Working Group I could readily attract leading scientists from all over the world and could also build on past assessments. But Working Group III, in particular, had to fight a tough battle to keep scientific issues separate from political issues. The working group structure was modified after the first round of assessments leading up to the UNCED in 1992. Working Group II was given a more precise mandate: scientific-technical assessments of impacts and policy options, and the task of Working Group III was changed to consider available scientific analyses of the socio-economic implications of climate change.

3.2.2. Organization of the IPCC Assessment Work

No single scientist can be an expert on the entire field of scientific research relevant to the global issue of climate change. An assessment will therefore require a breakdown of the subject matter into several separate fields of research. In fact, in the course of organizing their work for the Second Assessment Report, the three IPCC working groups found it necessary to define altogether 50 such fields to deal with the assessment satisfactorily. These are presented as separate chapters of the three working group reports.[3] A team of scientists ("lead authors") was selected for each of the chapters in the assessment reports and the composition of the group has to reflect different views to the extent this can be foreseen. The chapters summarize the state of knowledge as available in the

---

3 The report of Working Group I contains 11 chapters, that of Working Group II, 28 chapters, and Working Group III, 11 chapters.

published scientific literature. Furthermore, controversial views have to be considered, assessed and described. The drafts of the chapters were circulated to scientists in the respective fields for review, and the lead authors were obliged to consider and respond to comments that were scientifically well founded.

The second step of the assessment is a review of the draft chapters by government experts, at which time a short "Summary for Policy-Makers" is also circulated for consideration. The latter is prepared by the respective Working Group Bureaus and the lead authors of relevant chapters. This summary is to be based on the chapters and extract information of particular relevance to policy-makers. This, of course, is not exclusively a scientific task, but a first draft of such a summary is meant to serve as a basis for discussion in the Working Groups. The Summary for Policy-Makers is approved in a plenary session of the respective working groups, which are attended by government experts. On those occasions some lead authors are available as experts, with the obligation to point out possible discrepancies between scientific findings as described in the chapters and the conclusions in the Summary for Policy-Makers that is emerging in the course of the approval process. Neither the chapters nor the summaries to policy-makers should recommend specific actions, since this necessarily requires political considerations. Conclusions should instead be expressed in terms of consequences of alternative assumptions about future changes.

On the other hand, the plenary sessions do not consider the individual chapters of the report for approval. These remain under the authorship of the lead authors and are published as such. This procedure aims at ascertaining the scientific credibility of the assessment reports. It is also meant to prevent, as far as possible, political considerations from influencing the scientific assessment.

The Working Group plenary sessions become the occasions when the scientists engaged by the IPCC are confronted with political views and even attempts to clothe special interests in scientific terms. It is essential that the line between factual information and political judgement is upheld at that time. Not surprisingly, approval of the Summaries for Policy-Makers can become difficult and takes time. It is for others to judge how well the IPCC has succeeded in producing balanced assessments that are well founded in the available scientific literature. Here let us simply note that the IPCC process enjoys the informal approval of a large part of the scientific community that has contributed to the research relevant for the climate change issues.

3.2.3. Participation in the IPCC Process

Climate change is a genuinely global issue. Possible future actions to mitigate or adapt to climate change and issues of burden-sharing will engage all countries of the world. Political negotiations should build on factual presentations of available knowledge. Such presentations are likely to gain political recognition the more scientists, from many countries all over the world, are given the opportunity to participate as members of the lead author teams and to serve as scientific/technical experts in the review process. However, there are major differences between countries with regard to their support of research relevant to the global climate change issue, particularly between the developing and the industrialized countries. The IPCC recognized early the need to engage scientists from developing countries, not least in order to provide opportunities for them to work with key scientists from the industrialized world and to become better acquainted with the relevant scientific literature. An IPCC committee was formed already at the second IPCC session in 1989 to provide a forum for discussions of this kind.

The modifications of the IPCC rules of procedure agreed in 1993 in preparation for the work towards a Second Assessment Report formalized further the procedures to ensure broad participation of scientists from developing countries. According to the new rules, there should be at least one expert from a developing country in each lead-author's team. The IPCC has received some support from the Global Environmental Fund GEF of the World Bank to improve such broad participation. Still, it will take time to achieve geographically well-balanced lead author team. This requires long-term capacity-building in developing countries.

The IPCC process of assessment is also open to non-governmental organizations (NGOs). On a few occasions some of their scientists have served as lead authors, but above all they have been involved in the review process. As is the requirement for other expert reviewers, their comments and proposals for modifications are to concern factual issues based on the published scientific/technical literature. There is, however, a clear danger that non-scientific arguments may be brought forward. My personal judgement is that the final outcome of the assessment process has been influenced only marginally by political considerations.

3.2.4. Can the IPCC Assessment Process Be Further Improved?

The view has been expressed that the IPCC process is cumbersome and that it cannot respond quickly when necessary, for example in serving the process of negotiations within the FCCC. This may be true, since a full assessment can hardly be completed in less than 18 months. This, however, may be the price we have to pay in order to secure accurate information well anchored in the scientific community. This time can hardly be much reduced without compromising the integrity of the final reports.

The IPCC is, however, considering the possibility of initiating special assessments when the need arises to assess the reality and implications of some recent and remarkable scientific findings. The scope of such reports would then instead correspond to a single chapter in the full assessment reports and might be completed in perhaps a year.

Another need might be more urgent. The full assessments are sometimes difficult to understand for those who need the information. Moreover, they may not be given in a form directly useful in political discussions, since they are primarily supposed to serve as authoritative scientific overviews that maintain the credibility of IPCC efforts in the eyes of the scientific world. The Summaries for Policy-Makers, on the other hand, are often too condensed to provide answers to specific questions that may arise in the negotiating process.

It has therefore become desirable for the IPCC to initiate the preparation of reports in which specific scientific questions of interest for the ongoing negotiations would be defined jointly by the IPCC and the Subsidiary Body for Scientific and Technological Advice (SBSTA) of the Convention, and analysed by a comparatively small team of scientists commissioned by the IPCC. The resulting reports would be based on comprehensive assessments that have been peer-reviewed. The review process is thereby simplified and the time required to complete such reports reduced substantially. Even more importantly, direct cooperation would be established between a central group representing the political process (of the Conference of the Parties) and a small group of scientists from the IPCC – both with their respective roles clearly defined.

3.2.5. IPCC Assessment and International Research Planning

It was agreed as early as in 1989 that international planning and co-ordination of research should not be a task for the IPCC. Several international organizations were already in existence for this purpose – for instance, the World Climate Research Programme (WCRP), the International Geosphere Biosphere Programme (IGBP) and the Human Dimensions Programme on Global Change (HDP). Of course, some scientists have been engaged in both kinds of activities, but still a clear separation between the two objectives has been maintained. The work of the IPCC is aimed at providing information into the political process of negotiating international agreements, while on the other hand international research planning and coordination should provide a basis for national funding of global research efforts, in which coordination between funding agencies in different countries is often desirable.

3.3. Lessons From the IPCC Assessment Process

The IPCC assessment process is a novel approach intended to provide scientific information in an unbiased form to serve international political negotiations. It has been successful in the sense that, only some two years after the IPCC was formed, did the UN General Assembly agree on the basis of the IPCC First Assessment Report to create an Intergovernmental Negotiating Committee (INC). The FCCC was proposed by the INC to UNCED in Rio in June 1992 and signed by 154 countries on that occasion. Similarly, ratification of the Convention was remarkably quick; it could enter into force less than two years later. A first session of the Conference of the Parties (CoP) of the Convention was held in Berlin in March–April 1995. However, a more relevant question to ask in this context may be: How well anchored in the scientific community are the conclusions reached by the IPCC?

3.3.1. Scientific Controversy

Scientists do not easily agree on scientific matters. And, indeed, controversy is fundamental for scientific progress. To be constructive, debate should be based on views and results that are well documented in the scientific literature. Various views have been brought forward in the course of the IPCC assessments: some were rejected because they were not based on careful scientific analyses, while others have been analysed and well justified views have been accepted. For practical reasons, the teams of lead authors have been limited, but one or several dozens of reviewers have been real contributors to each IPCC assessment chapter. A few examples may be of interest in the present context.

Firstly, there is the question of the magnitude of the positive feedback mechanism due to an increase of water vapour when climate becomes warmer. This fundamental issue has attracted considerable attention (Sun, De Zheng and Lindzen 1993; IPCC 1992, 1994). Our knowledge is still quite incomplete here, which is reflected in the large range that the IPCC has assigned to the sensitivity of the climate system to changing concentrations of greenhouse gases. A low sensitivity implies a slight feedback as advocated by some, but recognition of such a value does not exclude the possibility that the feedback could be quite large, as shown by many model experiments. The possibility of a negative feedback, as advocated by Sun et al. (1993), was considered unlikely by the IPCC, however, and rejected. Very few have disputed this IPCC conclusion. Continued research will most likely resolve this issue some day, but for the time being considerable uncertainty remains.

Another controversial issue is the reliability and interpretation of the global mean temperature curve since the mid-1800s. It should first of all be clear that a change due to increasing greenhouse gas concentrations in the atmosphere cannot have been more than a tenth of a degree, until well into the twentieth century. About 70% of the present radiative forcing due to human emissions has actually occurred since 1950. Earlier temperature variations as seen in the record must therefore have been primarily of natural origin, not least due to volcanic eruptions. A recent and well-documented example could be noted for the period 1991–1994, caused by the June 1991 eruption of Pinatubo in the Philippines. Furthermore, some of the variability as shown in the record well into the present century may well not be reliable because of the paucity of the data. This is a serious dilemma, since our ability to distinguish a human-induced signal in the record depends on the magnitude of the "noise" due to natural variability. At any rate, a rather long time-series is required to establish a trend with any reliability. This is the reason why a trend derived from satellite data from 1979 to 1990 cannot be reliably interpreted as determining the magnitude of a human-induced trend.

Many other controversies have been carefully analysed, but not always resolved because of inadequate understanding. They may then be reflected in the IPCC reports as uncertainties. In general, we may say that *hardly any articles published in scientific journals which apply a peer review procedure reject the IPCC's conclusion that continued atmospheric emissions of greenhouse gases will lead to a significant increase of the global temperature and associated changes in regional climate.* The rather recent finding that sulphate aerosols partly counteract greenhouse-gas warming was quickly accepted because of the careful studies on which the conclusions were based, although considerable uncertainty remains about the magnitude of this forcing. The most recent IPCC assessment (IPCC 1996) actually concludes that taking aerosols into consideration has improved the agreement between model calculations and observations of climate changes that have occurred during the past few decades.

3.3.2. Is the Scientific Assessment Process Politicized?

The view has been expressed that some scientists may have been influenced in their scientific work by the wish to ensure financial support for their research (see Boehmer-Christiansen 1994). This is a very serious accusation, and I know of no solid evidence to support such a conclusion in the field of climate research. Anyone with the opportunity to attend meetings of lead authors discussing the content of an IPCC report would wonder at such an accusation. The frank and factual discussions are generally of high quality, and biased views cannot survive long in such debates. The outcome of the IPCC assessments cannot have been influenced in any significant way by dishonest scientific behaviour because of the broad participation by scientists and the openness of the process.

Another aspect of the assessment process concerns the widely differing attitudes to the scientific process shown by representatives from various NGOs. Environmental activists, for example, seize eagerly on the occurrence of extreme events (hurricanes, floods, droughts, etc.) as signs of an ongoing change of climate. Even though extreme events may be harbingers of change, there is still as yet little scientific evidence to prove this, nor can we as yet ascribe such changes to human interference. Much more careful analyses are warranted here.

On the other hand, countries whose economy is highly dependent on their supplying the world market with fossil fuels (coal, oil and gas), as well as industrial representatives in this field, and those representing energy-intensive industry (steel mills, smelters, etc.), are reluctant to accept that climate change might be a significant risk. One way for them to reinforce their arguments can be to discredit the scientific assessment process. Their policy would be to wait for more accurate projections of what might happen in the future, fully aware that projection of the future can never be very accurate. This does indeed complicate negotiations aimed at reducing emissions.

It is indicative that the number of lobbying organizations engaged in the climate change issue has increased markedly in recent years, particularly in the United States. Those representing industrial interests have criticized the IPCC for not being objective in carrying out assessments, but this view has not been supported by careful analyses of the scientific literature. A few organizations have engaged scientists to support their aims, but seldom can they show records of more extensive publications in the relevant scientific literature. A closer analysis of their activities, and not least how such activities are financed, might be of interest.

Any individual, whether a scientist or not, seldom expresses his or her views exclusively on the basis of factual evidence. All of us evaluate any kind of issue as human beings, biased because of our special interests. The IPCC process aims to bring in many key scientists in the assessment process so as to eliminate, as far as possible, non-scientific value judgements and to define the scientific task carefully. Applying scientific information in the political process is of course not a scientific task, but there is a need to express available knowledge in a simple form relevant for the non-controversial as well as the controversial political issues that are bound to arise. For that reason, establishing forms of interaction between the scientific community and politicians has also become an important task for the IPCC. If this is to succeed, there must be mutual respect for the different roles that scientists and politicians play.

3.3.3. Climate Change and the Press

Over the past decade, the media have given increased attention to the climate change issue. Still, it is not very well understood outside a rather small group of people primarily in the industrialized countries, although environmentally concerned politicians around the world are reasonably well informed. Climate change is being recognized as an important environmental issue, but its complexity is not yet adequately understood to engage a broader section of society.

The uncertainty of present knowledge is a difficult obstacle to get round when explaining the current situation to the public and clarifying the differences between acting and not acting. In journalism, the traditional approach in dealing with such issues has been to choose two scientists who have different views, have them confront one another, and in this way present a "balanced" view. However, individual scientists are seldom able to judge the climate change issue as a whole unless they have access to broadly based assessments of the kind produced by the IPCC. Controversies as depicted in the press are therefore often misleading, since they are usually presented on the basis of limited factual information. Rarely is the scope and the significance of the IPCC assessment adequately known and understood outside a rather limited group of people.

4. Prospects for the Future

The climate change issue is a long-term issue which will be with us for decades. The IPCC has successfully contributed to its scientific assessment in the sense that its reports have gained recognition as summaries of current knowledge, generally accepted by the scientific community. The careful analyses and the full recognition of the uncertainty of projections into the future have been essential in achieving this. Also, new findings have seldom appeared as "surprises", because surprises have in principle been expected. This has also meant that no attempt has been made to foresee extreme events that are important for assessing impacts of climate change, because of their fundamental random occurrence and because trustworthy weather statistics from climate models are not yet available.

It seems likely that we now are entering another phase in the climate change issue. Soon we should be reasonably sure that a change of climate is on the way, and the most important task will then become to determine how sensitive the climate system is to the forcing that mankind is causing by emissions of greenhouse gases. Running the climate models in real time, and attempting to validate them against observed changes, will gradually improve the reliability of their projections into the future, although also here the possibility of surprises should not be discounted. Sometimes these may make us less concerned about what might happen in the future; on other occasions, unexpected changes may call for more far-reaching measures.

Future work will require more stringent analyses of different options for mitigation and adaptation. There must be general recognition that industrialized and developing countries will have to collaborate far more closely, since there is no room for free-riders. This will imply addressing difficult equity issues which may well block such cooperation. Again, carefully summarized knowledge about the fundamental characteristics of both the natural and human systems as well as their interplay will gain increasing importance. A gradual change in how such information is gathered and made available to politicians may be required. The IPCC mode of operation will have to be continuously modified so as to keep pace with the new demands that most certainly will arise.

Finally, the climate change issue must be seen in a broader context. Linked to other environmental issues, it must be viewed as one of many complicating factors for the sustainable development of our world.

Chapter 4

Uncertainty in the Service of Science: Between Science Policy and the Politics of Power[1]

Sonja Boehmer-Christiansen

1. Introduction

This chapter has two themes: scientific institutions as political actors involved in making climate policy, and the internal politics of the research enterprise. Both themes are explored for their relevance to environmental policy. They emerged rather unexpectedly during a study of the Intergovernmental Panel on Climate Change (IPCC) and proved to be of particular interest to political science.[2]

The first theme deals with the politics which were inspired by the scientific consensus on the global warming hypothesis as negotiated among research institutions in 1985. It will be argued that in a world of competing industries, governments and fields of knowledge, the research enterprise itself constitutes a powerful political actor which deserves scrutiny because it will affect policy as well as being affected by the decisions made by policy-makers. Are the relationships between science and government arranged as well as they might be? Whereas uncertainty advances the enterprise of science, certainty is allegedly sought by policy-makers, and assorted vested interests either stand to gain or lose from lack of knowledge. Controversy over the "truth" of knowledge and its quality tends to become a formalistic battle behind which vested interest, fears and expectations hide, including those of the research enterprise itself.

The second theme concerns the widening gulf between policy analysis as an academic discipline and policy-making in practical terms. The

1 This chapter is based on a three-year project funded by the British Economic and Social Research Council contract Y 320 25 3030 "The formulation and impact of scientific advice on climate change". Support by IPCC scientists, especially in the UK, is gratefully acknowledged.
2 While scientific bodies do not like to be studied from a political perspective, more openness, sensitivity to non-scientific knowledge and their own shortcomings are likely to improve the public understanding of science.

former advocates rational decision-making principles in conditions of scientific uncertainty and claims to ensure that policy-makers have the best possible scientific advice. In reality, and from empirical observation, this academic approach usually excludes the "policy-maker" as part of the process, and policy-making institutions are rarely examined. The approach is a-political because it deliberately excludes the use, or application, of knowledge in real political contexts. This also means that the political processes of competition and cooperation between distinct political entities with specific interests are largely ignored. The conventional academic approach therefore remains unsatisfactory to those who seek to understand how the world really works and for those unsatisfied with preparing prescriptions for bureaucracies under the guise of their alleged "policy relevance".

The chapter is divided into four sections: (i) a brief examination of the significance of science – and the IPCC – for political science and environmental policy, including basic assumptions and questions, (ii) a history of scientific advice-giving on climate change, seeing the IPCC in a historical and political context, (iii) a detailed discussion of the implication of scientific and socio-economic uncertainties for policy and "consensual knowledge", and (iv) some conclusions about how global policy-making might be improved, without weakening science and the contributions it can make.

Warnings of climate change and what ought to be done about it have been made for over a decade, but little has been written on why this advice is rarely heeded. This gulf should worry political scientists as well as policy-makers and concerned citizens, who, through their taxes, pay for the advice and the "learning" that is supposed to be taking place.

2. The Significance of Science in Global Warming: Basic Assumptions and Questions

Science, by its very nature, cannot deliver a clear verdict on a complex environmental subject until change has already occurred. By the time a new "eco-system" has established itself in response to new inputs or removals, it is likely to be too late for human intervention that is likely to have predictable effects. Nevertheless, human beings may still intervene for "other" reasons and at any stage of scientific certainty. It may be foolish to wait too long. This, however, is a decision which science itself cannot make. Indeed, science, as an activity always in search of funding, has little incentive to say we know enough, because diagnostic

knowledge will itself become increasingly redundant as the policy process moves from assessing a problem towards a solution. Even if it were
possible to remove all uncertainties, that would not be in the interest of
research, and that in turn raises the question of whether the research
community is the best judge of what ought to be done.

The possibility of "unholy" alliances between knowledge-makers,
reluctant policy-makers and potential losers arises, as does that of strong
alliances among winners – some of whom may no longer care about
"the facts" and scientific opinion, once a potential threat has been found
politically useful. Policy-making in the real world is not so much the
result of the outcome of scientific debates, as it is the result of battles
between alliances using knowledge as their major weapon.[3] Climate
change politics raise all these issues and invite far-reaching questions
about environmental policy and the roles played in its formation by scientific institutions, commercial and industrial vested interests as well as
environmental lobbies dedicated to "saving the planet".[4]

The strong alliance, whether open or hidden, between institutions
engaged in studying a global problem – in this case climate variability
and possible long-term warming as predicted by general circulation models – and those who define such warming as dangerous and advocate
solutions, provides a fascinating case-study of the interaction between
the politics of science and the formation of environmental policy.
Solutions are usually targeted at the reduction of emissions from the
combustion of fossil fuels using a wide range of technological changes
and economic instruments. Many of these were in fact developed in
response to high energy prices during the 1970s and are now solutions
looking for a problem. While "politics" is mainly about these solutions,
appeals to scientific evidence and the need for further research are an
essential part of the political processes shaping our "globalizing" planet.

It is both easy and exciting for anyone to think and make proposals

---

3 This field of analysis is vast, and the view presented here cannot but be personal. It
is, however, based on several years of research, close contact with the scientific community, a major review of the literature, close involvement with the world of energy
policy analysis, hundreds of interviews and observation of many conferences addressing "global warming". The policy process was followed, with varying degrees of detail,
in Great Britain, Germany, the United States, Australia and Norway. The author also
admits to some sympathy for the coal mining industry and coal miners, who have
already emerged as "victims".

4 In terms of political analysis, environmental pressure groups become vested interests
in the sense that they must act to maintain a market for their ideas and an income
from their supporters. As a result, there is an on-going search for new environmental
threats, while a tendency to exaggeration is built into their functioning. This makes the
need for uncommitted science more urgent.

about how to respond to anthropogenic global warming as "predicted" and quantified by scientists. Indeed, there has been a most enthusiastic response to the problem among organized groups: ranging from Greenpeace, most churches, to the nuclear industry and the energy-saving lobby. The United Nations bureaucracy, the World Bank and donor organizations have been drawn into the issue. In doing this, they must of course trust "science" and accept the problem, quantification plus numerical uncertainties, as given. This allows them to concentrate their efforts on suggesting solutions "in society" and in the process attract converts or customers. For global warming (but not cooling, an earlier "concern") the solutions generally put forward and already being negotiated fall very squarely within the realm of energy policy, and hence the realm of power.

On the solutions side of the issue, another large research agenda suggests itself. It is one which is clearly "relevant" to governments (and hence fundable), as well as to environmental lobbies and many industries now obliged to implement the Framework Convention on Climate Change (FCCC). This Convention is widely acclaimed as an achievement of science. It should be noted at the outset, however, that climate and climate change remain undefined in both this treaty and the documents of the IPCC. This Panel was set up jointly by the World Meteorological Organization (WMO) and the United Nations Environment Programme (UNEP) in 1987, with the International Council of Scientific Unions (ICSU) in the background. The ICSU and WMO already then possessed very strong research interests in climate variability and climate change as scientific subjects, with UNEP adding a powerful policy dimension: planning for global sustainability. These bodies gave the IPCC the task of advising mankind on the science and impacts of climate change, as well as on "realistic" response strategies – a truly formidable task.

The political and public acceptability of solutions, particularly of difficult and costly ones, increases if there is agreement on what the problem is, and if there is "consensual knowledge" – which is too readily equated with scientific consensus. I argue that the IPCC was formed to generate such a consensus between scientists, and that in return major and exciting research programmes were being funded. However, agreement about what? In fulfilling its brief, the IPCC found it had to serve many masters.

From the IPCC story arise a number of philosophical and political questions about science policy and the behaviour of scientific institutions in world politics. In the model presented here, science ceases to be viewed as the provider of objective knowledge, arbiter of disputes and source of authority. It may well do all this, but science is also an independent political actor with interests, institutions and strategies. These are applied, according to this model of science, in response to growing political pressures to do "policy-relevant" research that is useful to government or industry. Scientific institutions must compete for resources and status. They also provide ladders to political influence for ambitious individuals.

From this analysis I conclude that politicians (acting – one hopes – for the benefit of "society") have a greater responsibility in evaluating the funding and selection of research priorities than appears to be the case in most countries today. Leaving these all-important tasks to the research enterprise inside and outside government will necessarily mean that subjects with the greatest research interest (and needing the latest research tools and toys) will be given priority, rather than those most in demand for society. By adding to global change research dimensions which are not readily considered "relevant" – those of socio-political analysis – the politics of climate change become complicated and more ambivalent.

The research enterprise, for which the IPCC acted and continues to act as a successful mouthpiece, is a globally organized and influential industry largely dependent on taxpayers' money. It is in its interests to persuade the rest of us that its desires are those of all humanity, and that the promised future knowledge which only it can provide will serve mankind's well-being and progress, not just powerful elites. A second strategy is to discover and promise to resolve dangerous threats. Without arguing that bodies like the IPCC are deliberately misleading the public or acting conspiratorially, I hope to show that the IPCC should be studied as a political institution acting in the interests of "science" and research bodies rather than environmental protection. Environmentalism and the threats upon which it thrives have proven a boon to science and especially predictive modelling of complex eco-systems and natural cycles – global warming being the example par excellence.

In contemplating the task given to the IPCC, we should note that there is a world of difference in terms of costs, interests and actors (and hence politics) between scientifically understanding a problem – such as the generation, emission, subsequent fate and wide-ranging physical impacts of "extra" greenhouse gases in the atmosphere – and doing something effective about it. Giving expert "policy advice" is also considerably easier than implementing it in the real world. Readers of the global change literature written by IPCC authors (and most others) will therefore note the avoidance of historical, political and cultural dimensions, indeed of most qualitative issues that cannot be modelled mathematically. The issue is largely dehumanized; human beings, when taken into account at all, are seen largely as the problem, the emitters and ravagers of nature. Later they enter again as planners, regulators and technologists, able to persuade "their" societies to make major changes in public and private spending for the sake of the future and a presumed common global interest. Given this basic neglect of the human dimension, why then did so many research bodies and a large number of governments respond so enthusiastically to "global warming"? This is the underlying question asked in this chapter.

One answer seems, indeed, too obvious. When problem and solution together are included in the political agenda, the effect cannot but be greater than if a problem is raised alone.[5] This in turn raises the question of what came first: the problem or the solution? Can science supply problems such that its own interests are advanced by the solution it advocates? Furthermore, if the problem remains plagued by uncertainty, then decades of research may be needed while solutions are being prepared on paper and in laboratories. Costly implementation can be avoided and governments maintain what may well be an illusion: that it is in control of the world of economic competition and enormous inequalities – in wealth, access to resources and political power, including scientific knowledge. The transformation of a scientific subject into a political and even legal issue is therefore likely only as a response to a specific, enabling context.

Such enabling conditions can be ignored by global modellers. Firstly, the world of knowledge must become tied to the world of interest, for "climate policy can only be successful if it does not merely protect abstract environmental goods but serves concrete interests and pursues several goals" (Fischer 1992: 33). In addition "shortages of money and the awareness of unsolved problems are usually the main reason for scientists to ask for a new science policy, though more noble motives are normally added" (Thorsrud 1972). Combining these conditions implies that three sets of questions need to be asked in addition to those concerning the nature of the environmental problem and its possible solutions: (i) who drew attention to the issue, (ii) who turned a "scientific experiment with mankind as actor" into a problem, and (iii) why did this happen in the mid-1980s rather than earlier?

The scientific debate about global warming goes back many decades. The scientific arguments have been around for a very long time and were certainly raised internationally at the 1972 Stockholm Conference. Yet despite the ensuing major research effort, uncertainty still surrounds

---

5 Poverty and unemployment have not generated vast global or even national research effort because science cannot suggest politically attractive solutions. Solving the AIDS/HIV problem, on the other hand, promises wealth to the discoverer of a cure.

the issue and, in my judgement, is still growing. Underlying the more obvious gaps in scientific knowledge clearly stated in IPCC reports are, however, two more fundamental questions. They do not derive from "science" as defined in English-speaking countries, but arise from within the technical sciences and the humanities.

(1) Does mankind in the abstract (or in aggregate) possess the means to solve the climate problem, assuming that we can understand it sufficiently for rational policy responses? In other words, is it technically feasible to reduce the emission of greenhouse gases sufficiently to have a desired impact on atmospheric concentration and predicted climate change?

(2) Do human societies as they exist and function today possess the political, administrative and economic capacities to design and implement effectively – without risking political instability – the technological solutions advocated by technical experts?

While the answer to the first question, to judge from the expert literature, must be yes, the answer to the second may well be no and might explain the reluctance with which politics is now dealing with the issue. My own scepticism derives from knowledge gathered through political enquiries and a "ground-truthed" approach to knowledge: history, geography and the humanities. Is it wise to try to interfere with the man–land– climate relationships that have evolved over millions of years? May not the prevailing rationalism, and the scientific predictions upon which it is based, be misused to legitimate political struggles about the distribution of global wealth, making ecological survival even less likely because of the political consequences? If implementation is driven by politics and not science, then science is really irrelevant to any outcomes, except to scientists themselves. What really matters to policy may not be long-term predictions, but their immediate economic, political and social impacts.

Having raised some really difficult questions, let us in the next section see how and by whom scientific advice on climate change – a much more neutral term, though less so than climatic variability – has been given.

3. Scientific Advice on Climate Change in the Context of Energy Politics

3.1. Background

The following is a brief history of these warnings and recommendations emerging from the institutions of the natural sciences with the promise that "Earth systems science" can provide causal understanding good enough for governments to intervene with preventive or mitigating consequences (*Ambio* 1994). For the political scientist, the question arises: how genuinely did scientists believe in the threat of dangerous warming and their ability to predict it "in time"? Or, was the threat used, if not directly by them, then by its many allies taking science as the source of authority, as a means for obtaining funding for prestigious natural science research areas in atmospheric, oceanic and space physics and chemistry, supported by high-cost technological developments in telematics? Faith in predictability and tactical behaviour to attract support can coexist. Hence even if the answer to the second question is yes, the question remains about whether there is a problem. However, the credibility of the natural sciences and all those relying on it for authority will be weakened, and the organization of seeking advice might have to be re-examined.

The political "pick up" of the threat of climate change in the mid-1980s is probably best explained not by the activities of a handful of research scientists (see below), but by the impact of their pronouncements on other actors and agents, especially changes in the prices of energy. After a peak decade, fossil fuel prices began to fall rapidly, leading the energy winners of the 1970s to search for arguments to persuade governments to intervene on their behalf. The price of crude oil collapsed in 1986, when the Chernobyl accident also increased fears that the world would turn away from nuclear power towards coal.[6] Continued investments in nuclear power, energy efficiency and renewables needed justification, for what was needed was an increase in the price of fossil fuels and hence a world-wide reduction in the demand for coal and oil (Boehmer-Christiansen 1993). Climate change, like acid rain, promised to improve the competitiveness not only of nuclear power, but also of fuels and technologies adopted to beat high oil prices.

The nuclear science lobby, closely associated with climate research in the United States since the 1960s, became particularly active. A politically very influential alliance of industrial R&D and commercial interests was being created as the rapid rise of energy prices during the 1970s came to an end. Countries with pro-nuclear, if disputed, energy policies were among the first to recognize the opportunity. For example, the working group on energy of the German Physikalische Gesellschaft warned the public of a threatening climate catastrophe in December

---

6 The German "Greens" at that time were all in favour of coal, provided acid emissions were removed.

1985. It called upon German politicians, the science and business communities and all citizens to reconsider energy policy and to support international action being taken to conserve energy and develop new energy
technologies. These physicists were quite certain about the causation of
global warming: they blamed the burning of fossil fuels, including natural gas (an argument heard much less in the UK, Norway and the United
States), as well as deforestation and various industrial activities. A mean
rise of sea level between 5 and 10 meters in less than one hundred years
was predicted, though quantitatively certain predictions about extent
and timing would not be available in the foreseeable future. Industry was
called upon to supply low emission solutions that were optimal for the
national economy (Deutsche Physikalische Gesellschaft 1985).

Fossil fuels have remained cheap since the mid-1980s, and the
demand for transport is soaring as is the demand for coal outside Europe.
This has made the costs, to governments, of intervention high at a time
when all governments are desperately trying to cut expenditures for electoral and ideological reasons. Few governments therefore have an intrin-
sic interest in climate change unless they are prepared to impose energy
tax increases to increase their own revenues, or want to use global
warming "responsibility" as a stick to "greenmail" the wealthy "North".
The number of non-scientific reasons for "wanting" climate chamge is
therefore large, with the market (or rather industry) emerging as bene-
ficiary only for those fuels and technologies which reduce greenhouse
gas emissions profitably. The owners of natural gas are therefore on the
winning side – probably the main reason for their relative silence.[7]

3.2. The Evolving Global Change Research Agenda

Attempts by scientists to draw attention to climate change as a policy-
relevant research issue date back, in the United States at least, to the
1950s (Hart and Victor 1993: 643–668; Weart 1992: 19–27). Claims
by climatologists that five to ten years of research would sufficiently
narrow uncertainties and tell international policy-makers "enough to

---

7 Switching to natural gas in electricity generation leads to a large one-off reduction
in CO_2 emissions which benefits countries possessing or importing a fuel owned mainly
by oil companies. The other way of reducing emissions cost-free is deindustrialization
as observed in the former USSR and East Germany where energy demand has declined
sharply. Britain has benefited from both; the United States faces major problems
because of its heavy reliance on fossil fuels at home and as exporter. Carbon taxes will
alter the price of all fuels, but especially of fossil fuels, the demand for which can then
be expected to decline.

justify changes in energy policy" also have a considerable history. Michael McCracken, a major US research manager within climate science, revealed that in 1965 "the President's Science Advisory Council proclaimed that within two or three years super computers would allow scientists to make useful predictions down to the regional level on CO_2 and climate change" (note in Veggeberg 1992: 13). Almost 30 years later, the same author commented, no such fine detail was possible.[8]

The issue was discussed at some length at the 1972 Stockholm Conference on the Human Environment when climate change was accepted as a research issue and a UN institute for planetary survival was proposed by the United States after a considerable domestic debate which brought German and US scientists together (Kellog and Schware 1981).[9] At the preparatory First International Conference on Environmental Futures held in 1971 in Finland, the climatologist R. A. Bryson gave a keynote paper on "Climatic Modification by Air Pollution" which considered the cooling role of aerosols in climate, as well as threats to the ozone layer (Polunin 1972: 167).[10] In the subsequent discussion, a research scientist from Finland, E. Halopainen, stated that the development of numerical general circulation models to simulate the present climate and the behaviour of the atmosphere in long time-scales "is a natural first step in attempts to predict what happens to the atmosphere as a result of man's activities". He claimed that this was the ultimate goal of the Global Atmospheric Research Programme (GARP) which was the predecessor of the current programme and also a joint effort of the ICSU and WMO. The research director of GARP during the 1960s was the Swedish meteorologist Bert Bolin, who has been the chairman of the IPCC since 1988.[11] Halopainen also believed that "the fundamental importance of developing these numerical models should

---

8 By 1993 many scientists were objecting to excessive demands made on climate modellers by policy advisors in government. McCracken has argued that it might be wiser to refuse but maintain the integrity and credibility of science, as did a small number of IPCC representatives to the cited SPRU questionnaire.

9 American–German Aspen Institute workshops addressed climate change, producing forecasts not very different from those of the IPCC.

10 Bryson also argued that the time had come to do real interdisciplinary work at the international level "rather than leave the science to the scientists and the social stuff to the social people: put them together and find a solution". This is being attempted now, but is proving extremely difficult.

11 Professor Bolin has worked on the carbon cycle and is a major advocate of earth systems science. He has published on climate change and energy policy since 1971. In 1980 he addressed the WMO on the subject of "climatic changes and their effects on the biosphere", in which he concluded that "it may take decades or centuries" before changes brought about by "human interference" may be noted (1980: 47). He has

also be recognized and stressed" and told those present that the development of these models had been "the most strongly emphasized matters in the MIT meeting last year" (Polunin 1972: 168).

However, scientific institutions failed to draw political attention to the climate threat until the mid-1980s. By then, a rhetoric of danger and concern for mankind created by globally organized research institutions and assorted allies, including the environmental movement, had prepared the issue and disseminated it, especially via the Brundtland Report by the World Commission on Environment and Development (WCED), world-wide to governments and academe.[12] By the mid-1980s not only had the global energy scene changed fundamentally, but resources were now needed for the successor to GARP and the ambitious research plans of the WMO.

New research plans developed since the mid-1970s promised to provide all the knowledge needed to respond effectively to global warming. This research agenda, collectively called "Global Change", was initially drafted in the United States (NRC 1983).[13] In 1986, the ICSU adopted the global International Geosphere Biosphere Programme (IGBP) based on this draft and now described as complementary by the current US Global Research Program. The IGBP was expanded by the ICSU and adopted by it in 1986. The IGBP also supplements the WMO's own climate change research plans which are part of the World Climate Research Programme (WCRP) (Ferguson and Jäger 1991).[14] IGBP fund-

directed the Swedish International Meteorological Institute and was research director at the European Space Agency and collaborated closely with the Stockholm Environment Institute. He and John Houghton addressed the World Energy Council on global warming and energy policy in 1993 (WEC 1993). Bolin addressed the energy lobby in December 1996 in London. The conference was supported, *inter alia*, by MITI, UN Conference on Trade and Development, US Department of Energy and USEPA on Controlling Carbon and Sulphur International Investment and Trading Initiatives.
12 The WCED is a non-governmental organization sponsored by senior politicians but provided with information and advice by the scientific community and environmental pressure groups. Its connection with a major gas producer may not be accidental. Countries with major fossil energy resources were not among its sponsors, though the North American environmentalist lobby was heavily involved.
13 Implementation was recommended by the 1985 Villach Conference. For an outline of the "supplementary" US effort, see US National Research Council, 1990, Committee on Global Change (US National Committee for the IGBP), Research Strategies for the US Global Change Research Program, National Academy Press, Washington 1990. US IPCC participants are closely involved and the impression is given that the IPCC early on could be understood as an attempt by the "rest of the world" to tap into this effort. This is not inconsistent with the view that the IPCC acts as the mouthpiece for the IGBP and WCRP.

ing did not begin until 1990. The WCRP was started in 1979 as the successor to the never fully completed GARP, but experienced funding difficulties. Both GARP and the WCRP were explained to be what we would now call "consumers" of research findings as means towards advancing climate forecasting and the understanding of climate variability, or, at most, climate change. Political attention clearly needed to be attracted if both programmes were to be implemented globally, which was essential for obtaining full data-sets.

The aim of global change research, according to US sources, is to understand, monitor, and predict global change. According to the ICSU, it is "to describe and understand the interactive physical, chemical and biological processes that regulate the total earth system, the unique environment that it provides for life, the changes that are occurring in this system, and the manner in which they are influenced by human activities" (ICSU/IGBP 1994: 9). According to my reading of ICSU documents, these place little emphasis on prediction for policy guidance, though echoes of "relevance" are clearly there and are immediately linked to uncertainty. According to the ICSU, the means of achieving scientific understanding was to develop a "fully coupled, dynamic model of the earth system useful for projecting changes over multidecadal time scale ... (which) demands that changes in the biosphere and biogeochemical feedbacks ... be incorporated into earth system models". Human beings were to be studied as a source of perturbations, and "the system's subsequent responses" were viewed as "an ongoing biogeochemical experiment at the global level", the consequences of which were "far from clear". The IGBP was described as supplementary not only to the WCRP, but also to the more recently set up Global Climate Observing System (GCOS), which now acts as a lobby for the space sciences and related technology interests involved as suppliers of observational data. Public and privately managed research efforts were seen as intimately linked because:

> Understanding the nature of the link between the biogeochemical cycles and the physical climate system represents a fundamental goal of the IGBP. This understanding bears directly on key scientific questions concerning the co-evolution of different components of the earth system including life, as well as on the most pressing environmental questions of our time. Present understanding of these issues is very incomplete; the attack on the problem will require extensive interdisciplinary collaboration. This attack will employ a hierarchy of models: it will include interdisciplinary problem analysis and the synthesis, interpretation, and

---

14 Ferguson and Jäger provide an outline of this international programme and its presentation at a political forum together with the first IPCC Report.

application of global-scale data sets, especially those obtained by continuous monitoring from space; and it will also require the active participation of the WCRP (Moore and Braswell 1993: 2).[15]

At the very heart of IGBP/GCOS research is the Global Analysis, Interpretation and Modelling (GAIM) project. It is to "assist the IPCC process by conducting timely studies that focus upon elucidating important unresolved scientific issues associated with the changing biogeochemical cycles of the planet ... and to maintain scientific liaison with the WCRP Steering Group on Global Climate Modelling" (IGBP 1994: 98).[16] This group is an extension of the US Earth Observation System in which NASA is closely involved, NASA being a major actor in the IPCC process, having been intimately involved in its management and currently acting as the home of the secretariat of one of its new working groups (see below). Leading IPCC leaders, including chairman Bert Bolin and co-chairman of the group on scientific assessment, Sir John Houghton, are involved in GAIM and have participated in IGBP/WCRP planning activities.[17] Global change research is now a major market for the products of remote-sensing. Many sections of the natural science enterprise themselves have agreed that considerable progress has been made since the mid-1980s (Boehmer-Christiansen and Skea 1994).[18]

This achievement has not entirely been the outcome of IPCC activities.

---

15 The military language may be noted, as well as the claim that it is science that defines what are the most important environmental questions.

16 *Ambio Special Issue* (1993) disseminates the plan and reports first findings. It contains contributions from leading IPCC figures, including Bert Bolin, who writes on policy-making, viewing policy as a product of facts and values, with scientists providing the former.

17 Sir John Houghton is a British meteorologist and former director of the UK Meteorological Office. He has served as chairman of IPCC Working Group I and been actively involved in designing the Global Climate Observing System. Through his links with the Rutherford Appleton Laboratory he is well connected with the European space science community. Both Bolin and Houghton have been very active in the environmental politics of their own countries, as well as at the international level.

18 While only 10% of all replies to this questionnaire to IPCC participants supported vigorous remedial action and 73% saw any type of action (no-regret, high-cost, research only) justified by precaution rather than compelling scientific evidence, only 6% had observed very little impact on the growth of knowledge. There was wide agreement that the concern about climate change had advanced knowledge; further, that this advance had been greatest for climate prediction, climatology and oceanography, but least for socio-economic causes and impacts of response strategies. Most respondents expected uncertainties to be significantly reduced in 10–15 years, though much less so for socio-economic issues.

A small non-governmental body comprising "Northern" research managers had called the first shots in 1985. It could be argued that this group, the Advisory Group on Greenhouse Gases (AGGG), rather than the IPCC got "global warming" onto the political agenda, and much more rapidly than many scientists had expected.

3.3. Villach 1985: University Research Initiates

Any history of the global change research and the IPCC would be incomplete without reference to the brief but important impact of the AGGG, which had its beginnings in the autumn of 1985 at the Villach Conference, the Second International Conference on the "Assessment of Carbon Dioxide and other Greenhouse Gases in Climate Variations and Associated Impacts". The scientific agreement that the world should listen to science and needed to be frightened about global warming, as well as about its many uncertainties, was reached at this conference held in an Austrian village and sponsored by the ICSU, WMO and UNEP. Here global warming was endowed with its first "scientific consensus" (WMO 1986).

Direct involvement of scientific institutions in the organization and policy-direction of the international effort to "combat global warming" that began in 1985 can be traced through the activities of the AGGG up to its disappearance at the Second World Climate Conference in 1990 and the assumption of its tasks by the IPCC. The AGGG consisted of a mixture of environmental and energy policy researchers also active in the science policy domain as research managers for institutions like IIASA, Harvard University and the Stockholm Environment Institute (SEI). The organizations represented by individuals tended to be active at both the national and international levels, many having close links with "their" government and its foreign policy goals.[19]

The organizers had in advance commissioned a book from SCOPE, ICSU's Scientific Committee on Problems of the Environment. Many subsequent IPCC participants made major contributions to a book which one major IPCC coordinator has called the Panel's Bible (Bolin 1986). Under the leadership of Professor Gordon Goodman from the SEI, the AGGG was to organize both the Toronto Conference on the Atmosphere

19 The SEI (formerly Beijer Institute) and IIASA in Austria had been welcoming homes for energy and environment modellers for about a decade. Both had good connections with US institutions; IIASA with the USSR. Important individuals were Bill Clark of Harvard/IIASA, Tom Malone (US meteorologist and international science leader), Gordon Goodman (SEI) and Martin Parry (UNEP/UK).

in 1989 and the Second World Climate Conference in Geneva in 1990. Both paid considerable political attention to climate change research, with leading scientists showing little reluctance to tell governments what to do. Their positive responses, especially from the United Kingdom, surprised many. Australian scientists, for example, were surprised, for "almost overnight, greenhouse-induced climate change moved from being the business of specialised bodies ... to become the centrepiece of a major political happening" (Tegart and Zillman 1990). At Geneva and even more so at subsequent conferences, the specific threat of global warming rather than simply climatic change became linked with the strong advocacy of certain energy policies.

After Villach, governments had to decide by whom they wanted to be advised on the climate threat and what should be done about it. The AGGG wanted "independent" scientists, i.e. itself, to do just this. As "independents" they would be funded by universities, environmental and charitable organizations (including UNEP) and be close to the ICSU. WMO members, however, had a different agenda. Worried that UNEP would exploit the climate problems without science, they wanted to set up an intergovernmental body that could reach "the heart of government" and presumably its pockets more effectively. The big national laboratories with the greatest computing power were making their bid.

3.4. WMO and Its Capture of the IPCC

The idea of an intergovernmental panel under WMO guidance was first mooted in the early 1970s in response to the efforts which led to the 1985 Villach Conference. By 1987 an impressive, knowledge-based alliance promoted the threat of global warming, not only through WMO and the Brundtland Report, but by now with the support of various influential national and international environmental bureaucracies eager to expand their briefs and competencies. Demands for a global treaty underpinned by global research were becoming louder. The WMO's World Climate Programme also needed support. This programme consists of two components: the small World Climate Impact Assessment and Response Strategies Programme (WCIRP) co-ordinated by UNEP, and the above-mentioned WCRP. In 1990 the programme was described as designed "to gain improved knowledge of climate mechanisms which bring about climate change so as to determine to what extent climate can be predicted and the extent of man's influence on climate". Studies were needed of the global atmosphere, oceans, sea and land ice, and the land surface which together constitute the Earth's climate system. Claims to policy relevance were now very clearly stated. The programme was "to introduce climate considerations into the analyses of rational policy alternatives and to develop the capability to warn governments of the economic and social impacts of climate change (WMO/ICSU 1990; also Ferguson and Jäger 1991: 163–169).

The WMO was also directly involved in the close observation of the global climate system and in climate modelling, calling this exercise "numerical experimentation" (WMO 1990; 1991).[20] In 1990 a Steering Group on Global Climate Modelling was set up to deal with coupled, as distinct from merely atmospheric, models. It was reported that model predictions had changed little since 1985, though their scientific foundations had become firmer. Atmospheric models were increasing in resolution, but this alone would not cure climate-model errors. Better physics was needed as well as bigger computers. More research had to be funded. The IPCC was available, for it had been set up in 1988 very much in response to governmental pressure exerted via the WMO. In early 1987 the WMO invited its member governments to set up a scientific advisory panel and appoint their WMO representatives to it; in the autumn of 1988 the IPCC was formally set up. As a non-governmental body, the ICSU had been left out. The WMO Executive Committee made its proposal in consultation with major governments, especially their meteorological offices, in response to the 1985 Villach Conference. The US State Department had apparently become suspicious of the AGGG and exerted influence within WMO.

In May 1991 at the 11th Congress of the WMO, the links between the IPCC and WMO were very explicitly stated when it was decided that WMO research should provide support to various intergovernmental activities, including the IPCC and of the Intergovernmental Negotiating Committee (INC) for the FCCC, as well as the process of implementation of the Convention. Member governments were asked to include members of the climate-science community in INC. UNEP was invited to continue its lead role in the WCIRP. Indeed, WMO research was to be reconstituted so as to "provide an interagency interdisciplinary framework to address the full range of climate and climate change issues including research into the economic and social consequences of climate and climate" (WMO 1991: 83). WCRPs coordinating committee was to invite the chairmen of the IPCC and the IGBP and continuing support was expressed for IPCC scientific, technical, and socio-economic analyses underlying policy

---

20 Observation is fundamental to most WCRP projects, e.g. the World Oceanic Circulation Experiment, where robots are replacing ships. The meeting was attended by a number of senior IPCC people, including the chairmen of IPCC working groups. Robert Watson (NASA) also attended. He is now advising the US president.

options. Governments and others were urged to increase their cash contributions and the Executive Council decided to change WMO's Research and Development Programme to "Atmospheric Research and Environment Programme" in response to an "overwhelming increase of international concern with environmental issues during the 1980s". The programme was to be implemented in cooperation with UNEP and ICSU.

4. The IPCC: Function and Achievement

At its first meeting the Panel defined its brief, apparently after considerable debate about the rights and wrongs of giving governments policy advice, as distinct from scientific guidance. It was decided to set up three working groups meeting concurrently: on scientific assessment (UK-based), impacts (USSR/Australia-based) and response strategies (US-based). A fourth group was to deal with problems faced by developing countries. These were guided by a Bureau consisting of the chairmen of the groups. The IPCC would be financed by voluntary contributions, and its underlying research efforts would remain primarily national responsibilities. Formally, IPCC would not sponsor, much less conduct, research. Its task was to gather and assess the findings of research underway. Thus, it does seem odd that virtually all its chairmen were well-known research managers.

The Panel first met in the autumn of 1988 to elect the chairmen of its working groups, who were indeed government-appointed individuals of high status in the WMO, with great political experience in the fields of science policy.[21] This structure did not suit policy-making, as scientists and officials hardly ever met, but it protected the independence of the natural scientists as much as possible. The scientific assessment of climate change was the task of Working Group I (WG I), which would interact with WCRP/IGBP/GCOS and produce the well-known assessments of climate change for the 1990 World Climate Conference, for UNCED and for the 1995 Berlin Conference.

The results of WG I were to feed into WG II for its assessment of the impacts of climate change – as mentioned a subject close to the heart of UNEP, which had been active in creating North–South links with refer-

---

21 For example, the British and Canadians sent their permanent WMO representatives; the United States, research managers and EPA officials with little direct policy influence; the Russians, a major policy advisor until their virtual disappearance; the Australians, an experienced science policy bureaucrat as well as research managers; the Germans, senior "independent" scientists and more junior people from the Environment Ministry. Linkages between the IPCC and national governments have therefore varied considerably.

ence to the "common" threat of climatic catastrophe. These two groups were, in logic at least, meant to supply data and ideas for the formulation of "realistic" policy-response options among which governments were free to select (old WG III). In practice, there was very little contact. In 1993 the IPCC was reorganized by combining working groups II and III into a huge but fragmented group coordinated from NASA and co-chaired by an atmospheric physicist known for his previous involvement with WG I and ozone research. The old WG I was altered but little, and was left to get on with its central task of assessing the outcomes of modelling experiments without "political interference", as originally intended by the WMO.

The new WG III has largely become the home of modellers who deal with sanitized, i.e. largely value-free, matters loosely described as "socio-economic" and "cross-cutting". The new WG III was given a Canadian chairman with impeccable WMO credentials. It considers efficiency and equity issues as being separable.[22] The old WG III was attended by "policy-makers" who were at most junior advisors in environment and foreign affairs ministries. This working group attracted not only governments, but also representatives from industry and assorted NGOs. It was the most volatile, unstable and hence least academically productive group, but presented governments (and participants) with a most useful learning exercise about the kind of conflicts that the INC and the FCCC would have to face. If the first two groups were seen to supply the rationale for action, the old WG III became a forum for those advocating changes, *inter alia*, in national energy systems.

IPCC-connected research networks now involve several thousands of researchers, mainly in the natural sciences, but they have been joined since 1992 by a growing number of social scientists who add "global data sets" on emissions, economic (GNP) impacts of carbon taxes, population growth (related to energy demand), health and agricultural statistics and speculative impacts of temperature changes, as well as equity arguments. They are all expected to present their results in "value-free" form. While it is far too early for a full evaluation of this global experiment in interdisciplinarity, genuine collaboration has proved difficult and is often resisted.[23]

An update of the 1990 IPCC report was produced for UNCED in 1992

---

22 The brief of this group is now widening rapidly. However, mankind is still being treated largely as a single emitting and reproducing species that alters the nature of the terrestrial surface, i.e. as it is for the HD programme.

23 The number of individuals, institutions and countries involved varies with the subject being studied – ranging from climate modelling (the dominant group), hydrology, to agriculture, economics, energy-demand modelling and public health.

(Houghton 1992). A shorter version of this report, published by WMO and UNEP, concluded with five key uncertainties: sources and sinks of greenhouse gases, clouds, oceans, polar ice sheets, and land surfaces. These uncertainties were to be reduced by improved model experiments, improved observation, better understanding of processes (climate-related, socio-economic and technological), national inventories, more knowledge of past climate changes and improved exchange of data. Between 1990 and 1992 (when the FCCC was being drafted) there was a strong perception, within the IPCC and elsewhere, that policy-makers needed a "consensus" view. From personal observation, I can add that this consensus was reached by careful drafting and rewriting under the guidance of government officials. Model results were usually reported to be "in broad agreement" when scientists themselves were most interested in their differences. Given the nature of the knowledge that was being condensed, consensus was most readily achieved on those subjects that were least well understood.

Once agreed by small groups of draft authors who had full discretion to take note of "peer review" comments, these reports were published by a few governments, with Britain probably being most generous to "its" group, WG 1. IPCC products were widely disseminated by interested governments and UNEP/WMO in the name of "the scientific community", usually accompanied by claims about how many scientists out of how many countries had contributed to the effort. In fact, the reports were written by a handful of people on the basis of drafts provided, selected and evaluated by "lead authors" appointed by WG chairmen in consultation with national scientific elites. Each chairman would be responsible for the prompt delivery of the report of his group, which was drafted by 10 or so lead authors meeting occasionally, often begging governments for money to allow their Third World colleagues to attend.

The efforts of working groups were coordinated, indeed enabled, by small secretariats provided (or not) by national governments. Many thousands of pages of scientific research, already compressed into hundreds of summaries, were reduced by these small teams into several tens of pages called "Policy-Makers' Summaries" and subjected, like the report chapters, to a review process. Reviewers were selected by the working groups, again with the approval of government officials. Small drafting teams would have the choice of accepting or rejecting any amendments that were received (Skodvin and Underdal 1994).[24] The politically most important summaries were negotiated among lead authors under the

---

24 These authors recognize that policy-makers prefer expert consensus because it relieves them of having to make difficult choices. However, science progresses less by consensus than by debate and dissent.

guidance of the working group chairmen, whose personality would define how any conflicts would be resolved.

Behind the IPCC, largely government-funded research flourished, especially in North America and Europe, but Japan, India and China were increasingly drawn into research networks. Knowledge was to be collected and assessed to suit political needs and time-tables, a task which became frustrating if not impossible to comply with honestly. In particular, scientists repeatedly argued that policy-makers wanted them to deliver quickly on two issues which would make policy-making so much easier: to provide a definition of a single global warming potential to which all greenhouse gases would be included in a single index,[25] and to predict climate change at the regional and subregional levels.[26]

The first would allow trade-offs between the various gases and sinks and permit the implementation of the comprehensive approach favoured by the United States, which was having profound difficulties with reducing CO_2. This instrument is now enshrined in the FCCC and will require major research efforts for its "objective" definition. The second would allow governments and regional bodies to actually "plan" for expected changes in various climatic indicators, thus providing for activities which need not – ever – lead to implementation. However, neither of these two issues is of immediate interest to earth systems science.

4.1. Critical Views

The IPCC has been criticized, but only a few scientists have published their reservations, and limited organized opposition has been observed only in the United States, from within the world coal industry and very recently in Europe.[27] Not surprisingly, scientific opposition in the United

---

25 At a special workshop held in Boulder, Colorado, in late 1990 experts agreed that the science for GWPs was preliminary at best. Science could at best provide guidelines, but using CO_2 as the reference gas would introduce further uncertainties. "Policy-makers must be aware of the need for revision ... and that a truly comprehensive index could, to some degree, reflect aspects of a compound's contribution to broader impacts. However, such cross-issue concept would involve value judgements, hence might prove very difficult to quantify and even more difficult to negotiate" (Draft Report of Workshop, 31 May 1991).

26 In practice, scientists appear more interested in developing their global models and do not readily appreciate policy needs. An observer from the Climate Impact Centre (Maquarie University, Sydney) who attended the 1993 regional modelling workshop of WG I noted growing frustration and tension between modellers and policy-people (interview, May 1994).

States has tended to come from the political Right within universities and from individuals concerned about academic freedom, rather than from government laboratories. The Marshall Institute argued in 1992 that "the fact that the expected greenhouse signal is missing from the records suggests that the computer models have considerably exaggerated the size of the greenhouse effect", refers to solar activity as an important factor and suggests that warming of half a degree spread over many decades would have but a small effect "lost in the noise of natural climatic fluctuations" (Marshall Institute 1992: 25–26). There is enough time to wait for better results before taking action, argue the scientists associated with this body, so influential during the Reagan and Bush Administrations.[28] The IPCC has of course acknowledged cooling due to "acid rain", a subject which in the 1970s raised concern about climatic cooling; others again are critical of the emission scenarios used by the IPCC and the many political assumptions built into these.[29]

The perhaps most outspoken critic of the IPCC consensus and a member of the Marshall Institute, Professor Richard Lindzen of the Massachusetts Institute of Technology, addressed the Organization of Petroleum Exporting Countries (OPEC) in 1992, arguing that;

> The notion of scientific unanimity is currently intimately tied to the WG I report of the IPCC ... [on which] university representation from the United States was relatively small ... since funds and time needed for participation are not available to most university scientists ... the report is deeply committed to reliance on large models ... [which] are largely verified by comparison with other models. Given that models are known to agree more with each other than with nature [even after "tuning"], this approach [of creating consensus] does not seem promising (Lindzen 1992: 126).

He further alleged that pressure had been brought to bear "to emphasize results supportive of the current scenario and to suppress other results"

---

27 For example by the George C. Marshall Institute, Washington DC, the *World Climate Review* published by the University of Virginia; and the European Science and Environment Forum has published a book on the subject of politics and climate change research (Emsley 1995).

28 It may be noted that, despite a very different rhetoric, the Clinton Administration has not been able to deliver much action. Like the UK, the United States now hopes to reduce emissions somewhat by switching to gas and allowing energy efficiency to take its natural course.

29 This subject is too complicated to go into here. However, IPCC WG I experienced major difficulties, not wanting to accept scenarios handed to it by two very "interested" governments. In 1992 it therefore put forward "scientific" scenarios. Such scenarios depend on assumptions about fuel prices, economic growth and technological change, as well as regulatory measures implemented.

– a conclusion which is confirmed to some extent by a questionnaire (Boehmer-Christiansen and Skea 1994).[30] As IPCC emission scenarios were most damaging to the prospects of fossil fuels, some saw the hand of the nuclear power lobby behind their adoption, others the green "soft" lobby urging society to switch to wind, wave and pedal power. Both clearly stood to benefit. However, national analyses of climate policies suggest that the direction and nature of these pressures vary significantly between countries depending on national energy situations and policies. The political pressures on WG I to make "neutral" policy pronouncements was therefore considerable and coincided with the Panel's own interests, though not necessarily with the views of individual scientists.

Scepticism has by no means disappeared. The very notion that there is an IPCC consensus has been questioned (Böttcher 1992, Bates et al. 1996). Sceptiscism even appears to be growing, especially as far as the products of very costly model experiments are concerned, as it now takes months to run a single simulation. Yet the need for accurate forecasting and causal understanding is now enshrined in the FCCC. Its implementation has been made dependent on warming being "true", dangerous and human-induced. Yet it has also been claimed that the numerical experiment performed on general circulation models which underpin the warming hypothesis is one that "fails on all counts", produces data "by the bucket load" but may well not really fit with the promise of averting catastrophe (*Economist*, Nov. 5, 1994: 117).[31] Global average temperatures are more easily calculated from physical theories or complex models than actually measured "on the ground" and then generalized, when they become quite meaningless for policy purposes. Climate is a far more complex phenomenon than either temperature or rainfall.

4.2. The Social Sciences: A Cinderella

The International Social Science Research Council (ISSC) realized only too late that the WMO was trespassing on its territory. A small Human

---

30 The IPCC strives indeed for consensus and goes to great lengths to market this effort. It may be noted that the German Enquete Commission on the same issue was willing to submit dissenting opinions. Political conflicts over interpretations and solutions were openly admitted and accepted as "normal". The possibility of such reports was discussed by IPCC in 1994, but does not appear to have borne fruit. Ambivalence reigns instead.

31 The failures listed include the lack of controls and the inability to repeat the experiment. Human-made and natural changes cannot be distinguished, and results will be available when it is too late. Yet governments are spending billions of dollars on this research. Other explanations are surely called for.

Dimensions of Global Change Programme (HD) was therefore added to the IGBP in 1991 after considerable debate and several competing attempts. This programme is now running from its base at Michigan University's Consortium for International Earth Science Information Network, which views itself as a response to "the need for research collaboration between natural and social sciences ... as social science data, research and models will be necessary both to understand the dynamics of natural processes of global environmental change and to deal with the social and economic implications of these changes".[32] The HD now seems dedicated to the exploration and measurement of the concept of "sustainability", a notion which appears to have replaced the "limits to growth" in the jargon of international research.

The idea appears to legitimate highly quantitative and data intensive research by both natural scientists and economists – with equity added as an abstract political value that might be achieved by economic instruments, given political decisions that, in 1993 at least, "had not been addressed" (Frankhauser 1993: 36). Both macroeconomic and climate modelling are now "predicting" a dangerous future and hence recommending various degrees of government "intervention" in markets, indeed almost acting as justification for a return to public policy as a worthy issue. The HD community became heavily involved in the writing of the second IPCC report, to be published in early 1996. One well-known economist and participant of the IPCC WG III, Professor David Pearce, has expressed the view that this social science contribution will be largely an academic exercise with scant policy relevance (Pearce 1995: 5). Nevertheless, much HD expertise and knowledge will be needed for complying with the reporting procedures already enshrined in the FCCC. This can be interpreted as providing a degree of legal compulsion for the funding of global change research in the natural, technical and social sciences, selected sections of which have come to report to governments via the IPCC and largely under the guidance of national research councils or similar bodies.

With this condensed and selective history of scientific advice in mind, let us turn to an analysis of the links between scientific uncertainty (or research agendas), climate policy and the politics of energy in particular.

---

32 Roberta B. Miller, President and Chief Executive Officer of the Consortium, Interaction and Collaboration in Global Change, addressed the IGBP in January 1993, predicting increasing uncertainty arising from this collaboration. The Consortium publishes *Human Dimension Quarterly*, which is written almost entirely by North American social scientists.

5. Policy Implications

5.1. Cascading Uncertainties: Do Policies Need Scientific Predictions?

The uncertainties of climate change predictions include incomplete understanding of how climate works and changes over time, of the carbon cycle and how it is affected by natural and human-induced changes in land-use. Is climate really predictable from past trends, or does it behave in a chaotic fashion? Might feedbacks from changes in the water vapour content of the atmosphere maintain some satisfactory temperature balance? Are human-induced changes really significant enough to alter climate over long time-scales, when at least 18 factors influence climate, most completely beyond the control of mankind (Turner et al. 1990)? There cannot but be huge doubts about the assumptions which underlie the emission scenarios used by models to achieve a doubling of carbon dioxide concentrations sometime next century and which are based on estimates of future emission of greenhouse gases from a large variety of sources, some quite beyond the control of society. Why should greenhouse gas concentrations double during the coming decades, something all global models assume?

Uncertainties therefore increase sharply as the policy relevance of the knowledge needed for rational decisions improves, because the uncertainties of one field of knowledge (or model) feed directly into the next one. These were indeed a major theme in discussions among scientists in the 1980s when climatologists, ecologists and energy demand forecasters made their political statements at Villach. Scientific uncertainty has remained a permanent theme in all IPCC reports and is, in my judgement, growing stronger rather than weaker. For example, in its Updated Opinion for the Rio Conference in 1992, the IPCC WG I weakened its earlier emphasis on certainty and predictive capacity and told governments that: (i) there are many uncertainties in our predictions; (ii) the size of the (observed) warming is broadly consistent with predictions of climate models (broadly being a very vague term, meaning that the issue is debated); and (iii) the observed increase could be largely due to this natural variability; alternatively, this variability and other human factors could have offset a still larger human-induced greenhouse warming (IPCC 1992: 5).

Virtually each introduction to the chapters of the Second IPCC draft report begins with statements about uncertainty; for instance, "climate change presents the analyst with a set of formidable complications: large uncertainties, the potential for irreversible damages and costs, a very long planning horizon, long time-lags between emissions and effects, wide regional variations, and multiple greenhouse gases of concern ... the presence of strong uncertainty, however, does not mean inaction. It means filtering paths of action and choosing which strategy with which outcome would be best".[33] The intellectual challenge is exciting, especially as computational powers are increasing rapidly. However, for impact studies of climate change the situation is even worse as the possible "cascade" of errors is even larger. Here "uncertainty pervades all levels of climate impact assessment, including the projection of future greenhouse gas emissions, atmospheric greenhouse gas concentrations, changes in climate, their potential impacts and the evaluation of adjustments". All is not lost, however, the experts continue: "There are two methods which attempt to account for these uncertainties: uncertainty analysis and risk analysis" (IPCC 1994: ix).

Even a superficial reading of scientific reports in popular science journals will indicate a degree of debate and uncertainty about climate change science not revealed by IPCC summary reports, not to mention pronouncements from UN bodies dedicated to global warming.[34] The extreme position of denying any policy relevance and describing their work as purely "academic" has not, however, been taken by natural scientists – only by economists modelling global emission abatement costs and their impacts on global GNP. Ironically, these are the very experts from whom many natural scientists had hoped to obtain support for their cause. The 1995 Berlin Conference, when the Parties to the FCCC met for the first time, therefore faced serious problems. Research seemed to have revealed greater ignorance and more complexity, though luckily a large piece of Antarctic ice did break off most conveniently, and observations by lay people and scientists in many parts of the world suggested that warming had already taken place. The science debate therefore continues, with emission reduction now as the holy grail of some,

33 Citations are from two draft chapters of this report reviewed by the author. The IPCC lead writers recommend portfolio and international measures, such as joint implementation, on efficiency grounds. They also argue that efficiency and equity can be separated. The realism of this is surely doubtful. The political implications of any recommendation remain unexplored, as the IPPC is instructed to remain "value-free".

34 The debate about further research needs can be followed in the *New Scientist*, as well as in the full IPCC reports and relevant conference reports, e.g. of the 1990 World Climate Conference. UN publications are a less reliable source because of the huge (over)commitment of the UN bureaucracy to climate change as a justification for a large range of initiatives, most of which are research-intensive and indeed as a battleground for North–South relations.

but a serious mistake or even conspiracy, to others (Victor and Salt 1994: 2–30).[35]

Newspaper reports in 1995 were full of scientists still eagerly awaiting observational proof of their theoretical predictions, while others, including certain governments, appear convinced. Some US government scientists have claimed that it is still "too early to draw firm conclusions" about sea level changes observed by satellites, and in 1996 several authors discussed the possibility of human-induced aerosols in addition to those already modelled, such as mineral and carbon particles, were producing a net cooling effect (*Nature*, April 4, 1996). Indeed, a variety of competing views dared publication even in Europe (Bate et al. 1996). The IPCC concluded in November 1995, after heated debate that "our ability to quantify the human influence on global climate is currently limited because the expected signal is still emerging from the noise of natural variability, and because there are uncertainties in key factors. These include ... long-term natural variability ... changes in concentrations of greenhouse gases and aerosols, and land surface changes. Nevertheless, the balance of evidence suggests that there is a discernible human influence on global climate" (Callender 1996). Combatants in the debate may pick either sentence. The second one does not refer to warming. The British scientific civil servant in charge of communicating IPCC climate change science to the public is reported as saying (at the very moment when the British press was claiming that "Britain sets the pace in war on global warming") that "we are trying to draw out of the scientific community a consensus recognising that policy-makers need a balance of probabilities. Many computer modellers say that the results of their climate models are still very uncertain. But the chance of nothing happening is very unlikely" (Clover 1993: 12). Knowing that something happening is not enough for passing laws and regulations, and far less for implementation: but it is sufficient for planning, report-writing and drafting, activities which point towards the most interested parties in the current political process. There is no ready basis for rational decisions based entirely on an environmental cost-benefit calculus. Uncertainty thus provides justification for further research, data collection and advice given from a rapidly growing set of professionals concerned with "the environment", its assessment, regulation and management – with reference less to current generations and problems (which cannot but be political), but more to future scenarios. This is likely to suit politics.[36]

---

35 These authors argue that there is now a need to experiment and learn, rather than set unreasonable targets in conditions of distrust.

36 An example of the utility of scenario-building is the revision of UK predictions for British CO_2 emissions just prior to the Berlin Conference. In 1990, when the UK

As the observed and predicted warming range now seems to be declining, aerosols, which once were feared would generate global cooling, are seen as the cause. The implications for energy politics are exciting. If the burning of fossil fuel maintains some balance between warming and cooling, albeit at the cost of acid rain damage, some may argue that this damage is preferable to the risks associated with the nuclear route, or to the costs of a large-scale shift to renewables. These uncertainties have three broad impacts on the policy process: (i) They justify the search for more knowledge concerning the nature of the problem. As a minimum, science will strive to measure uncertainty. (ii) They allow socio-economic and technological research to solve a postulated problem and proceed irrespective of whether this problem is real or not. For example, engineers, economists and ecologists can do endless research about whether, which and how emissions ought to be reduced. (iii) They provide the various stakeholders in the policy formation process, the potential political and economic winners and losers of regulation with arguments in the process of persuasion.

Persuasion, not expertise and rationality, becomes the primary agent in the political process in areas where knowledge is insufficient and yet is claimed to be the major instrument for policy design. Uncertainty therefore becomes a part of two distinct, even competing, realms of politics: science policy and environmental policy formation. Both in turn have major impacts on real policies, here the complex mix of interventions in energy markets by which emission reduction is to be implemented.

Science policy is a public struggle by scientific institutions for resources and attention. It involves a united front ("scientific consensus") towards outsiders, but also a more private, fiercely competitive battle between various branches of knowledge. Science politics are organized globally but directed primarily at national governments as funders and sponsors (Boehmer-Christiansen 1995). Research will claim environmental policy relevance and political neutrality. In its extreme form, environmental policy will be transformed into science policy. The environmental politics shaping policy are, however, much more open and widely reported than science politics as the contending parties seek out public opinion as an ally. In order to create "winning coalitions", appeals to saving the

---

wanted to ensure that the United States signed the FCCC, a prediction of a 30% increase in emission by 2000 made stabilization a tolerable commitment. When reality produced a decline and the expectation of further CO_2 reductions, the UK could act as climate leader in Berlin. The decline was brought about by a switch from coal to gas and nuclear power in electricity generation. An attempt to increase government revenues from raising energy taxes failed, however.

planet and other "hegemonic myths" are fundamental (Rayner 1994).[37] The links between "hidden" science and "open" environmentalism need to be better understood. Unstable alliances between the two underlie the politics of climate change.

5.2. Responses of Vested Interests

Facing uncertainty means having the choice of whether to belief or disbelief. Pronouncements made by scientists are accepted as truth by some and rejected by others. From observation, this confusing response is best explained by the vested interests a party has in a particular regulatory debate, and more knowledge is indeed generated as such a debate moves toward resolution. As such resolutions cover timetables, targets, technology transfer, energy taxes, joint implementation, withdrawal of subsidies or public support for clean energy sources, the expected impacts of policy intervention are revealed and create very immediate losers and winners in most countries and among many organizations. The politics of regulatory capture are therefore upon us.

Insurance premiums are already being increased with reference to more natural catastrophes rather than past errors. Several governments have raised taxes, allegedly as part of their climate protection policy. While there are not enough data on impacts, costs and benefits for rational policy-making at the national level, more than enough is known to justify global recommendations, research, planning and report writing, the drafting of legal and other instruments of regulation, as well as the setting up of new national and international institutions or new sections in existing ones. Especially international bureaucratic interests associated with the climate agenda cannot admit to ignorance, or refuse to propose policy options because of insufficient science. Too much political capital has been invested, and the claims of the prophets of doom must be taken seriously. Electors and markets need to be persuaded that problems can be solved better because the solution is already available. For example, the electricity sectors of industrializing countries are expected to grow rapidly and remain highly profitable. Profits are more likely to flow from the "South" to the "North" – if new technologies and cleaner fuels can be sold to rapidly industrializing economies, especially if joint implementation becomes a practical policy instrument for reducing greenhouse gas emissions. The economic goal "technology transfer", the battle about selling the cleanest (or most research-intensive) rather than

---

37 Rayner distinguishes five such myths: global vulnerability, per capita equality, historical obligation, voluntary frugality and market solutions.

the cheapest products or options to the "South". The implications which such policies can have for the distribution of wealth and power have not been investigated by the IPCC, nor by most environmental lobbies.

Only time will tell whether the new knowledge created in response to the threat of global warming – a cooling prediction would produce very different coalition of interests – is "valid" as well as effective in what it claims to be able to prevent. "If we act now, then we will never know whether we were right", a senior climate scientist was heard to complain in private. The current generation will not know whether dangerous anthropogenic climate change will be prevented, but this change is already part of current politics being played with the future. Yet threats of future catastrophes are ancient tool of politics. Threats serve to create group cohesion, and future threats may be used to counter present ones.[38] Heaven and hell have always been part of politics; what is new is the deep involvement of computer predictions derived from complex yet still highly simplifying mathematical models, incomplete theory and scenario building based on cascading errors. It is a part of contemporary environmentalism and possesses an apparent "rational" ideology to underpin economic globalization and the political battles for access to minds, markets and resources so characteristic of post-Cold War international relations.

So far, the practical politics of climate change have largely turned into international conflicts over the regulation of energy demand, prices or technologies: how to replace cheap, widely available fuels employing many people (but few researchers) with more expensive and technology-intensive sources of power; how to replace simple but polluting technologies with more complex and expensive ones? In economic terms, the objective of climate change politics is to reverse the return to a period of cheap energy by "internalizing externalities". Science has provided the justification and economics the means. IPCC leaders have supported both, repeatedly linking the alleged problem to energy policy solutions with major political implications (Bolin 1993: 42; Houghton 1993: 47). Politics has already revealed winners and losers.

The main industrial beneficiaries of global warming are, with "likely certainty":

---

38 Powerful organizations have always used threats to advance their own interests while claiming to protect society. If politicians knew that global warming threats were being exaggerated, and there is evidence that some were told in private that this was the case, then their reluctance to do no more than fund more research is revealed as a rational response to knowledge. However, this leaves environmentalists and "socio-economic" research in a difficult position.

- the nuclear industry and R&D establishments, with global warming arguably the only strong argument remaining in their favour. By the mid-1980s the industry had become seriously threatened by falling fossil energy prices and Chernobyl. It had long been a major funder of climate change research in the United States, Britain, Germany and France

- the distributors and owners of natural gas, because of widespread switching from coal to gas in the generation of electricity, especially if efficiency improvements due to combined cycle turbine technology are included, can yield up to 30% reductions in CO_2 emissions

- insurance companies, who would be able to justify increased premiums to pay for past losses, on the expectation that future claims would be even higher

- a whole complex of industries with investment in making more energy-efficient appliances, buildings or materials. Selling "clean" technology (and gas) to industrializing countries, including joint implementation mechanisms, would be good for business, as would the privatization of energy resources and power stations. This, in turn, allows much-needed private investment to flow "South" into sectors already highly profitable

- small, near market producers of renewable energy technologies, such as wind and water, who could plead for subsidies. Subsidies have also benefited the incinerators of rubbish, who practice "energy recovery" rather than waste minimization.

These interests were wooed to varying degrees by organizations like the Greenpeace and the Environmental Defence Fund, who saw them as providing solutions to a problem that confirmed their view of nature and humanity. In turn, they became effective mouthpieces demanding global changes in energy policy. The "climate change" interests in government are equally complex and reflect not only industrial and sectoral responsibilities, but also foreign policy goals and the direct interests of environmental bureaucracies. In the climate debate, governments will therefore rarely remain neutral, simply awaiting expert advice; rather, they will seek to shape global developments to suit their domestic capacities and expectations.

Among the losers we find the coal and oil producers, who face a growing commercial threat from the new technologies and fuels created in the era of high oil prices and now seeking subsidization, natural gas being the important exception. The fossil-fuel lobby took some time to realize the threat to its interests, but is now better organized and has made extensive use of scientific uncertainty in its arguments against hasty action. Its venom has been concentrated on IPCC emission scenarios, which do not predict, but simply assume, a doubling of carbon dioxide concentrations. Deliberate regulation to reduce coal consumption would seriously undermine the economies of major coal exporters (Australia, the United States, South Africa, Colombia)[39] and of a number of rapidly industrializing countries, especially China. These countries have extremely good economic and political reasons for worrying about international demands for reductions in their emissions according to targets and time-tables. They have generally been portrayed as villains by environmentalists, who tend to ignore the fact that the greenest governments tend to be those to whom emission reduction comes cost-free.

Inside science itself, other battles have been raging. The essence of science politics is the competition between the various branches and institutions of science in the making of public science policy. Competition is about status, access to policy-makers and resources. Will global change research be done in the big national laboratories, in less dependent and less readily managed universities, or by "private" consultancies delivering neatly packaged products to funders with little time to read? Will this research be managed by scientists, government departments or national research councils? What will be the main beneficiaries – space technology or space physics, the study of the carbon cycle or of past climates? Will the British Meteorological Office remain able to "outmodel" the German Max-Planck Gesellschaft and will national space agencies support European satellites to provide the observational data needed by their general circulation models? Should not "my" government have its "own" science base before being able to join the negotiations as an equal? Is UNEP to be given new tasks? Central to global warming science politics was the friendly competition between the big climate models found in a handful of laboratories and, less friendly, between governments that needed markets for the expensive products of their space agencies.

Global warming consensus has been negotiated between scientific policy elites in order to maintain the claim to relevance of "earth system science" research. As long as this research is an open process subject to political scrutiny, there may be little to worry about, any biases being self-correcting over time. In the meantime, however, uncertainty fuels political debates and conflicts about response strategies. Rationalism and causality also create much larger problems when the prediction of future societal behaviour is required: for example, mainstream economic theory

---

39 Australia did not agree with the CoP-2 declaration "taken note of" by the parties to the FCCC in Geneva, 1996. The United States agreed to binding limits, but none were agreed.

tends to advocate narrow efficiency optimization not easily reconciled with demand for public action to protect common goods. Consensus among socio-economic theorists attracted to the IPCC was therefore more about the assumptions of their theories than about what ought to be done.

5.3. Six Hypotheses

For political science (and for reflective policy-makers) all this is an exciting area of investigation. Learning is opened up which includes observations of the use and misuse of knowledge in the political process. Six hypotheses have emerged from my work, all highly relevant to global warming politics.

5.3.1. Science Policy As the Cheap Option of Environmental Policy

Research viewed as an industry creating products that promise future utility needs justification. It thrives on ignorance and uncertainty. Once all is known, science ceases to be interesting. Certainty is therefore a danger to the research networks upon which the IPCC represents and from whom it must draw information and judgements.

Once an environmental problem is "known" (or believed), policy will go elsewhere for advice. Solving a problem that natural science has identified is a very different issue, taking societies into the realm of practical implementation. However, the expectations of what this implementation might mean for polluters, investors, regulators and those offering solutions in future, is already becoming part of the policy formation process today. In this process, actors on the side of "solutions" have made much use of the concept of scientific consensus.

A political study of the behaviour of one of the largest global industries, the research enterprise, assumes strategic behaviour. Scientific institutions cannot but act to maintain and enhance their interests. Their enterprise remains firmly based in a handful of industrialized countries, but needs a global information base and is firmly tied to developments in telematics – information technology and electronics. The research enterprise must be considered as a major actor in climate change politics, in addition to industrial and bureaucratic organizations.

The list of political issues related to climate change science seems endless, though short compared to those affecting industry and environmental regulators, at least potentially. Yet very little systematic research has been done on these politics and its implications for environmental policy. A major dimension of each set of politics is the utility of uncertainty to actors wishing to shape policy. Uncertainty itself is a political actor because so few people are able to judge the accuracy of claims to "scientific consensus" once science leaders have delivered this as a product in the name of the "scientific community".[40] Such a "consensus" is negotiated at the very top of scientific hierarchies by individuals with a firm belief in "science" and their own worldview, and whose task it is increasingly to protect the research enterprise in a harsh world of "steady state" funding and growing competition. Inside the IPCC the competition has been tough and has involved the use of UN organizations, especially the WMO and UNEP as described above. Large areas of relevant, above all socio-political research were initially neglected (and later emasculated) because of the superior ability of the natural sciences to organize, disseminate and attract global interests to their "neutral" but strongly R&D related research agendas.

The implications for environmental policy need to be observed and discussed. Once the production of knowledge is shaped by the globalization of research agendas justified with reference to global environmental problems, concern for implementability of environmental policy emerges on three grounds: (i) neglect of non-(natural) scientific knowledge in its design or making; (ii) stifling of scientific debates in order to protect the policy relevance of already funded research; (iii) impact on the overall direction and objectives of policy and hence priorities selected for action.

Research objectives rather than societal need may direct policy and weaken the ability of societies to deal with existing environmental (and social) problems. Instead, what is promoted is curiosity-driven research as defined by a small number of individuals carefully selected from the global pool of ability.[41]

Science policy cannot replace environmental protection. While science can predict or estimate change, it cannot value these changes except as deviations from something "natural". The natural rather than the created or constructed therefore tends to become the norm. The human and social are devalued. Wilderness becomes superior (because purer) to gardens, and environmental fundamentalism is indeed implicitly advanced.

---

40 Is the 1994 IPCC working group I report on "Radiative Forcing of Climate Change" any more true because it has been compiled by 25 lead authors from 11 countries, based on contributions from 120 contributing authors from 15 countries?

41 A social analysis of IPCC participants indicates that all were English speakers; about 70% of the SPRU Questionnaire respondents gave English as their native language. Very few women participated at the research or policy level, and promising researchers from for example China would soon be attracted to the major laboratories in the "North".

Such a development would deny centuries of interaction between man and nature. A simplistic naturalism has been replacing socio-cultural judgements in some sections of the environmental movement.[42] With society and politics out of the way, environmental policy would be shaped by green ideology or science politics and mathematics – given that environmental research, especially that relating to climate change, is increasingly mathematical. Such a development might be contemplated in the light of thoughts by A. N. Whitehead on the nature of mathematics:

> Let us grant that the pursuit of mathematics is a divine madness of the human spirit, a refuge from the goading urgency of contingent happenings (quoted in Barrow 1991: 172).[43]

Can divine madness serve environmental contingency?

5.3.2. "Rational" Policies Serve Political Purposes

Deducing action guidelines from theories which treat mankind as a single species pursuing its self-interest – be this as "the world community" or an aggregate of optimizing individuals – assumes not only that a common or aggregate interest can be defined, but also that it can be effectively pursued by social engineering, by a consistent policy-mix implemented over a very long time. History, geography and politics do not share such assumptions and raise serious reservations about the achievability of such a project. Even at the local level, the plan rarely becomes reality. There is no single society or environment and many political questions arise about the widely recommended "technology transfer" from "North" to "South". Such transfers may be prescriptions for selling more expensive technology to rich elites in poor countries who use up most of the available national energy.

The debate on the practical achievability and political impacts of climate policies seems to have a rather optimistic bias: Most experts seem

---

42 If bodies like the IPCC were asked to define "dangerous" under the FCCC, for example, a political judgement would be handed to a small, unelected group with specific interests.

43 Barrow deals in some detail with the profound differences between words and numbers. He states that modern science is founded almost entirely upon mathematics which is acultural and apolitical; it discovers rather than invents, but restricts its powers to the natural world. However, he makes a very sharp distinction between the natural world, about which everything might be knowable, and the worlds of society and culture, for which he makes no such claims.

to agree that the technical means (including economic instruments) for turning the energy "juggernaut" around do exist provided mankind behaves rationally (Dornbush and Poterba 1991). However, most of the recommended technical solutions will require major new investments, which in turn will have profound political as well as economic impacts. The envisaged investments and regulations are expected to advance technology and competitiveness in the "North" although their climatological impact must remain speculative at best. The impact on poor countries is also difficult to predict. If they accept the climate threat as "real" and blame the North, will the pursuit of a strategy of "green mail" – that is, technology transfer – actually benefit them (Jordan 1994)? Considerations of the secondary impacts of climate policies form the essence of political judgement. Political judgements are responsible to specific constituencies and not to humanity as a single entity.

The implementation of major environmental agreements suggests that the political power to intervene successfully to reshape economies and man–nature relationships is limited and carries with it considerable political implications. Societies do not learn very quickly or even permanently; social cohesion is rapidly destroyed and political stability threatened. Exploitation comes more easily than improving welfare, myth more readily than truth. Without stable institutions and functioning political systems, sustainability, however well defined and measured, is not likely to happen.

5.3.3. Scientific Consensus: Insurance Policy for Government or Foundation for Socio-Economic Modelling?

The political utility of scientific consensus, i.e. agreement among scientists about what is known, what is not and what might be given sufficient funding, relates to questions of authority and legitimation, not merely the provision of information for policy design. The belief among academics that politicians need scientific consensus before they can act is, however, quite surprising, even naive. It may reflect either or both the successful insertion of scientific institutions into the administration of environmental policy, or derive from the Anglo-Saxon two-party system. In the latter case advisory bodies set up, for example by the British Parliament, are considered to have an impact on policy only if their recommendations are bipartisan, i.e. when representatives of both parties have been able to reach consensus on the basis of the advice they are given. This, however, is clearly achieved by bargaining and not "science".

A third explanatory factor concerns the need of politicians to protect themselves against policy failure. By demanding scientific consensus even on issues at the frontiers of research, politicians and their advisors "delegate" responsibility "down" to what are in fact paid servants of power: the experts (Cole 1992). However, at the frontiers of knowledge, there cannot be consensus about the "facts", but only a political agreement negotiated between individuals supported by major scientific institutions. Pushed too hard towards constructing consensus when debate comes more naturally, science will deny all policy relevance. Once it is accepted, as it often appears to be, that environmental policy must be derived from science, then the pressure for a policy dedicated to learning more rather than doing is strong, and one that raises questions of learning by whom and for whom.

When "science" is certain, on the other hand, politicians and bureaucrats have an excuse for avoiding thinking or accepting responsibility for their judgements. This "need" for the facts advances the usefulness or relevance of research, especially when "science" confirms what politicians want to do anyway. When it does not, scientific knowledge in its most rigorous sense is readily "deconstructed", especially by those who do not understand its fragility. This potential power has long worried students of public policy. From the perspective of policy-makers, who often have scant direct experience of how science works and are indeed in need of "objective" decision-criteria, scientific consensus can have profound attractions even if it is not needed when there are non-scientific reasons for taking, or promising to take, action.

Studying the fields of knowledge attracted to the IPCC reveals another group of people who need scientific consensus, and probably need it much more than government. Agreement on specific measurements of risk or danger are needed most by social scientists who want to quantify policy advice and build their own intellectual edifices or numerical models on the objective results of the natural sciences. This brings us back to the problem of cascading uncertainties. Whereas politicians are experienced in making decisions on the basis of belief, prejudice, ideology, fashion or bargaining, social scientists interested in prescriptive policy analysis argue that consensual knowledge is essential (though not sufficient) for a rational environmental policy (Underdal 1989). It is therefore argued that scientific consensus is a product of an interdisciplinary science policy that tries to shape environmental policy to its own advantage. This may not be entirely voluntary because the pressure to remain policy-relevant requires the institutions of science to walk a tightrope between creating sufficient certainty to initiate political demands for action, but also uncertain enough to justify the search for more knowledge in their particular disciplines. They therefore make unreliable allies for policy-makers who want to act, and more useful ones for those who do not.

5.3.4. Environmental Politics: Science as Tool of Persuasion, Regulation and Planning

If environmental politics is seen as the art of using assertions, facts, opinions and evidence about "the environment" in order to persuade markets or government, consumers or taxpayers to change their behaviour, then all "groups" or institutions with a climate policy are "interested political parties". Science has become their favourite tool of persuasion, creating the above-mentioned opportunities and problems for scientists. The political stakes of policy are far higher than those of science – high enough to involve not only national governments, but virtually all intergovernmental organizations. Not a single ministry, but several will be involved, with major interdepartmental conflicts typical rather than rare. Climate policy is never made in a vacuum, but must adjust to an existing and strongly patterned policy context which includes macroeconomic and foreign policy objectives. When these politics become too complex to generate decisions, then simply funding more research stands out as a cheaper and more convenient option than adopting new legislation, taxes and enforcement strategies. Via technological change, the market will occasionally offer solutions which government can then present as policy achievements.

With commercial competition often a part of climate change politics, the interests to be considered include not only industry (investors, companies, employees) and governments (competing administrations as well as politicians manoeuvring for advantage at domestic and intergovernmental levels), but also a great many environmental and development organizations. It is they who have largely defined, with the help of scientific evidence, environmental change as degradation or threatening to life. Visions of the future, especially worst-case scenarios, become tools of persuasion. A great deal of professional expertise now lives off these scenarios by providing the "facts" as well as the solutions.

Climate politics therefore consists of debates about environmental robustness versus catastrophe, about the credibility of model predictions and the reliability of observational evidence, of the seeking of competitive advantage through regulation (or deregulation) and of subsidies or their abolition. Arguments with potential losers will be serious. Businessmen responsible for real jobs and products clash with green activists who feel a responsibility not for the unemployed or poor living today, but for the planet's future.

Vested interests, it must be noted, have not defined the problem: this has largely been done by the research enterprise, supported by environmentalists. The political power of those defining the problem is very considerable, however. As has been argued for the evolution of decision-making in the EU, "... organizations will only have a marginal effect ... More important will be who can give the conceptual discourse direction and thereby shape ideological hegemony... Those possessing the capacity to develop effective strategies will also have a great say in the process of conceptualization. By producing 'leading conceptions' they have an influence on the choice of action and objectives" (Kohler-Koch 1994: 21). These small groups, I have argued, are not bench scientists, but scientific managers with direct access to government as a major employer of "expertise". Once the problem has been defined and solutions proposed, they are tested for their practicality and broader impacts in debates by the real stakeholders: politicians, investors, taxpayers, employees and employers: that is, all the individuals and groups that must "pay the price" of any solution or benefit from it. Cost and benefits will be allocated or redistributed through electoral systems, taxation and regulatory regimes, as well as the penalties or rewards of markets.

5.3.5. Climate Policy is not Decided, it Emerges as Resultant of Many Forces

The scientists' policy-model underlying the IPCC can be deduced from many documents as well as the formal descriptions provided by its chairman (Bolin 1994: 27). It envisages natural science (and branches of value-free "socio-economic research") as the only source of advice needed by policy. Politicians merely add value to fact. This is not how policy is made in the real world, but an elegant simplification which exaggerates the role of science and the difference between fact and value. Although it is not fully clear how and by whom climate policy has been made in recent years, the range of actors and arguments seem similar.

At least as regards the policy process in Britain and Germany it appears that climate policy proposals emerge from small sections of the environmental bureaucracy and reflect their current expertise. Agreed between invited experts and middle-rank civil servants, the latter have to "sell" their proposals to senior administrators and finally politicians with very little knowledge of the subject and more pressing immediate concerns. Climate policy must be attached to these pressing concerns and it will be these, rather than science, that will determine the positions taken. In Britain, policy emerged from a small "policy network" that could rely on early and strong support from the very top. The experts and behind them, the research enterprise, played a major initiating role. Indeed, climate policy became research policy, and research policy foreign policy, each made possible by developments in the field of energy which promised and delivered emission reductions at very little cost to government.

Policy in general seems to emerge from complex bargaining at several interconnected levels. Thus, no single decision-makers can be defined: public opinion with politicians (party politics); government departments with each other (bureaucratic politics), commercial competition (for market shares, government subsidies or regulatory help); ideological competition (for concern and membership of NGOs), as well as alliance-building (for or against government intervention). Such a process of policy formation directly links environmental policy to good government, raising questions about who best represents the public interest, both global and national, in climate negotiations. Environmentalists and scientists have tended to ignore this dimension. What should be the roles of unelected NGOs, national parliaments and political parties? How much power should society invest in allegedly apolitical "experts" who may well act to advance knowledge rather than its implementation? And what are the likely world political consequences of widespread intervention in energy markets? At this point we enter a world of speculation which can only be explored through observation, interpretation and retrospective analysis, not quantitative model building. Motives, judgements and investments are the real factors that shape the future.

5.3.6. Impacts on the Global Political Economy May be Harmful

Given its widespread use in the current research and management, it is difficult to envisage climate policy disappearing from world politics irrespective of the fate of remaining scientific uncertainties. Rather, we might expect predictions to adjust to what is desired by authorities and current fashion. Political analysis therefore invites the interpretation of climate change negotiations in the broad context of the forces which shape the global political economy: power struggles involving trade, technology transfer and investment flows. Changes in energy prices and fuel competition are major agents of change, with direct but unpredictable effects on emissions. By controlling emissions, the major measure advocated in climate policy, governments may seek some control over energy prices and markets. This involves major interventions by public authorities, the political and economic consequences of which are very difficult to predict. From the perspective of political economy it is of interest to "guess" how the climate-change debate might have run for the various actor groups had the research lobby closed ranks around a prediction of "global cooling" – as seemed possible in the late 1970s. Probably very little political energy would have been generated as no research-intensive solutions, other than burning more fossil fuels, would have presented themselves (Weart 1992).

On the basis of speculations about the potential political impacts of proposed emission reduction strategies, it can be argued that the policies sought by the most powerful actors, the winners in the game of climate politics, are likely to strengthen rich communities, wherever they are found, at the expense of the poorer ones – unless (and this is highly unlikely) major redistributive efforts are undertaken in advance. Environmentalism, once combined with contempt for humanity, is not likely to provide the conceptual incentives for such developments. Rather, spreading global inequality and poverty may attract ethical justifications from environmentalist doctrines that put the blame not on "structures" but on individual failure, including uncontrolled "breeding" by the poor and their refusal to restrict their mobility and desire to consume. It will be forgotten that the new, high-technology and information-based economies will require fewer people. Human labour per se will decline further in value, adding to the negative view of human society of many environmentalists. The much lower economic evaluation of a peasants' life compared to that of a Northern city dweller by IPCC authors has already caused resentment (Pearce 1995: 5).[44] Rather than "healing" North–South rifts, global warming seems to be intensifying the confrontation.

The global warming threat may thus come to justify both reaction and modernization, the centralization of governmental power or its wider dispersal among societal actors. In some countries, it may justify a new Malthusianism, in others the greening of Christianity by offering a modern interpretation of paradise, hell and sin, or a green Marxism by providing new reasons for the battles against capitalist exploitation, not of Man but of Nature. While rural romantics may tilt against urbanization and technological change, postmodern thinkers may use the environment to reject the welfare state and its faith in science and progress. To modernists, the environmental imperative would have to

---

44 IPCC economists have calculated that a Bengali labourer ($150,00) is worth much less than a "Northern" citizen ($1.5 million). Calculations about the number of people likely to die (200,000 per year within 50 years) seem not only unwarranted by the "science base" but might be compared with millions dying now because of inaction over TB and poor nutrition, branches of policy that are not research-intensive.

serve as justification for major efforts in public investment and public policy planning: for "ecological modernization".

6. Conclusions: Open Science Politics and Humanist Knowledge

Stabilizing or reducing greenhouse gas emissions globally will require a far greater effort than just doing more research. What will be necessary is a transformation of the governance of the world and its energy supply and demand system, indeed of the financial and political principles upon which the global political economy is based. The nature of this transformation is now an academic battleground with important resonance in the world of Realpolitik. The outcome, if there is any, will not be left to environmentalists and their self-interested allies. Climate change negotiations are likely to run for a very long time indeed and will do so as part of the "sustainability" debate, a slogan which may yet provide a cover for the myths behind which political forces compete for global hegemony. Or it may disappear like the concept of "limits to growth", after having served a similar purpose of making research "relevant".

The policy agenda is still malleable. To prevent malign developments, the politics of climate change need to be understood together with the ambivalent role of scientific institutions. Tactical behaviour by scientific bodies should be recognized but not necessarily condemned as unethical. Science policy elites defend their constituencies by giving advice which is politically useful. This motive is best developed at the international level, where the protection of sovereignty and national choice remain overriding goals and ecological interdependence can be rhetorically recognized with minimal concessions to political independence. Those who do not understand "science", i.e. do not possess general circulation models, can be excluded from the negotiating table – or be trained so as to gain "capacity".

Research is now treated as an industry "bidding" for missions and marketing the relevance of its products. It will behave like a commercial entity. Adjustments to "truth" are the responsibility of the buyer, who will need to seek full information from more than one source. As consumers of resources, scientific experts are unable to direct resources to solutions. Their greatest power lies in their ability to raise concern and direct resources to learning – no more. However, being engaged more than ever before in the creation of markets for data and knowledge means that the research enterprise should not also carry, or take upon itself, the responsibility for directing policy. The wise, fair and appropriate use of knowledge should not lie with the institutions of science, but with those of politics. The search for more knowledge about climate change and its impacts would become redundant if the search for "real" solutions were to predominate. This is not likely to happen unless millions of people have the desire and confidence and in their combined ability to determine their own future. In the meantime, the truth of climate change remains a promise in the hands of space agencies, the WMO, the ICSU and Meteorological Offices. Can we trust them?

As prophesy remains part of climate policy advice, the political price of yielding to those who propose that we should manage our economies in the interests of the future may well be too high. Future societies may have to adjust to the change their predecessors have caused, as we have had to adjust to the legacies of our forefathers. It could even be argued that, transferring people away from areas to be flooded in a century or so may do less damage to the planet-societies system than the conflicts which, for example, a significant carbon tax is likely to provoke. The dangers of political misuse of the future seem too great. That is why no clear policy recommendations will emerge from this exploration of the use of knowledge as a tool of environmental and energy politics.

For policy analysts the question arises of how to encourage "structures" and attitudes which can allow knowledge to enter the political process without politically or commercially motivated selection. Without integrity, knowledge cannot fulfil its major global responsibility: to legitimate joint actions which serve some common interest. Knowledge should not be funded because it claims to be "policy-relevant". Rather, governments must protect its development and openness so that policy relevance may be judged in the democratic process. Science needs to be tested by disputes, not funded to be consensual. Science is too readily "deconstructed" to supply the motivation for major societal changes. Only human attitudes, interests and beliefs can create a coalition in favour of big causes and it behooves the academic community, including the natural sciences, to look at global causes sceptically and critically.

Underlying this argument is the assumption that better policy outcomes are likely to result from the joint application of many types of knowledge generated not by complex computer models but in the political process involving very many human minds and views, many undoubtedly assisted by such models. There is little social value in prescribing "perfect" solutions if effective implementation first requires the creation of a "perfect" world to satisfy the assumptions required by mathematical logic. This may require a clearer separation between science as knowledge and scientific institutions, between science policy as resources and goals dedicated to research and environmental policy as a set of goals and instruments applied to the protection of the physical world. Above all, theoretical knowledge needs to be tested against observation, experience and cultural values. It is these realms which will, in the end, decide between uncertainties that matter and those that do not.

Chapter 5

Decision-Theoretical Approaches to Climate Change

Sven Ove Hansson and Mikael Johannesson

1. Introduction: An Extraordinary Complex Decision-Problem

Climate change is a complex process that involves a multitude of both natural and social causal factors, and a large amount of recalcitrant scientific uncertainty. As a consequence of this, decision-making on measures against global warming and its effects poses unusually difficult decision-theoretical problems. In this chapter we review the major issues involved and discuss some of the approaches proposed in the literature.

The ultimate energy source for weather and climate is radiation from the sun. Some of the radiation that is not reflected back to space is absorbed by the atmosphere, but most of it is absorbed by the land, ocean, and ice surface. Incoming solar radiation is balanced by outgoing infrared radiation at the top of the atmosphere. Some of the outgoing radiation is trapped by naturally occurring greenhouse gases, principally water vapour (H_2O), but also carbon dioxide (CO_2), ozone (O_3), methane (CH_4) and nitrous oxide (N_2O). This natural greenhouse effect keeps the surface and troposphere about 33°C warmer than what they would otherwise have been. Various human activities add more greenhouse gases to the atmosphere. It is generally agreed by climatologists that this will lead to a further increase in temperature. This is the anthropogenic part of the greenhouse effect, often somewhat misleadingly called just "the greenhouse effect" (IPCC 1994a: 15). There are strong reasons for believing that the anthropogenic greenhouse effect may have severe consequences for human life on our planet, by leading to higher sea levels, and worsened conditions for agriculture, among other things.[1]

---

1 See Chapter 3.

Three major practical approaches or strategies with respect to managing the anthropogenic greenhouse gas effect seem to be available. The first of these is *prevention or mitigation*. In order to prevent or reduce the anthropogenic greenhouse effect we can reduce the concentration of greenhouse gases in the atmosphere. Emissions of greenhouse gases can be cut back. Furthermore, the carbon sinks can be increased, or prevented from being decreased, through measures such as reforestation, decreased deforestation, or increased content of organic carbon in the soil through modified agricultural methods (Arrhenius 1992).

According to mainstream scientific models, in order to achieve a stabilization of atmospheric CO_2 concentration at all considered levels between one and two times today's concentration (that is to say, between 350 and 750 ppmv), anthropogenic emissions must be reduced to levels substantially below those of 1990 (IPCC 1994b: 4). In order to avoid any increase above today's concentrations of greenhouse gases, current anthropogenic emissions of long-lived gases (CO_2, N_2O and CFCs) need to be reduced immediately by more than 60% (IPCC 1990b: 2).

The second strategy is *compensation*. Various proposals for "climatic engineering" have been put forward (Wiman 1995). In order to compensate for greenhouse gas emissions, large mirrors could be placed in space to reflect sunlight, dust could be injected into the atmosphere to screen out sunlight, or iron compounds could be spread to fertilize polar water in order to stimulate the growth of phytoplanktons and thus withdraw CO_2 from the atmosphere.[2]

The third strategy is *adaptation*. Human civilization can adapt in various ways to the effects of global warming. Walls can be built to protect against sea-level rise. Irrigation systems can be constructed to cope with changing patterns of precipitation. New crops can be grown that are more suited to the new climate, etc. At least in one case, such adaptation has already taken place. In anticipation of future rises in sea level, Royal Dutch Shell has decided to increase the height of its gas platforms in the North Sea from 30 m to 31–32 m above the present sea level, ironically, in order to facilitate energy production that will contribute to further aggravation of the greenhouse effect (Wiman 1995).

The decisions that need to be taken in response to the anthropogenic greenhouse gas effect include a choice between these options and various

---

2 The limit between prevention/mitigation and compensation depends on exactly how the problem is defined. If the problem is described as "global warming" rather than "global warming due to anthropogenic greenhouse gas emissions", then most of the strategies described here as compensation should instead be counted as prevention/mitigation.

combinations of them. We must also decide when, and based on what type of evidence, various types of action should be taken. Furthermore, the distribution between nations of the costs and other burdens imposed by these actions has to be decided.[3]

Climate change is a difficult decision problem for several reasons. Perhaps the most obvious of these is the mere magnitude of the problem. A major change in climate may, in the long run, have world-wide catastrophic consequences. Measures to prevent, mitigate, compensate or adapt may turn out to be extremely costly and to require changes in strongly entrenched habits and life-styles. Decision-making on climate change is also complicated by a series of problems that are caused by the intricate natural and social phenomena that are involved. These problems can be divided into three major categories: scientific uncertainty, co-ordination problems, and problems connected with the time factor.

Obviously, *values* have an important and far from unproblematic role in decision-making about climate change. What are our obligations to posterity? What are the obligations of industrialized countries to developing countries? How important is biodiversity? Value issues like these have often been confused with both scientific and decision-theoretical issues. We believe that a systematized account of the decision-theoretical issues can facilitate clarification of the value issues by making them more isolatable.

The following three sections of this chapter are devoted to each of the three above mentioned problem-areas. In the first section we discuss the scientific uncertainty. The possibilities for reducing this uncertainty or making it accessible are outlined, and we comment on different decision-rules for coping with it. In the second section we discuss some of the coordination problems that are involved in climate policies. These problems are related to the great differences between countries and peoples concerning emissions of greenhouse gases, vulnerability to the greenhouse effect, etc. Finally, in the third section we discuss problems related to the time factor.

The three categories are partly overlapping; in particular, the third category overlaps substantially with the other two. We have chosen this classification since it is helpful in systematizing and comparing, from a decision-theoretical point of view, the various strategies that have been proposed in the literature.

---

3 See Chapter 6.

2. Scientific Uncertainty

In spite of extensive research, there are still many and important unanswered questions concerning the anthropogenic greenhouse effect. This applies especially to the climatological issues. Our knowledge is still incomplete about what determines the climate and how it is influenced by greenhouse gas emissions and other human activities. It also applies to the effects of potential climate change on the natural environment and on human society. It is essential for the policy-making process to be aware of the nature of this scientific uncertainty and to realistically assess the chances of reducing it. To the extent that it cannot be eliminated, we need decision-making procedures and criteria to cope with it in a rational manner.

2.1. The Nature of Scientific Uncertainty

A vast number of factors and potential mechanisms need to be taken into account when determining the effects of increased greenhouse gas concentrations in the atmosphere. Complex natural systems such as seas, forests, and soils are influenced by a change in the climate, as are human activities such as food production, international trade, and energy production. Social factors such as health, population, migration, and life-styles also need to be analysed. The degree of uncertainty varies in

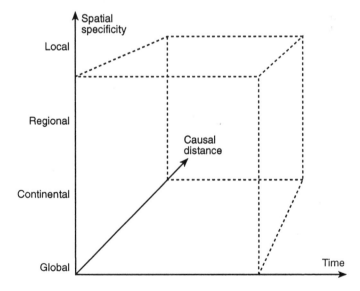

Figure 1: Dimensions of Scientific Uncertainty

different subfields of climate studies. Three dimensions can be identified that roughly determine the degree of uncertainty (see Figure 1).

Time: Our knowledge about emissions, greenhouse gas concentrations caused by the emissions, and temperature changes caused by these concentrations is better for the immediate future than for the more remote future. The same applies to our ability to judge how human societies will react to various possible effects of climate change. Uncertainty increases as the time horizon moves into the future.[4]

Spatial specificity: Uncertainty also increases as we move from global to regional and local effects. The models employed in calculations of future effects of greenhouse gas emissions are generally believed to be reasonably reliable in their prognoses of changes in the *average* temperature. They are much less reliable in their predictions of how changes in temperatures and precipitation patterns will be distributed over various regions on earth.

Causal distance: Uncertainty increases rapidly with the length of cause/effect chains (see Figure 2 for examples of causal chains). A useful summary of the degrees of uncertainty connected with various scientific judgements on climate change was provided in the 1990 report by the Intergovernmental Panel on Climate Change (IPCC) (1990b: 2–3). The IPCC's conclusions were divided into four categories with respect to scientific reliability or certainty. First, the IPCC were *certain* that some statements had been scientifically validated. Secondly, they *calculated with confidence* that certain other statements were true. About a third group of statements, the IPCC stated that "*based on current model results, we predict*", and for a fourth group they said that "*our judgement is that*".

The highest degree of reliability ("certainty") was assigned to the following statements:

- There is a natural greenhouse effect which already keeps the earth warmer than it would otherwise have been.
- Emissions from human activities are substantially increasing the atmospheric concentrations of greenhouse gases.
- The increase of greenhouse gas concentrations will enhance the greenhouse effect, resulting on average in an additional warming of the earth's surface.
- The main greenhouse gas, water vapour, will increase in response to global warming and further enhance it.

---

4 In the same way, but to a smaller extent, uncertainty increases if we move backwards in time. On the time-scale of the history of science, older measurements are fewer and less reliable. On a geological time-scale, judgements of palaeoclimate in older geological periods are usually more uncertain than judgements of younger periods.

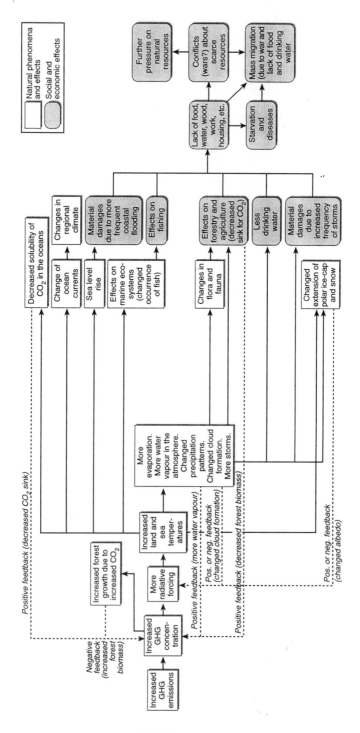

Figure 2: Climate Change – Examples of Causal Chains

The second degree of reliability ("calculated with confidence") was assigned to the following statements:

- CO_2 has been responsible for over half the enhanced (anthropogenic) greenhouse effect in the past and is likely to remain so in the future.
- Atmospheric concentrations of the long-lived gases (CO_2, N_2O, and the CFCs) adjust only slowly to changes in emissions.
- The long-lived gases would require immediate reductions in emissions from human activities of over 60% if their concentrations are to be stabilized at today's levels.

Important examples of the third degree of reliability ("predictions based on current models") are the quantitative predictions of the temperature increase and sea-level rise at different emission levels.

One example of the fourth category ("our judgement") was the appraisal that global mean surface air temperature has increased by 0.3°C to 0.6°C over the last 100 years. Another example is the statement that the effect of warming on biological processes may increase the atmospheric concentrations of natural greenhouse gases.

Scientific uncertainty adds up to two possibilities: That the effects of greenhouse gas emissions may be much less serious, and that they will be much more serious than has been concluded by the IPCC in their major scenarios.

As was observed by Bolin (1994: 27), the claim that the IPCC has overestimated the effects of increasing greenhouse gas concentration in the atmosphere is most commonly found in unrefereed report series and in the popular press. There are few papers in the refereed scientific literature that support this claim. Among the few researchers that challenge the conclusions of the IPCC from this side are Sun and Lindzen (1993) who have outlined how more intense convection in the tropics could change the distribution of clouds and water vapour in a way that would compensate for increased greenhouse gas emissions. However, they do not seem to have been able to show that this is likely to occur (Bolin 1993: 3). Another minority opinion with respect to the effects of increasing concentrations of greenhouse gases is represented by Mikhail Budyko (cited in Miller et al. 1989: 24), who acknowledges that they will lead to increased temperature and climate change, but claims that this will be a change for the better, since he is convinced that increased precipitation will make agriculture possible in previously arid areas in Africa and Asia.

On the other side, some scenarios have been proposed that will lead to more serious effects than those projected by the IPCC. Usually these scenarios are based on feedback mechanisms that could increase the rate of greenhouse gas build-up in the atmosphere.[5] The worst-case scenarios include a run-away greenhouse effect. This means that positive feedbacks make the climate warmer at such a pace that it cannot be stopped by any measures available to humanity. Although not much discussed, a runaway greenhouse effect cannot be definitely ruled out. In January and February 1992, Greenpeace International polled 400 climate scientists, including those that were involved in the 1990 IPCC study and others who had published on issues relevant to climate change in *Science* and *Nature* during 1991. Of the 113 scientists who responded to the questionnaire, 45% said that a runaway greenhouse effect is possible if no action is taken to reduce greenhouse gas emissions; 12% believed such a scenario to be probable (Leggett 1992: 30).

2.2 How Far Can Scientific Uncertainty Be Reduced?

Scientific uncertainty is a matter of concern to both scientists and decision-makers. It is useful to distinguish between what scientists *qua* scientists can do about scientific uncertainty and what decision-makers can do.

The most obvious approach that scientists can have to scientific uncertainty is to try to reduce it by producing new knowledge. Indeed, new knowledge about global climate change is accumulating at a rapid rate. Unfortunately, however, scientific uncertainty does not always decrease with increased knowledge. New research may result in three types of new information that influence the certainty/uncertainty of previous conclusions.

1. *Decreased uncertainty through new results that confirm or corroborate previous assessments:* As an example of this, in a reassessment in 1992 of their report from 1990, the IPCC wrote that since 1990 some progress had been made in quantifying the cooling effect of aerosols (IPCC 1992a: 4).

---

5 The following are three examples of such mechanisms: Increased concentration of water vapour in the atmosphere leads to higher temperature, which leads to increased evaporation and increased concentration of water vapour since warmer air can contain more water vapour. Increased sea temperature leads to decreased solubility of CO_2 in the sea, which leads to more CO_2 in the atmosphere and higher temperature. Massive quantities of CH_4 are trapped in permafrost onshore and offshore in polar regions. Significantly increased temperature in high latitudes might cause release of huge quantities of CH_4. Increased concentration of CH_4 in the atmosphere will lead to increased temperature and more CH_4 could be released. The first two examples are included in Figure 2.

2. *Increased uncertainty due to new results that falsify or cast doubt on previous assessments:* This, too, can be exemplified with the 1992 reassessment. The IPCC wrote: "we now recognize that there is increased uncertainty in the calculation of GWPs (Global Warming Potentials), particularly in the indirect components" (IPCC 1992a: 5).[6]

3. *Increased uncertainty due to new results that reveal mechanisms or phenomena that were previously unknown or so little known that they had not been taken into account:* In 1992, the IPCC referred to "a number of significant new findings and conclusions" since 1990. An important example was the information that "depletion of ozone in the lower stratosphere in the middle and high latitudes results in a decrease in radiative forcing which is believed to be comparable in magnitude to the radiative forcing contribution of CFCs globally-averaged over the last decade or so" (IPCC 1992a: 4). The discovery of this mechanism led to a complete reappraisal of the net contribution of CFCs to the greenhouse effect.

New knowledge of all three categories, perhaps in particular the second and third, force us to make new assessments of scientific conclusions that influence decisions and policies. The global system is so complex that no guarantee seems to be attainable, now or in the near future, that all relevant mechanisms have been identified, let alone their impact correctly calculated. In particular, physical, biological, and social feedback mechanisms are difficult to identify and appraise. The climate system is in principle a chaotic system, and it can therefore be expected to be predictable only to a limited degree.[7] It does not seem plausible that surprises can ever be ruled out in this field of science (IPCC 1990b: 3 and 19; Bolin 1994: 28).

2.3 Making Scientific Uncertainty Accessible

Besides reducing scientific uncertainty, scientists can also attempt to *describe* it in a way that is accessible to decision-makers. The need for this has repeatedly been emphasized by the chairman of the IPCC, Bert Bolin. In a speech in 1991, he emphasized that "if sound scientific/

---

6 The GWP index is defined as the time-integrated warming effect due to an instantaneous release of a unit mass (1 kg) of a given greenhouse gas in today's atmosphere relative to that of CO_2 (IPCC 1990b: 12).

7 In a chaotic system semi-linear changes may be discontinued quite abruptly. Some climatologists are of the opinion that the fairly sudden return to much colder climate during 500 years at the end of the latest glacial period (about 10,000 years ago) in at least parts of the northern hemisphere is an example of this (Bolin 1993: 5).

technical arguments led to different results in an open exchange of ideas, then this was an indication of genuine uncertainty; that uncertainty should be recorded and this principle must be upheld very strongly" (quoted from Bernson 1993: 20). More recently, he said that "in the past, Working Group I carefully distinguished between: what we know with certainty; what we are able to calculate with confidence and what the certainty of such analyses might be; what we predict with current models; and what our judgement will be, based on available data. A similar approach will be used by all three working groups in the next assessment" (Bolin 1993: 27).[8]

In science, the burden of proof resides decidedly with the person who claims that a previously unproven effect exists, rather than with the person who doubts it. This must be so in order to prevent scientific progress from being blocked by the pursuit of a large number of blind alleys. In practical decision-making based on (uncertain) scientific knowledge, there are often reasons why the burden of proof should be distributed differently. Such a view of the burden of proof has often been promoted in applied toxicology in particular (Rudner 1953; Jellinek 1981; Hansson 1991). In our opinion, the global climate issue is a case in point, i.e. due to the potentially devastating and irreversible effects it is morally indefensible to defer action until all scientific doubts have been put to rest. This view concurs with the precautionary principle: "Where there are threats of serious or irreversible damage, lack of full scientific certainty shall not be used as a reason for postponing cost-effective measures to prevent environmental degradation" (Rio Declaration 1992: Principle 15).

Hence, the burden of proof for decision-making purposes cannot be assumed to be the same as that for scientific purposes. This makes it necessary for scientists to present their material in a way that clarifies the uncertainties involved. Ideally, decision-makers with different opinions in burden-of-proof issues should be able to use the same scientific material. The nature, extent, and possible consequences of scientific uncertainty should be clarified as fully as possible. In particular, close attention should be paid to less probable but more serious effect scenarios, such as the runaway greenhouse effect, that diverge substantially from the most probable scenario.

In spite of the unusual awareness of the IPCC's chairman of the problems of scientific uncertainty, scientific disagreement does not have a prominent role in the publications of the IPCC. The reason for this may

---

8 The task for IPCC Working Group I is to "assess available scientific information on climate change" (IPCC 1990b).

be a tendency among the participants in the IPCC process to seek consensus and to opt for compromises whenever possible.

In many, perhaps most, cases, the search for scientific consensus is both reasonable and useful. In relation to the science of climate change, however, a too consensus-oriented approach may hide away one of the most important aspects that decision-makers need to take into account, namely scientific uncertainty. In order to provide decision-makers with as adequate information as possible, diverging opinions as well as majority opinions should be presented. Committee procedures for the IPCC and other similar scientific bodies remain to be developed that encourage rather than discourage the public presentation of minority opinions. This is particularly important in the case of the IPCC, since the vast majority of the world's leading experts in the relevant fields are involved in the IPCC process.

In conclusion, it seems to be fairly obvious how scientists as a group should react to scientific uncertainty, in order to be as helpful as possible to decision-makers: They should try to reduce the uncertainty, describe any remaining substantial uncertainty accurately, and avoid overly consensus-oriented procedures that may lead to the neglect of possible but less probable cases.

2.4 Decision-Rules for Scientific Uncertainty

We can now turn to the decision-maker's perspective on scientific uncertainty. Decision-making under imperfect knowledge is the central issue in most decision theory. Although most of the discussion on climate change has not referred to decision-theoretical concepts, several of the traditional approaches in decision theory have reappeared under new names in this discussion.

Expected utility is the dominant approach to decision-making under risk. It could, more precisely, be called "probability-weighted utility theory". In expected utility theory, each alternative is assigned a weighted average of its utility values under different states of nature, and the probabilities of these states are used as weights. This is the common procedure in risk analysis. Risk analysts multiply "the probability of a risk with its severity, ... calls that the expectation value, and ... uses this expectation value to compare risks" (Bondi 1985: 9).

In order to apply this approach to global warming, numerical probabilities would have to be assigned to the various scenarios. We are not aware that this has been done in a systematic fashion. Nor is it to be expected that meaningful probabilistic estimates will be available in the near future.

By far the most common approach in economic and decision-analytical studies of climate change is instead to base one's calculations on the "most probable" case presented by a scientific body such as the IPCC. All scenarios other than the most probable ones are disregarded.[9] This approach may be called the maxiprobability method. It does not have an established name, since it seems only to have practitioners, no explicit proponents.

The difference between the expected utility and the maxiprobability methods does not seem to have been sufficiently understood. To illustrate it in a schematic way, let us make the thought experiment that probabilities and utilities are as follows:

Table 1: Climate Change – Hypothetical Outcomes

case	probability	utility
A	0.01	−10000
B	0.04	−1000
C	0.90	−100
D	0.05	0

Here, C represents the most probable (IPCC) case, D the "best case" in which the greenhouse effect is completely outbalanced by other mechanisms, and A and B represent cases with very serious consequences, such as could occur if positive feedback mechanisms were much stronger than what is now believed. Under the assumptions given in Table 1, the maxiprobability method will estimate the utility at −100, whereas the expected utility method estimates it at −230.[10] More generally, the major difference between the two methods is that the maxiprobability method is insensitive to scenarios with a small probability but a high disutility (or utility). Hence, in the presence of very serious (but improbable) scenarios, the method used in all economic decision-studies that we are aware of is more risk-taking (more optimistic) than expected utility theory, i.e. it is more risk-taking than the standard method used in risk analysis.

9 The IPCC has presented several scenarios, but they differ only in the rate of emissions. None of them is a worst-case scenario where all factors that have been taken into account are maximally conducive to increased greenhouse gas concentration. Furthermore, several important factors have not been taken into account, such as biological feedback mechanisms (IPCC 1992a: 18). The calculated effects are usually based on greenhouse gas concentrations that are equivalent to a doubling of the CO_2 concentration. The effects due to concentrations higher than those corresponding to doubled CO_2 concentrations have not been put in focus by the IPCC.

10 Note that case A, although least probable, contributes most to aggregated expected disutility.

This should be seen against the background that the expected utility rule itself has been criticized for being a too risk-taking (too optimistic) decision-rule. The strength of this decision-rule comes out when it is applied to a long series of similar events. If reliable probabilistic information about these events is available, then the expected utility rule is a fairly safe method to maximize the outcome in the long run. Suppose, for instance, that the expected number of deaths in traffic accidents in a region will be 300 per year if safety belts are compulsory and 400 per year if they are optional. Then, if these calculations are correct, about 100 more persons per year will actually be killed in the latter case than in the former. If we aim at reducing the number of traffic casualties, then, due to the law of large numbers, this can safely be achieved by maximizing the expected utility (i.e. minimizing the expected number of deaths). This argument depends, however, on the large number of road accidents, which levels out random effects in the long run. It is not valid for case-by-case decisions on unique or very rare events (Hansson 1993). In particular, this applies to major catastrophic events such as a nuclear war or a major ecological threat to human life. A strong argument can be made that the expected utility rule takes too big risks, i.e., is too optimistic, in decisions with very high stakes, such as those that relate to global warming. This makes the maxiprobability rule still more problematic.

Another well-known decision-theoretical method or rule is the *maximax* method. The maximax strategy may be roughly summarized as: "Behave as if we live in the best of all possible worlds". More precisely, it can be expressed as follows:

> *The maximax rule:* Consider only the best possible outcome associated with each option. Choose the option (one of the options) whose best outcome is better than that of all other options.

In decision theory, maximax is commonly regarded as an irrational or at least highly problematic strategy (Rapoport 1989: 57). In the debate on global climate change, some authors have maintained that nothing should be done until the scientific uncertainty has been considerably reduced. This comes closer to a maximax strategy than any other proposal in climate policies that we are aware of.

Another decision-theoretical method is the *maximin* strategy. This strategy can be summarized as "behave as if we live in the worst of all possible worlds". More precisely, it can be formulated as follows:

> *The maximin rule:* Consider only the security level (worst possible outcome) associated with each option. Choose the option (one of the options) whose security level is better than that of all other options.

The maximin rule is commonly recognized as a major decision-rule under uncertainty.

A serious problem in the application of the maximin rule is that in principle, *any* decision may have catastrophic unforeseen consequences (Hansson 1995). In principle, any decision with respect to global warming may lead to the extinction of all human beings. For all that we know, a strong preventive strategy that eliminates most of the anthropogenic greenhouse effect may leave us unprotected against some (unknown) mechanism that drastically cools down the climate, and which would otherwise have been balanced by the greenhouse effect. If extremely improbable possibilities like these are taken into account, then the maximin rule cannot distinguish between the decision alternatives, since they all have the same worst outcome.

There are various ways of coping with this drawback of the maximin rule. One is to completely disregard extremely implausible hypotheses. In that case the phrase "possible outcome" in the above definition of the maximin rule may be replaced, e.g. by "possible and not extremely improbable". Another is to supplement the maximin rule with an additional rule that distinguishes between alternatives that have equally bad worst cases. Such a rule can make use of the (imperfect) probabilistic information that is available:

> *The modified maximin rule:* Consider only the security level (worst possible outcome) associated with each option. If there is a single option whose worst outcome is better than that of all other options, choose it. If the best security level is obtained for several options, choose among those of these options for which the security level (worst outcome) is least likely.

Given the present state of scientific knowledge, the modified maximin rule will urge us to choose an alternative that minimizes the probability of global disaster, i.e. an alternative with strong preventive measures against global warming.

Which decision-rule should we apply in decision-making on global warming? Clearly, this is a value-laden question that cannot be answered by an argument that refers to scientific or decision-theoretical rationality alone.

Owing to the high stakes involved, we believe that any decision-rule that is more risk-taking (more optimistic) than the "standard" method of expected utility is unsuitable. This rules out not only the maximax rule, which few have seriously proposed, but also the maxiprobability method, which is the one most commonly used in the literature on climate change.

In our opinion there are good reasons for choosing a decision-method that is less risk-taking (less optimistic) than the expected utility rule, such as some variant of the maximin rule. However, the importance of the choice of an exact decision-method should not be exaggerated. The recommendations, at least for the immediate future, of various decision-rules may very well be the same. In particular, we suspect that expected utility maximization and perhaps even the maxiprobability method will make the same practical recommendations as less risk-taking methods such as the maximin rule. This, however, is a hypothesis that remains to be investigated.

2.5. No-Regret Measures

Surprisingly many of the measures that have been proposed to prevent or adapt to climate change are "no-regret" measures. There are other reasons in favour of them that are quite independent of global warming. Some examples:

- Reduced use of fossilized fuels not only decreases emissions of CO_2; it also reduces emissions and concentrations of sulphur dioxide, nitrogen oxides, hydrocarbons, toxic metals, tropospheric ozone and other substances with negative effects on human health and the environment. Some of these substances are also major causes of weathering of stone buildings and artefacts and of corrosion on metal constructions.

- Some ways of reducing the use of fossil fuels, such as the introduction of more efficient engines and better heat insulation of buildings, will in many cases be advantageous from a purely economic point of view (WCED 1987: 196–200; Mills et al. 1991).

- Irrigation systems that make agriculture less sensitive to "normal" variations in weather also prepare it for effects of global warming.

- The development and use of crops that are less sensitive to variations of the weather can also contribute to our ability to cope with climate change.

- By planting trees, the CO_2 uptake is increased, and at the same time erosion is prevented and the local/micro climate in arid areas is ameliorated.

- Dams and other constructions that protect against flooding in coastal areas are also a contribution to our chances of coping with a rise in sea level due to climate change.

Each of these types of measure is motivated even if, contrary to what there are strong reasons for believing, the threat of climate change should turn out to have been a false alarm. It is an obviously sound strategy for promoting measures against climate change that have reasonable concomitant motivations. Therefore, these are "no-regret" strategies.

3. The Coordination Problem

Global warming and the anthropogenic greenhouse effect is a matter of concern to all countries. No country has it in its power to avert the threat of a future global disaster with measures taken only within its own borders. This is a truly international problem that can only be solved with concerted action in which many nations take part.

3.1 The Nature of the Coordination Problem

Nations differ widely in their vulnerability to the consequences of an enhanced greenhouse effect. Countries with large land areas at or below sea level are seriously affected, as are countries and regions with bio-physically fragile resource zones such as mountains and arid and semi-arid tropical areas, which are already faced with crisis situations with the present climate (Jodha and Maunder 1991: 411). In such fragile zones even a minor climate change can make the conditions for agri-culture even worse and result in serious negative effects. On the other hand, some countries and regions may, at least in the short run, even be "winners" and for instance get a better climate for cultivating crops as a result of the greenhouse effect.

Economic factors are another source of important differences between countries. Generally speaking, less developed countries are much more vulnerable than industrialized countries, because of their lack of resources to implement adaptive measures and to repair material dam-ages. The present cost to the Netherlands of protecting its land below the sea level is less than 0.1% of its GNP (Bolin 1993: 12). Arguably, rich countries can cope with many problems created by climate change with-out major sacrifices, whereas the costs may be insurmountable to a developing country.

On the other hand, emissions that give rise to the greenhouse effect originate, to a highly disproportionate degree, in the richer and indus-trialized countries of the world. Hence, advantages and disadvantages of the processes that lead to greenhouse gas emissions largely accrue to different countries.

Since the emissions, sinks and effects of greenhouse gases are global, a single country's preventive measures are not enough. Neither are the compensatory measures undertaken by a single country, since the results of such measures are uncertain and will affect other countries. Adaptation is the only strategy type that can be applied on a national level. Prevention or compensation has to be global. By reducing its emissions, a country makes only a small contribution to the global situation.[11] This effect is all but negligible, particularly for small countries. Only if other countries do the same, can a substantial effect be achieved.

In the absence of prevention, the need for adaptive measures will continue to increase. In the long run however, adaptive measures will not be sufficient although in the short run they might be both necessary and the most cost-effective measure for a single country.[12]

All this sums up to the greenhouse gas issue being a difficult problem of *international justice*. In addition, it also creates problems of *implementation*: How can countries be made to coordinate their efforts and achieve, through concerted action, what none of them can achieve alone? In particular, how can the free-rider effect be avoided, i.e. how can we avoid the situation where countries gain from measures taken by others without contributing to these measures themselves (see also Chapter 1, Section 7)?

3.2 Handling the Coordination Problem

It is important to distinguish between the two perspectives on the co-ordination problem that were mentioned in the previous section. In the perspective of *justice*, the issue is to find the best possible outcome of the collective decision process. In this perspective, practical feasibility is not required. The fairest distribution of costs and sacrifices between countries may very well be unrealistic, since it requires most concessions from those in a position to dictate the outcome of international negotiations.

---

11 The United States is an exception. This country alone, with its less than 5% of the world's population, is responsible for about 25% of the global emissions of CO_2 from fossil fuels.

12 One example: If the only serious short-term threat to the Netherlands is the rise in sea level, it would, to begin with, be more cost-effective for the country to build higher dams than to reduce emissions of CO_2, since the former measures are much more specific. But the sea-level rise will continue so long as the preventive measures taken are not enough or until the sea level is stabilized at a much higher level (maybe 60–70 m higher than present). Building dams will then be more and more expensive and in the end it will probably be impossible to protect the land from sea-level rise by building dams.

But when we are searching for the most *just* solution, this should not concern us.[13]

In the perspective of *implementation*, the issue is instead how nations can and should find realistic ways of coordinating their effects to cope with the threat of climate change.

It seems to us that most discussions on implementation have been too little concerned with the issue of justice. Some authors seem to take for granted that the attainment of sufficiently strong measures against global warming is all that matters. The agenda that they set up (or rather take implicitly for granted) for the implementation problem is therefore: How can countries coordinate their efforts to achieve the most effective measures possible against the anthropogenic greenhouse effect?

The distribution of sacrifices between countries then becomes merely a practical issue. Any method that leads to the desired global outcome is taken to be acceptable. This attitude is not defensible upon further reflection. The sacrifices necessary if major climatic change is to be avoided are so large that their distribution is indisputably an important issue. Therefore, the agenda for the implementation should rather be: *How can countries coordinate their efforts to achieve as effective and just measures against the anthropogenic greenhouse effect as possible?*

With this formulation of the issue, international justice must have a central role. Without trying to develop a full theory of justice between nations, we wish to propose *methodological individualism* as a central feature of any such theory that is plausible. Basically, the concept of justice cannot be meaningfully applied, other than in an indirect sense, to non-sentient entities such as nations. Therefore, a just distribution between nations is essentially a distribution between nations that leads to justice between people.

As noted by Agarwal and Narain, sources of greenhouse gas emissions must be judged according to their dispensability:

> Can we really equate the carbon dioxide contributions of gas guzzling automobiles in Europe and North America or, for that matter, anywhere in the Third World with the methane emissions of draught cattle and rice fields of subsistence farmers in West Bengal and Thailand (1991: 5)?

From the viewpoint of individual justice, the constructions most common in *Realpolitik*, such as assigning to each nation a quota in proportion to its emissions of greenhouse gases in a reference year, are extremely inadequate. Any policy that assigns total quota to countries

---

13 See Chapter 6 for an elaboration on burden-sharing.

without paying attention to the needs of the inhabitants of those countries can only be a mockery of justice.

A proposal that is much more reasonable from the viewpoint of (individual) justice is to assign to every single country a permissible emission level for greenhouse gases that is proportionate to its part of the world population, and such that the sum of all national quotas is compatible with the principle of sustainable development. As an example, if we need to reduce the world's CO_2 emissions from fossil fuels by 70%, countries with high *per capita* emissions, such as the OECD countries, have to decrease their emissions by more than 70%. On the other hand, many non-industrialized countries will not have to reduce their emission at all, or could even increase them and still be below the permissible limit.

4. The Time Factor

Time plays a central role in our endeavours to understand and to cope with the threats of climate change. There are two major groups of temporal issues that require decision-theoretical treatment. The first of these is concerned with how much the future should matter to us, and how distant those future events are that we should at all be worried about. The second is to find a proper time for decision-making: What can be gained, or lost, from waiting before making a decision?

4.1. How Important is the Future?

The IPCC has chosen six different emission scenarios as a base for its calculations. All scenarios show an increase in concentration well above preindustrial levels by 2100 and none of them shows a stabilization of the concentration before 2100 (IPCC 1994b: 14). Despite this, none goes beyond the year 2100. Instead, all of them extend the time period from 1990 to 2100 (IPCC 1992a: 10–13). Apparently as a consequence of this, most policy discussions have focused on the same period of about 100 years. This is unfortunate, since global climate change is a long-time process, and today's emissions contribute to effects that may very well show up after much more than 100 years. From an ecological perspective, or even that of the history of human civilization, 100 years is an extremely short planning horizon. It would be difficult to defend the position that our responsibility for the future of this planet has a time limit of 100 years. To the extent that more serious problems can be expected after that period, these problems should be at the focus of decision-oriented discussions on climate change.

It is difficult, however, for policy-makers to plan for a future more distant than they have scientific information about. It should therefore be a high-priority task for climatologists, and in particular for the IPCC, to provide information pertaining to the more distant future, even at the price of a much higher uncertainty than is associated with prognostications for the next 100 years.

It is not sufficient just deciding that the distant future matters. We also need to have an opinion on *how* important the future is. Given that it matters to us now what happens 100 or 500 years ahead, how much does it matter? Would, to give just one example, a mass extinction of species be less serious if it happened after 500 years than it would if it happened now? If so, how much less?

The most common approach to this problem is that of the economist discounting future costs. The discount rate commonly used in calculations referring to climate change is a composite "social rate of time preference". It is the sum of "pure time preference" and a term reflecting prognosticated lower marginal utility due to a higher standard of living:

$$Y + Pg$$

Here, Y, the "pure time preference", simply reflects the degree to which we care less about what happens tomorrow than about what happens today. If $Y = 0.03$, then it is assumed that we value a loss 3% lower for each year into the future that it is postponed. P is the absolute value of the elasticity of marginal utility, and g is the growth rate of *per capita* consumption. Pg represents the lower marginal utility that future people are assumed to assign to various goods, as a consequence of being better off than we who are living today.

It has often been assumed that only the first of the two terms (Y) depends on moral judgements; the second (Pg) has often been taken to be uncontroversial and a "matter-of-fact". This becomes wrong, however, as soon as the elasticity of marginal utility is applied to "non-economic" goods such as a clean environment and biological diversity.[14] There is no reason to believe that future people will assign lower values to environmental assets than we do today. To the contrary, historical evidence indicates that environmental assets are being valued increasingly more highly. If the trend continues, then even negative discount rates may be adequate.

As an illustration, with a 5% discount rate, damage worth 1,000,000 dollars reduces to a mere trifle of 35 dollars if it occurs 200 years later.

---

14 For a discussion of the distinction between economic and non-economic values, see Andersen (1993).

There is (at least some) sense in using this as an argument for not paying more than 35 dollars today to avoid an economic cost of 1,000,000 dollars two centuries later. However, there is absolutely no sense in saying that if we are prepared to pay 1,000,000 dollars today to save a life, then we should not be prepared to pay more than 35 dollars today to prevent a death 200 years from now.

Conventional discount rate analysis, as applied in studies of environmental issues such as climate change, involves the application of economic concepts far outside the area for which they have a sound motivation. The effect of this misapplication is to promote a neglect of the future that few people would be prepared to defend from a moral point of view.

A major alternative to discounting future costs is the principle of *sustainable development*. According to the World Commission on Environment and Development (WCED) that was set up by the UN, sustainable development is "development that meets the needs of the present without compromising the ability of future generations to meet their own needs" and it requires both "meeting the basic needs of all" and "promotion of values that encourage consumption standards that are within the bounds of the ecologically possible and to which all can reasonably aspire" (WCED 1987: 42–43).

As we see it, it is an important consequence of the idea of sustainable development that decisions in environmental issues should not be dependent on limitations of the decision horizon. Something is wrong if a decision concerning preservation of the natural environment is based, e.g. on calculations with a time horizon of 100 years, but would be overthrown if the horizon is extended to 200 years. More generally, we propose the following principle for environmental decisions, as a corollary to the principle of sustainable development:

• Temporal extensibility: Decision-criteria should be insensitive to extensions of the time horizon. Taking the more distant future into account ought not to lead to drastic changes in the decision.

As we have already seen, much of the economic analysis of global climate change depends heavily on the choice of a time perspective of 100 years. Therefore, it does not satisfy the principle of temporal extensibility.

4.2. When Is It Time to Decide and to Act?

Implementation of preventive measures against climate change takes an unusually long time from decision to result. This time-lag is composed of the five major components summarized in Figure 3.

1. *The negotiation time-lag*: Several factors combine to make negotiations in the climate change issue difficult to conclude in a short interval of time:

- Many, indeed all, countries are involved, and many of them have special and diverging interests.

- Some countries may be winners, at least in the short run, from the enhanced greenhouse effect. This applies, in particular, to countries in cold climates with large agricultural sectors. Similarly, some countries may be losers from preventive measures against climate change. This applies most conspicuously to countries whose economy is based on the production and exportation of fossil fuel.

- The starting positions when reducing greenhouse gas emissions differ between countries. For some countries the best alternative is to reduce emissions by a certain percentage, for others energy efficiency standards are more favourable, and for still others an agreement based on greenhouse gas emissions *per capita* is the most favourable alternative.

- The gains obtained from preventive measures taken by a particular country do not accrue to that country any more than similar measures taken on the opposite side of the globe.

- There is no positive connection between the size of the emissions from different countries and the effect from climate change that might hit them. It is most likely that the non-industrialized countries responsible for relatively small emissions will suffer more than industrialized countries.

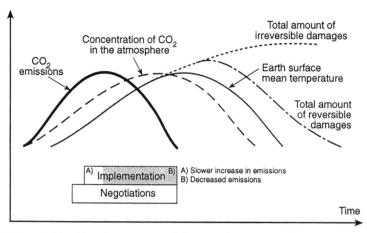

Figure 3: The Five Components of Climate Change Time-lag

- Measures against greenhouse gas emissions have great effects in most sectors of human society. Therefore, a large number of commercial interests are involved that will influence the negotiations. The scientific uncertainty is large and, as we saw in Section 2.2, we cannot expect it to be drastically reduced. Opponents of preventive measures can always claim, not without right, that we do not know for sure that the proposed measures are needed or that they will have the desired effect.

- Because of the time-lag caused by natural factors (see components 3, 4, and 5 below), the positive effects of sacrifices that we undertake will benefit later generations rather than ourselves.

All these factors combine to make negotiations difficult and consequently to protract them.

2. *The technological time-lag*: In order to reduce greenhouse gas emissions, major changes in technology and life-styles are required. Cars, household appliances, various machines and systems of transportation and heating must be replaced by new products and systems that make more efficient use of energy and/or use renewable energy sources. But that will probably not be enough. In order to reduce energy consumption we will also need a change in the life-styles of the industrialized countries; for instance, a reduced use of cars. In most cases the technological changes can only be expected to be practicable at the pace of the lifetime of these products and systems. For infrastructure systems such as the energy and transportation systems it will probably take about half a century. To change life-styles has been shown to be very difficult and to take a long time.

3. *The gas concentration time-lag*: Since the present emissions of greenhouse gases are far above the capacity of the sinks, greenhouse gas concentrations will continue to increase even if greenhouse gas emissions decrease. Because of the long atmospheric life-time of CO_2, if the emissions of that gas were "frozen" at the present level, its concentration in the atmosphere would not be stabilized until after several hundred years, and then at a considerably higher level than today's level (IPCC 1994: 14). Under pessimistic assumptions of positive feedbacks, the stabilization may – if it occurs at all – take considerably more time and take place at an even higher level.

4. *The temperature time-lag*: Increased greenhouse gas concentrations lead to an immediate corresponding increase in radiative forcing.[15] There

---

15 A change in radiative forcing is a change in average net radiation at the top of the troposphere because of a change in either solar or infrared radiation (IPCC 1994b: 7).

is, however, a time-lag between that effect and effects on the average temperature of the earth. This depends on the length of time taken to heat the sea. The sea has a cooling influence, until its temperature has reached an equilibrium with the radiative forcing. In the same way the sea will have a warming influence when and if the radiative forcing is decreasing. Calculations indicate that this time-lag is approximately 30 years (Klimatdelegationen 1994: 10).

5. *The reparations time-lag*: After a reduction in temperature has been achieved, it may take considerable time before the negative effects of the higher temperature and changed climate have been reversed. If and when the climate and sea level are back to normal, the recolonization of trees, plants and animals will take a long time, and the reparation and rebuilding of man-made constructions will take time too.

Some damage can never be repaired: deaths, extinction of species, loss of cultural heritage etc. In these cases, the reparations lag can be said to be infinite. Irreparable damage to the atmosphere cannot be excluded either.

These five components add up to making climate change an issue with a very long planning horizon. It takes many decades from the time decisions are made to take preventive actions to the time they can be expected to have full effect. Four of the five components (2–5) are in practice unavoidable. In our opinion, this is a strong argument for making the first component, the negotiation lag, as short as possible.

A possible argument against this standpoint is that by waiting we can obtain more information, and so be able to make a better informed decision. However, as shown in Section 2.2, we have no reason to expect that the scientific uncertainty can be substantially reduced in the foreseeable future. As has been observed by several authors, to wait for reliable knowledge could in this case mean to wait for disaster, since certain knowledge may very well come only with disaster (WCED 1987: 176).

5. Conclusion

We have identified three major problems-areas for decision-theoretical analyses of the climate change issue.

(i) *Scientific uncertainty*: There is not much hope that the scientific uncertainty will be sufficiently reduced to allow decisions to be made on a much safer base than at present. In important respects, new knowledge may even lead to increased scientific uncertainty. We have to live with scientific uncertainty.

It is essential that the nature and degree of scientific uncertainty is adequately presented to decision-makers. To provide more adequate information for policy-making purposes, consensus-oriented procedures in scientific committees such as the IPCC should be modified in order to bring out disagreement and uncertainties better. Not only the most probable scenario should be presented. It is at least as urgent to present worst-case scenarios even though they might be much less probable.

The most common evaluations of the global warming effect have made use of what may be called the *maxiprobability* criterion, i.e. attention has been paid only to the most probable state of nature. This method is more risk-taking than the expected utility rule that is otherwise the standard method in risk analysis. A strong case can be made that the maxiprobability method should be replaced either by expected utility maximization or by a still less risk-taking decision rule, such as some variant of the maximin rule.

(ii) *The coordination problem:* In our opinion, the agenda for the implementation should be to find how countries can coordinate their efforts towards achieving as effective *and just* measures against the anthropogenic greenhouse effect as possible. The criterion of justice has often been neglected in this context.

Basically, the concept of justice refers to people, not to nations. Therefore, a just distribution between nations is essentially a distribution between nations that leads to justice between people. From this point of view, any policy that assigns total quota to countries without paying attention to the needs of their inhabitants can only be a mockery of justice. This applies to some of the most commonly discussed constructions, such as assigning to each nation a quota in proportion to its emissions of greenhouse gases in a reference year. A proposal that is much more compatible with individual justice is to allocate nature's capacity to receive greenhouse gases to each nation in proportion to its population. These quantities can then constitute the permissible emissions for each country.

(iii) *The time factor:* We have found the standard method of discounting future costs to be inadequate and misleading. Instead, we favour the principle of sustainable development. A corollary to that principle is *temporal extensibility* for decision rules. By this is meant that decision-criteria should be as insensitive as possible to extensions of the time horizon. Taking the more distant future into account ought not lead to drastic changes in the decision. Unfortunately, most of the economic analyses of global climate change are heavily dependent on the choice of an arbitrary time horizon (mostly 100 years), and therefore do not satisfy the criterion of temporal extensibility.

The principle of sustainable development states that current development should not compromise the ability of future generations to meet their needs. But, as we have seen, increased concentrations of greenhouse gases in the atmosphere, as a result of anthropogenic emissions, may lead to very serious effects, indeed so serious that it will not be possible for future generations to fulfil their needs. Hence, a development with large greenhouse gas emissions may not be sustainable. Owing to scientific uncertainty we are not sure of the effects. But waiting for better knowledge is not a solution since most of the scientific uncertainty will probably remain until it is too late to avoid serious effects.

The principle of sustainable development, the scientific uncertainty, the risk for irreversible damages and the long time-lag from negotiations to the point in time when most of the reversible damages will be repaired, all together imply that we should reduce the anthropogenic emissions of greenhouse gases and that we should act as soon as possible. A good start would be to take no-regret measures, i.e. measures that are strongly motivated for other reasons as well.

Chapter 6

The Requirement of Political Legitimacy: Burden-Sharing Criteria and Competing Conceptions of Responsibility

Gunnar Fermann[1]

1. Introduction

There is now something close to a scientific consensus that global warming is a real threat and that it is caused by greenhouse gas emissions from human activities (IPCC 1990; 1992; 1995).[2] Given the broad scope and lasting nature of the problem, as well as the potentially huge costs involved in combating the threat, international cooperation based on the principles of cost-effectiveness and burden-sharing are viewed by most observers as essential preconditions for ensuring the success of abatement efforts. As an expression of international cooperation, the 1992 Framework Convention on Climate Change (FCCC) acknowledges both these principles, although in a rather general fashion.

One may question whether, to what extent, and under what circumstances the principles of cost-effectiveness and burden-sharing are compatible.[3] The task of the present chapter, however, is more limited. Leaving the detailed analysis of cost-effectiveness to Chapter 7, the present chapter is devoted to the question of how – and according to what moral principles – the costs involved in combating climate change may be distributed among various kinds of countries? Assuming that the climate change burden-sharing debate will intensify as the third conference of the parties (CoP3) in Kyoto, Japan approaches in December

1 This chapter is a result of my association with the Centre for Environment and Development at the Norwegian University of Science and Technology, and the Energy, Environment and Development Programme of the Fridtjof Nansen Institute. The chapter builds in part on previous work financed by the research programme SAMMEN, run by the Norwegian Research Council.
2 See Chapter 3 for a detailed exposition.
3 For contributions investigating the relationship between efficiency and fairness/equity in burden-sharing, see Fermann (1992: 17–24), Solomon and Ahuja (1991: 343–350), and Tarlock (1992: 871–900).

1997, six conceptions of burden-sharing with diverging implications for the distribution of abatement costs between various socio-economic regions of the world are first considered. Owing to the strong moral appeal of the argument that those causing a problem should take part in solving it, next a more committed attempt is made to sort out the confusion still surrounding the particular burden-sharing criterion of "responsibility". It is shown that emissions responsibility can be assigned different meanings depending on the time frame, range of greenhouse gases, sources, and scope of national responsibility applied.[4] In the concluding section an attempt is made to demonstrate that rational, egoistic, and utility-maximizing states may have incentives to accept emission responsibility as a premise for the design of a climate change burden-sharing regime. The underlying assumption of the present chapter is that a widely shared sense of fairness is at the core of political legitimacy, which, in turn, is a prerequisite for the general compliance with and effective implementation of internationally adopted regulations of greenhouse gas emissions.

2. Burden-Sharing Criteria

Even if an international climate change regime capable of curbing the emissions of greenhouse gases is designed according to the best advice of economists (cost-effective measures and arrangements), such a regime is nevertheless likely to impose heavy costs on the participants. Not suprisingly, therefore, delegates in the negotiations have invested considerable interest in the question concerning how these costs are to be shared and it is far from being resolved.

Among academics, the *equity issue* – the question of what a *fair* burden-sharing scheme should look like – has been recognized as possibly the most crucial question to be resolved if a strong climate change abatement regime is to materialize:[5]

---

4 It is acknowledged that a complete analysis of responsibility in the field of climate change would require discussing not only various conceptions of responsibility, but also the philosophical underpinnings, and the problem of measuring various conceptions empirically. This is beyond the scope of this chapter, however. It is left to the political philosopher to link the question of responsibility to a body of normative theory and to the natural scientist and statistician to circumvent the uncertainties involved in empirical measurement.

5 According to Kasperson and Dow (1991), equity is fairness of the process by which a particular decision or policy is enacted and of the associated outcome. The latter, distributional equity, is the concern of the present chapter.

If an agreement is not seen to be fair by countries, then they will not sign or participate in it.... The issue of equity becomes not only an ethical consideration, but a practical one in relation to the effectiveness of the agreement (Thomas 1992: 188).

Similar sentiments are echoed by Young and Osherenko, who hold that "institutional bargaining cannot succeed unless it produces an outcome that participants can accept as equitable, even when the adoption of an equitable formula requires some sacrifice in efficiency" (1993: 235), and, furthermore, by Burtraw and Toman, who argue that "it is almost axiomatic that an effective international agreement to limit emissions of CO_2 will not be undertaken unless the agreement is considered fair by the participants" (1992: 122).[6]

However, general terms like fairness, equity and burden-sharing have no unequivocal meaning and several criteria can be drawn upon to support divergent arguments about the allocation of abatement costs. Although responsibility – the notion that those causing a problem should be equally active in solving it – has an intuitively moral appeal, competing and complementing criteria for burden-sharing have been put forward and will undoubtedly be applied to justify positions on burden-sharing, thus influencing the debate on allocation of abatement costs. A distinction is made below between six classes of normative argument.[7]

2.1. Responsibility – Polluter Pays

The notion that those causing a problem should take part in solving it seems intuitively fair and is a strong incentive to *anticipate* environmental costs prior to damage occurring. At least two versions of the "responsibility" criterion can be identified.

The moderate version – the *polluter-pays-principle* (PPP) – emphasizes the responsibility of polluters to bear the costs of abatement efforts. PPP has some precedent, since it was adopted by the *Organization of Economic Cooperation and Development* (OECD) in 1974 as a guide to environmental policies (OECD Com. 223 1974; OECD 1991: 258).

The extended version – the *equal entitlements criterion* – is inspired by egalitarian doctrine and holds that "all human beings (are) entitled to an equal share in the atmospheric commons" (Grubb et al. 1992: 312). The equal entitlements criterion implies that the industrialized North will have to reduce its emissions to a level consistent not only with a stabi-

---

6 See also Albin (1993).
7 For alternative classifications, see Grubb et al. (1992) and Solomon and Ahuja (1991).

lization of anthropogenic emissions, but one that provides margins for developing countries to increase their emissions to a level consistent with continuous economic growth.[8]

2.2. Performance – Rewarding Efficiency

Countries commanding energy-efficient economies tend to suffer relatively higher marginal costs on (further) abatement efforts. The *efficient energy-use criterion* accounts for this fact by demanding that energy-efficient countries be compensated rather than punished for this achievement when allocating abatement costs. Two alternatives are conceivable: Firstly, energy-efficient economies suffering relatively higher abatement costs may be allowed to meet less demanding national abatement targets. However, it would be more cost-effective to allow these countries to gain credit for abatement efforts in less energy-efficient countries where abatement costs are considerably lower through so-called joint implementation measures (Fermann 1992: 19).

2.3. Concern – Victim Pays

The *willingness-to-pay criterion* implies it is fair that abatement costs be borne by those countries hardest hit by global warming. In a radical version, this argument of welfare economics implies that countries most severely affected by the impacts of global warming should be responsible for bearing the brunt of the abatement costs. While the logic of the willingness-to-pay criterion may be sound under other conditions, the cynicism of the argument becomes evident when pointing to the fact that the industrialized North is responsible for a majority of greenhouse gas emissions irrespective of what measurement of emission responsibility is applied (see Section 3), and that the less developed South probably will be the most severely affected by climate change. Thus, by implication, the willingness-to-pay argument suggests that the most severely affected and less reprehensible South should pay the costs of an abatement regime made necessary primarily by the North. Applying such an argument would ruin efforts toward persuading the developing South to participate actively in an international abatement regime (Grubb et al. 1992).

---

8 This argument is very much in line with Agarwal and Narain (1991). For a study calculating the future emission quota per capita for various regions of the world based on the equal entitlements criterion, see Den Elzen et al. (1992).

2.4. Capacity – Paying in Accordance with Solvency

Supporters of the *ability-to-pay criterion* argue that the costs of combating global warming should be distributed according to each nation's ability to pay. If the United Nations Scale of Assessment could be agreed upon as a distribution instrument, the ability-to-pay criterion seems likely to strike a middle ground between the polluter-pays principle and the equal-entitlements criterion on the one hand, and the willingness-to-pay criterion on the other. The ability-to-pay criterion may, in one form or another, prove acceptable to both the industrialized North and the developing South.

2.5. Full Compensation for the Poor

While the ability-to-pay criterion distributes the costs of an abatement regime according to an assessment of problem-solving capacity, the *distributional implications* criterion holds that an abatement regime should be capable of mitigating the international distribution of wealth, that is, improving the situation of the least well-off (Rawls 1971; Rose 1990: 927–935). At a minimum this would suggest that "developing countries be left at least as well-off under an emission control regime as they would be in its absence" (Grubb et al. 1992: 313). Such a scheme would almost certainly include provisions of technology transfers to developing countries and thus reawaken old controversies between the developing South and the industrialized North. One could expect the North to be sceptical of any abatement regime that would imply a "new economic world order".

2.6. Securing Basic Needs

Reflecting the "basic human needs" school of thought, the *reasonable emissions criterion* suggests that countries emitting more than what is deemed "reasonable" to support a consistent, modest standard of living, must accept far higher abatement costs than countries still facing poverty. This morally sound argument is unlikely to be rejected out of hand by the industrialized North. However, we may expect the North to be sceptical of any criteria which explicitly link the distribution of abatement costs to an obligation to ensure that basic human needs are met in the developing South. In the eyes of the industrialized North, such an acceptance could be seen as creating a costly precedent.

3. Diverging Conceptions of Responsibility

Having briefly reviewed several partly overlapping burden-sharing criteria, let us consider one of them more thoroughly. Although responsibility as a criterion for allocation of abatement costs (burden-sharing) has a strong moral basis, its specific meaning is ambiguous. In fact, competing conceptions of responsibility have been put forward with quite divergent implications for burden-sharing. The typology presented below is an attempt to organize this conceptual pluralism.

Accounting for the responsibility of various countries on the issue of climate change is to describe each country's contribution to the extended greenhouse effect caused by anthropogenic emissions. The task of assigning responsibility among countries and among socio-economic clusters of countries for greenhouse gas emissions is a complex venture, however. From a polluter-pays perspective, at least four factors need to be considered: the time frame applied; the sources accounted for; the greenhouse gases included; and the interpretation of "national responsibility".

3.1. Time Frame

As for the *time frame*, a distinction can be made between conceptions accounting for historically accumulated emissions and those accounting for current emissions only. Most statistics on emission responsibility are restricted to the reporting of current greenhouse gas emissions. As pointed out by Agarwal and Narain (1991), however, the validity of this operationalization of emission responsibility is limited since the problem of climate change is due to emissions of anthropogenic greenhouse gases *over time* (since the Industrial Revolution).[9] Applying the cumulative (historical) conception of responsibility implies that the burden of responsibility tilts even more to the early industrialized countries of the North than it would if only the current emissions of CO_2 are accounted for (Table 1).

Table 1: Energy-related CO_2 Emissions from Industrialized and Developing Countries (% of World Total)

	Industrialized countries	Developing countries
Cumulative[10]	86	14
Current	72	28

---

9 See also Mukherjee (1992: 95).
10 1860–1986 period.

3.2. Emission Sources: Energy-related and Biotic

A third distinction can be made between a conception of responsibility accounting only for emissions related to the production and consumption of energy (energy-related emissions) as opposed to a more encompassing conception also including emissions from biotic sources. Table 2 illustrates that the emission responsibility of developing countries increases considerably to the extent that biotic sources of CO_2 are included. It is evident, however, that even if biotic sources of CO_2 emissions are included, the industrialized countries are still responsible for a majority of CO_2 emissions.

Table 2: Cumulative Emissions of CO_2 from Industrialized and Developing Countries (% of World Total)

	Industrialized countries	Developing countries
Energy-related	86	14
Energy-related and biotic	58–80[11]	20–42

3.3. Range of Greenhouse Gases

With regard to the *range of greenhouse gases*, conceptions of responsibility vary according to whether one or more greenhouse gases are accounted for. Responsibility is also sensitive to what kind of greenhouse gases are measured and whether the differing radiative impact of various greenhouse gases is adjusted for. In Table 3, a distinction is made between a narrow and a wide conception of emission responsibility: By taking only the (current) CO_2 emissions into consideration, the industrialized countries are responsible for nearly three-quarters of world emissions (and more than four-fifths, if emissions *over time* are taken into account – see Table 1). However, if our conception of responsibility is broadened to also include the emissions of methane (CH_4), nitrous oxide (N_2O), and carbon monoxide (CO), the responsibility between industrialized and developing countries almost evens out.[12]

---

11 This range is based on alternative assumptions of historical land clearing rates (Subak 1993: 59).

12 It should be noted, however, that considerable uncertainty is related to the measurement of comprehensive greenhouse gas emissions, since CH_4 and N_2O emissions are difficult to inventory (Subak 1993: 58).

Table 3: Current Emissions of Greenhouse Gases from Industrialized and Developing Countries (% of World Total CO_2 Equivalent)

	Industrialized countries	Developing countries
CO_2 only	72	28
Comprehensive (CO_2, CH_4, N_2O, CO)	52	48

The above distinction is neatly illustrated by the practice of the International Energy Agency (IEA), which in its environmental reporting applies country estimates based on current emissions of energy-related CO2, whereas the World Resources Institute (WRI) in addition reports estimates of accumulated emissions on a broader range of greenhouse gases (CO_2, CH_4 and CFCs) including both energy-related and biotic sources.

3.4. Scope of National Reponsibility

From the above exposition it is clear that the developing countries will benefit to the extent that a conception of responsibility including cumulative emissions of energy-related CO_2 is applied. To the extent that a burden-sharing scheme is based on comprehensive emissions (including CO_2, CH_4, N_2O, and CO) from both energy-related and biotic sources, the developing countries will have to take a considerably larger share of the blame. The most important observation to be made, however, is that, irrespective of the measurement method used, the industrialized countries are responsible for a majority of the world's total greenhouse gas emissions (ranging from 52–86%). This is likely to change over time, however, as the developing countries are currently responsible for more than two-thirds of the *increase* in CO_2 emissions. This is the main reason why one observer concludes that the problem of climate change "is one of the very few on which the rich nations actually need the co-operation of major groups of the poor in the implementation of a solution" (Shue 1992: 377).

A reasonable objection to the various conceptions of responsibility outlined above is that they all ignore the fact that the developing countries are inhabited by three times more people than the industrialized countries. By linking responsibility to individual emissions, a *per capita related* conception of emission responsibility remedies this serious shortcoming – a measurement based on the compelling argument that all humans should have an equal right to the atmospheric commons (Subak 1993:

61). Table 4 shows that a *per capita* related conception of responsibility dramatically increases the part played by the industrialized countries in the problem of climate change. According to this mode of measurement, the industrialized countries are 6.5 times (3.26/0.50) more responsible for world CO_2 emissions than are the developing countries.

Common to most assignments of emission responsibility – including those of the IEA and the WRI reported above – is the application of a rather narrow *scope of national responsibility* based on a simple *territorial* criterion. Below, we intend to demonstrate that at least two other well-argued criteria based on *net investments abroad* and *consumption* should be taken into consideration, although this scarcely noticed distinction has yet to be accounted for in existing emissions statistics.

Table 4: Per Capita CO_2 Emissions from Industrialized and Developing Countries[13] (tons of carbon)

World average	Industrialized countries	Developing countries
1.12	3.26	0.50

3.4.1. Nation-based Emissions Approach

Existing emissions statistics – whether reporting current or accumulated emissions, CO_2 or other gases, biotic or energy-related sources – limit reporting to greenhouse emissions originating from within the borders of each country. This practice of linking (limiting) responsibility to each country's *territorial* emissions is for the present purposes conceptualized in the *nation-based emissions approach*. It is this approach which dominates the current debate, and which is reflected in the 1992 FCCC. It may be objected, however, that in the context of an integrated world economy, characterized by increasing interaction among countries, such an approach to responsibility provides too narrow a scope: It ignores the fact that some countries are responsible for emissions in *other* countries through their ownership abroad and consumption of imported goods and services.

3.4.2. Shadow Emissions Approach

The *shadow emissions approach* may be conceived of as an attempt to remedy some of the shortcomings of the nation-based emissions approach:

---

13 The category of "industrialized countries" includes high-income economies (states) with a GNP per capita of US$7,620 or more in 1990. The "developing countries" category includes countries with a GNP per capita less than US$7,620 (WB 1992: xi, Table A9).

By not linking responsibility to a fixed territory, but rather to *ownership and net investments abroad*, the shadow emissions approach accounts for the fact that countries with a long history of economic prosperity and international economic involvement emit more than is accounted for when the narrower nation-based emissions approach is applied. More specifically, the net extra-territorial shadow emissions of country A can be defined as the emissions resulting from A's investments, credits and activities abroad, minus the emissions resulting from other countries' financial and industrial activities in country A. Hence, whether the net shadow emissions of a country add to the nation-based emissions depends on the international economic standing of the country in question. It is clear, however, that due to considerable investments and ownership in developing countries, the industrialized countries are responsible for more emissions than are calculated by means of the territorial criterion for responsibility (nation-based emissions approach). The exact amount of CO_2 emissions resulting from the North's extra-territorial activities in the Third World cannot simply be inferred from net investment figures in developing countries. A proper assessment of extra-territorial CO_2 emissions requires detailed information on *how* their investments are *utilized* abroad: In what sectors and activities is external financing engaged? What is the ratio of CO_2 emissions per dollar in different sectors and countries? Unfortunately, our knowledge of these questions is incomplete; any serious attempt to quantify the net shadow emissions of relatively affluent countries depends on the rectification of this state of affairs.

3.4.3. Trade-Based Emissions Approach

While the shadow emissions approach represents an improvement in relation to the conventional nation-based conception of responsibility, it may be objected that neither of them provides a correct estimate of the distribution of responsibility among countries on the issue of climate change. By linking emissions responsibility to investment and ownership of means of production, the shadow emissions approach ignores the additional emissions related to *consumption*. The *trade-based emissions approach* accounts for emissions originating from the extraction and production of *imported* raw materials and goods not accounted for by the two other emissions approaches, thus allocating more responsibility to countries having a high consumption rate. The advantage of the trade-based emissions approach to responsibility is apparent. However, it may

---

14 See Table 5.

Table 5: Assessing Emission Responsibility – Approaches and Dimensions (total or per capita emissions)

A. TIME FRAME	B. RANGE OF GASES	C. SOURCES	D. SCOPE OF NATIONAL RESPONSIBILITY		
			NATION-BASED EMISSIONS APPROACH	SHADOW EMISSIONS APPROACH	TRADE-BASED EMISSIONS APPROACH
CLIMATE EMISSIONS	CO_2	ENERGY-RELATED	1. Historical emissions of energy-related CO_2 originating from the territory of a country	2. Historical emissions of energy-related CO_2 originating from activities conducted by a country at home and abroad	3. Historical emissions of energy-related CO_2 related to a country's consumption of home-produced and imported goods and services
		+ BIOTIC	4. Historical emissions of all anthropogenic CO_2 originating from the territory of each country	5. Historical emissions of all anthropogenic CO_2 originating from activities conducted by a country at home and abroad	6. Historical emissions of all anthropogenic CO_2 related to a country's consumption of home-produced and imported goods and services
	ALL GREENHOUSE GASES	ENERGY-RELATED	7. Historical emissions of energy-related CO_2, CH_4, N_2O etc. originating from the territory of a country	8. Historical emissions of energy-related CO_2, CH_4, N_2O etc. originating from activities conducted by a country at home and abroad	9. Historical emissions of energy-related CO_2, CH_4, N_2O etc. related to a country's consumption of home-produced and imported goods and services
		+ BIOTIC	10. Historical emissions of all anthropogenic CO_2, CH_4, N_2O etc. originating from the territory of a country	11. Historical emissions of all anthropogenic CO_2, CH_4, N_2O etc. originating from activities conducted by a country at home and abroad	12. Historical emissions of all anthropogenic CO_2, CH_4, N_2O etc. related to a country's consumption of home-produced and imported goods and services
CURRENT EMISSIONS	CO_2	ENERGY-RELATED	13. Current emissions of all anthropogenic CO_2, originating from the territory of a country	14. Current emissions of energy-related CO_2 originating from activities conducted by a country at home and abroad	15. Current emissions of energy-related CO_2 related to a country's consumption of home-produced and imported goods and services
		+ BIOTIC	16. Current emissions of all anthropogenic CO_2, originating from the territory of a country	17. Current emissions of all anthropogenic CO_2 originating from activities conducted by a country at home and abroad	18. Current emissions of all anthropogenic CO_2 related to a country's consumption of home-produced and imported goods and services
	ALL GREENHOUSE GASES	ENERGY-RELATED	19. Current emissions of energy-related CO_2, CH_4, N_2O etc. originating from the territory of a country	20. Current emissions of energy-related CO_2, CH_4, N_2O etc. originating from activities conducted by a country at home and abroad	21. Current emissions of energy-related CO_2, CH_4, N_2O etc. related to a country's consumption of home-produced and imported goods and services
		+ BIOTIC	22. Current emissions of all anthropogenic CO_2, CH_4, N_2O etc. originating from the territory of a country	20. Current emissions of all anthropogenic CO_2, CH_4, N_2O etc. originating from activities conducted by a country at home and abroad	24. Current emissions of all anthropogenic CO_2, CH_4, N_2O etc. related to a country's consumption of home-produced and imported goods and services

be criticized for weakening the incentive for producers and manufacturers to reduce their emissions, and for the overwhelming complexity of measuring emissions linked to consumption (Fermann 1992: 26–31; Subak 1993: 51–69).

3.5. Implications

The four dimensions considered above give rise to multiple conceptions of responsibility.[14] The specific time frame, combination of greenhouse gases, emission sources, and the approach to national responsibility applied will heavily influence the distribution of responsibility among various categories of countries – whether interpreted in terms of per capita emissions or not. Highly developed countries, with international economies and a long history of industrialization will probably be found most responsible if a mix of criteria including the accumulated and trade-based emissions of energy-related CO_2 are applied. The developing South would, on the other hand, probably be penalised to the extent that territorial and current emissions of CO_2 and CH_4 (from agriculture) were accounted for in the estimation of responsibility.

My aim here is not to argue in favour of one particular conception of responsibility, nor attempt to verify empirically the above assumptions about burden-sharing implications. My ambition has been more to expose the complexities involved in any attempt at deciding the relationship between burden-sharing and responsibility in the field of climate change.

4. Conclusion: Transforming Conceptions of Justice into Fair Burden-Sharing

The main conclusion to be drawn from the above exposition is that whatever conception of responsibility is applied, the industrialized and relatively affluent countries are responsible for a majority of the world's greenhouse emissions. This may place the North in a vulnerable position: Vulnerability is perhaps not so much linked to the direct impacts that global warming may have on the economies of these countries since the distribution of negative effects of climate change is probably negatively correlated to the share of greenhouse emissions (responsibility). The responsible North will most likely suffer less from climate change than the less responsible South in the short term. The vulnerability of the North is *political*: Because of its vastly superior economic resources and great responsibility for the greenhouse threat, the industrialized North is exposed to criticism. More than altruism or a sense of guilt, political pressure from the international community may provide an incentive for these countries to finance abatement costs according to some conception of responsibility. The burden of great responsibility may, by means of international criticism, induce these countries to take progressive action.

What are the incentives for taking this responsibility seriously? Of course, the willingness of each country to cover abatement costs cannot be inferred directly from its share of responsibility for global warming. It should not, for instance, be assumed that the United States government, being responsible for nearly one-quarter of energy-related world CO_2 emissions, would feel more obliged to reduce this share than Germany and Japan, which are responsible for "only" 5% each. The relationship between responsibility and burden-sharing is far more complex. Below, five *incentive mechanisms* are suggested which may be capable of translating responsibility into a corresponding allocation of abatement costs:[15]

Vulnerability (negative incentives):

a. *Direct environmental and socio-economic impact costs*: Fear of the *impact* that climate change may have on the environment and, ultimately, on their economies.

b. *Indirect impact costs*: Concern about the criticism which might arise from countries severely threatened by global warming and possible harmful effects on international migration and trade.

c. *Adaptation costs*: Fear of the costs involved in *adapting* to climate change.

Opportunity (positive incentives):

d. *Competitive edge*: The prospect that international abatement efforts may create new markets and business opportunities for technologically advanced countries, e.g. in the field of energy-conservation technology.

e. *Impression management*: The prospect that a progressive policy towards climate change may strengthen environmental image and enhance goodwill.

---

15 The list of incentive mechanisms is not assumed to be complete.

While these incentive mechanisms may have general relevance, their specific workings and roles will have to be decided empirically for each country. If, for instance, the United States government perceives that it has little to fear from global warming and is confident that it can deflect or absorb the criticism arising from a "policy of neglect", great responsibility in the field of climate change is unlikely to induce the United States government to cover a greater share of the abatement costs. But Japan, known to be sensitive to international criticism and a world leader in environmental technologies, may find that a progressive abatement strategy is in tune with the national interest, although it bears less responsibility for global warming than the United States.

The main implication of this brief discussion is that the influence of responsibility on the willingness to bear a corresponding share of abatement costs depends very much on each government's perception of vulnerability and its ability to take advantage of the challenge. This means that the actual relationship (correlation) between responsibility and burden-sharing (allocation of abatement costs) can be assessed only after these perceptions and capacity have been mapped. It should be emphasized, however, that perceptions on vulnerability and opportunity are not given once and for all. Most notably, new scientific evidence on the distribution of vulnerability – whether in terms of direct or indirect impact costs, or in terms of adaptation costs – may tilt perceptions in a completely new direction and become the basis of a new policy towards climate change. Technological breakthroughs in the field of energy conservation and renewable energies may work in the same direction by reducing the costs of abatement efforts.

Chapter 7

The Requirement of Cost-effectiveness: Climate Change and the Notion of an Effective Abatement Policy

Asbjørn Torvanger

1. Introduction

The Framework Convention on Climate Change (FCCC) focuses on aims and principles for climate policy cooperation between countries. In Article 3, where the principles of the FCCC are set out, it is stated that climate policy measures should be cost-effective to ensure global benefits at the lowest possible cost (Art. 3.3). Furthermore, all Parties are to implement measures to mitigate greenhouse gas emissions and facilitate adaptation to climate change (Art 4.1.b).

On this background, this chapter seeks to define cost-effectiveness in relation to climate policy, and discuss methods for calculating abatement costs and benefits in terms of reduced global warming, and the implications for choice of policy instruments. An important motivation for the focus on cost-effectiveness is the relation between the level of cost-effectiveness and the ambitions of the Parties to the FCCC with respect to reducing emissions of greenhouse gases. Greater cost-effectiveness means that more emissions abatement can take place for a given cost.

Even if the FCCC is weak with respect to legally binding targets, time schedules, and measures for mitigating greenhouse gas emissions, the Parties to the Convention are invited to engage in a climate process. Detailed targets and implementation mechanisms can be negotiated in later protocols to the FCCC. There can be a moral pressure on countries less willing to participate in this process. According to the FCCC, industrialized countries have a specific obligation to finance the additional costs of Third World participation in the climate process.

This chapter discusses characteristics of an efficient climate policy, without paying much attention to the short-term realism of some of the policy options. An efficient climate policy is defined in terms of social efficiency, thus minimizing the social cost of a given climate policy target, and in addition possibly finding the optimal scale of climate policy measures, comparing the total costs and benefits in terms of reduced global warming. Subsequently, some studies of climate change costs, climate policy measures costs, and model analyses of an optimal climate policy, are reviewed. Finally, short-term and long-term climate policy perspectives are discussed.

2. An Efficient Climate Policy

2.1. Conceptions and Definitions

An efficient climate policy can be defined at three levels of aggregation: (i) minimization of national costs, (ii) minimization of global costs, (iii) finding a global optimum.

In the process of minimizing costs of reaching a specific national emission abatement target, all abatement options must be considered. Abatement options differ according to: (i) project type (such as improving energy efficiency or switching from high-carbon fuels to low-carbon fuels, e.g. from coal to natural gas); (ii) different greenhouse gases abated (whether, e.g., carbon dioxide (CO_2) or methane (CH_4)); (iii) the economic sectors involved; and (iv) the policy measures available (such as taxes or emission quotas).

In the same way as reducing gross emissions, consideration must be given to measures to reduce net emissions through increased uptake of greenhouse gases in sinks.[1] The abatement effect of each policy option in relation to the cost must be found, so that options can be ranked according to increasing cost per unit of greenhouse gas abatement. An efficient climate policy is thus characterized by *an emissions abatement target being met by carrying out climate policy options according to increasing cost per unit greenhouse gas abated.* First to be implemented should be the options of lowest cost and highest abatement effect.[2] Such a procedure will lead to cost minimization and cost-effectiveness. For large-scale abatement projects or policy options, complicated calculations may be required, since implementation may influence the cost calculation of projects

---

1 Net emissions of greenhouse gases can be defined as gross emissions from combustion of fossil fuels and other sources minus removal of greenhouse gases by sinks. The most important sink category is sequestration of CO_2 in tree biomass through forestation.

2 As an illustration, assume that we can choose between policy option A at a cost of 2 mill. NOK that leads to a reduction of CO_2 emissions of 20,000 tons, and policy option B at a cost of 3 mill. NOK and reduction effect of 40,000 tons CO_2. Then the cost-effect quotient of policy option A is 100 NOK/ton CO_2 and 75 NOK/ton CO_2 for policy option B. Thus policy option B should be implemented before policy option A since the cost-effect quotient is lower for B.

implemented later. Ideally the effects and costs of all projects should be calculated simultaneously so as to capture all interactions, but this is less realistic due to the complexity involved and the insufficient information on possible abatement projects.[3]

To *minimize the global cost* of achieving a global emissions abatement target, policy options must likewise be implemented according to increasing cost-effect quotient irrespective of national borders.[4]

The most ambitious aim is to identify an optimal global climate policy. In addition to a minimization of global abatement costs, abatement projects should then be implemented until the cost of the next project equals the benefit in terms of reduced global damage from climate change. This necessitates a global cost-benefit analysis, and is more demanding than cost minimization since there is an uncertain relation between reduction of emissions and reduction of global warming, and between reduction of global warming and reduction in damage to the economy and the environment.

An optimal climate policy must also include an optimal adaptation to climate change in addition to an optimal reduction of net emissions of greenhouse gases. Future costs can be reduced if the planning of long-term investments in infrastructure, buildings, energy systems, etc., is carried out under the consideration of a possible future climate change likely to occur at a given rate of change. Some investments of this type will be optimal, since our society probably will have to adapt to some climate change in any case. Because of the inherent uncertainty in this field, investments that can increase future flexibility with respect to climate conditions should be of particular interest. Different adaptation options must be implemented according to an increasing cost-effect quotient, in the same manner as for emissions abatement projects.

An *optimal global climate policy* is therefore one that can reduce net emissions of greenhouse gases and invest in adaptation measures until the cost of the next policy option and investment option equals the benefit in terms of reduced damage from climate change.

2.2. Abatement of Greenhouse Gas Emissions

The government can influence and reduce the emissions of greenhouse gases by influencing the incentives of households and enterprises –

---

3 See Hanisch et al. (1993a).

4 This means that national borders ideally should play no role for choosing abatement projects, only the cost-effect quotient, since the climate effect is independent of the geographical location of the emissions. Consequently, referring to the example in note 4, project B should be undertaken first irrespective of its geographical location.

whether through carbon taxes, tradable quotas, direct regulation, or efficiency standards. Such measures can be undertaken at the local level, the national level, or at the international level, for example in the European Union (EU). Emissions can be reduced through increased energy efficiency, i.e. reduced energy use per unit produced. This means that technological progress and the introduction of new and more energy-efficient technologies to replace older technologies are important. Since different emissions are associated with different fuel types, emissions can also be reduced through fuel-switching – for example substituting oil for coal, gas for oil, and hydropower and renewable energy sources (e.g. biomass, solar heating systems and wind energy) for fossil fuels. Improved technologies may also mean that production factors like real capital and labor are substituted for energy. A more general effect is found in the substitution of energy- and emission-intensive goods and services for less energy- and emission-intensive goods and services. One example of this would be replacing private cars by public transportation.

One alternative to abatement of greenhouse gas emissions involves the use of sinks of greenhouse gases to reduce net emissions. If areas are available for forestation, carbon can be sequestrated in tree biomass. The increase in biomass stock (and the stock of long-lived products made of wood, such as houses) determines the volume of carbon removal from the atmosphere and net carbon fixation. Thus there will be a contribution towards reducing net emissions of CO_2 from forestation (and natural tree growth) if the resulting increase in biomass stock is greater than the loss from logging and other anthropogenic impacts (and natural tree decay).

2.3. Greenhouse Gases and Other Pollutants

Economic activities – fossil fuel use in particular – lead to the release of greenhouse gases and other pollutants to the atmosphere, such as sulphur oxides (SO_2), nitrogen oxides (NO_x), polynuclear aromatic hydrocarbons, and chlorofluorocarbons (CFC), where the last-mentioned are recognized as gases that deplete the stratospheric ozone layer. These pollutants may have detrimental effects on human health, cause damage to economic activities (such as corrosion of buildings and materials), and have wider environmental impacts. Climate policy measures often result in reduced emissions of both greenhouse gases and other air pollutants.[5]

---

5 One example is a carbon tax that reduces the consumption of fossil fuels, and thus also emissions of both carbon dioxide and gases like nitrogen oxides and sulphur oxides. The latter two gases can cause local or regional air pollution, but have no climate effect.

To calculate the net cost of climate measures, the benefits in terms of reduced pollution damage from such pollutants should be subtracted. This is required to find the efficient level of climate policy measures. Moreover, reduced emission of greenhouse gases may influence the impacts of other air pollutants through chemical processes in the atmosphere, and vice versa. Consequently, in determining an efficient climate and environmental policy the total effect of the policy measures should be considered: both the effect on greenhouse gases and the effect on other pollutants.[6]

2.4. Comparing the Climate Effect of Different Greenhouse Gases

There are interactions between different greenhouse gases, and between greenhouse gases and other gases in the atmosphere. These interactions are relatively well known, but there are difficulties in quantifying their effects. This is especially the case for the indirect climate effect of some gases, for example NOx, that may have a climate effect due to chemical interactions with greenhouse gases. The knowledge of sinks and reservoirs of greenhouse gases is also limited; for example, the role of the ocean as a reservoir for carbon. Thus it is a complicated task to calculate the atmospheric lifetime of carbon and changes in concentrations of atmospheric CO_2. For this purpose detailed greenhouse gas inventories are required. Hoel and Isaksen (1994) have calculated weights for greenhouse gases on the basis of atmospheric processes and lifetimes, and on the basis of the complicated relation between greenhouse gas concentrations and atmospheric temperature increase. Such weights are necessary to develop an efficient climate policy. One finding is that the weights will depend on future economic growth and the discount rate.

2.5. Uncertainty, Risk Attitude, Insurance

Many important types of uncertainty are associated with the climate change issue, relating both to physical–chemical conditions and the cost-effectiveness of climate policy measures (see Chapters 3 and 5). This uncertainty can, to some extent, be reduced through research and continuously updated knowledge. Manne and Richels (1990) have shown that such research can have a high yield, but will be sensitive to the knowledge-updating momentum, i.e. learning. This finding is supported

---

6 Aunan et al. (1993) present a model framework that can be applied to such integrated pollution control.

by Peck and Teisberg (1993) in a similar analysis, but they also find that the research yield is highest in a situation where the scale of climate measures is less than optimal, where research leading to reduced uncertainty increases the number and scale of climate policy measures. In any case, a climate policy must be developed in an environment of extensive uncertainty. In such a setting an important question is the level of risk we are willing to accept – our risk attitude or degree of risk aversion. This will largely be risk which we today accept on behalf of future generations, a point with implications for intergenerational equity. If we decide to carry out some climate policy measures in the next decades, the climate change process will be slowed down. Such measures can be interpreted as an inexpensive insurance against rapid climate change that could have major impacts on future economic activities and the natural environment. This type of insurance can save large future costs. Many climate measures – like increasing energy efficiency, saving scarce resources and reducing air pollution – make sense even if there should prove to be only minor climate change in the future. They are often referred to as "no-regrets" options.[7]

2.6. Discounting and Future Generations

Cost-benefit analysis employed for finding an optimal climate policy is primarily a tool for ensuring efficiency, i.e. allocating resources to maximize the discounted sum of net benefit over the time horizon. However, there is also an element of intergenerational equity associated with a climate policy option which is not covered by a cost-benefit analysis. One important issue in this relation is the choice of discount rate.[8] According to the opportunity cost principle the discount rate should be equal to the time preference if climate measures imply reduced consumption,[9] but be

---

7 No-regrets options can be defined as investments that are profitable under normal market conditions (including potential local environmental benefits). Thus such investments are profitable even if potential global climate benefits are not included.

8 Future costs and benefits are discounted at some rate due to time preference or the existence of a positive interest rate to make costs and benefits accruing in different years comparable. As an example USD 100 is more worth today to an individual than next year due to impatience or because USD 100 can be invested today to earn interest by next year. Thus if the interest rate is 10%, USD 100 today is worth USD 110 next year, and USD 100 next year is worth USD 100 divided by 1.1 today, which is equal to USD 90.91. Consequently the discount rate in this example is equal to the interest rate at 10%.

9 Since resources are limited and all wants cannot be satisfied, the opportunity cost of an action (e.g. investment option) is defined as the value of the foregone alternative action.

equal to the marginal product of real capital if climate measures imply reduced investments. Especially in the latter case the discount rate could become high enough to make the weight of future generations small. Because of effects of this type, cost-benefit analyses should be supplemented by ethical considerations regarding a fair and equitable welfare distribution between generations. One implication could be that more wealth should be transferred to future generations to compensate them for a relatively high rate of global warming and subsequent reduction in welfare if few climate policy measures are undertaken by our generation.[10] This transfer could take the form of improved technologies and increased real capital investments such as investments in buildings, machines, and infrastructure.

3. Climate Change Impact Costs

Climate change will affect both economic activities and the natural environment. The consequences will depend on the global climate change and local climate conditions. There will be impacts on agriculture, forestry, and the living conditions of animals and plants. In addition some coastal areas will be lost due to sea-level rise; the demand for heating or air-conditioning of buildings can change; an increase of pests and diseases can occur and affect human health, agriculture and forestry; the supply of fresh-water can be altered; and the frequency of hurricanes and extreme weather conditions may increase. The higher the rate and speed of climate change, the more vulnerable will economic activities and the natural environment be, with their limited adaptation capacity. Slowing down the rate is more realistic than stopping climate change, and could also be closer to the social optimum (Houghton et al. 1990; 1992). The evaluation of climate change also depends on the time horizon applied. Should the present scale of fossil-fuel use continue (in particular coal, where large deposits are available), atmospheric greenhouse gas concentrations can continue to increase for many decades. Owing to emissions from combustion of fossil fuels and other sources, the temperature increase may continue until the end of the next century, or even further into the future. Cline (1992) argues that a time horizon of 300 years should be applied for analyses to include the maximum concentration of CO_2 in the atmosphere, and consequently the maximum expected temperature increase and consequences for economic activities and the

---

10 This transfer would also depend on future economic and welfare growth in the absence of any climate policy measures, since a high growth rate would make future generations much better off than the present generation.

natural environment. Thus, a long time horizon in such analyses may imply higher climate impact costs.

Cline (1992 and 1994) has attempted to estimate the climate change impact for the United States measured as loss in Gross Domestic Product (GDP). Even if the estimates are subject to many uncertainties and relate to the United States only and only a limited number of impacts can be captured by GDP changes, his figures do indicate the likely extent of damages. Realistic estimates of damages are required to determine the optimal scale of policy measures and greenhouse gas abatement to mitigate climate change. Cline finds that the damage induced by a doubling of atmospheric CO_2 concentrations by the middle of the next century could reach 1–2% of GDP, based on model calculations where the expected temperature increase reaches 2.5°C. If the climate system is even more sensitive, the temperature increase and impact will be larger. Further into the future, the damage could equal 6% or more of GDP. Nordhaus (1991) has done similar studies employing other models and finds that the lower limit on damages is the same (1% of GDP), but with an upper limit of 2% of GDP. His figure is lower than Cline's due to different model assumptions. Nordhaus restricts his analysis to a doubling of atmospheric CO_2 concentration, which indicates that he is using a shorter time horizon than Cline.

These estimates are primarily relevant for the United States. Other countries can face smaller or larger climate change impacts, and some might even gain from a moderate change in climate. Those most vulnerable to climate change are developing countries that have large coastal areas prone to sea-level rise, densely populated, with an economy and natural environment vulnerable to climate change, and with political and administrative systems inadequate to handle such challenges (Hernes et al. 1995).

4. Costs of Climate Change Abatement Measures

Several studies have been undertaken of the social cost of taxes on CO_2 emissions (carbon taxes) from combustion of fossil fuels, in addition to those of Cline and Nordhaus. Most of these studies build on an assumption of a specific target for CO_2 abatement, find the required carbon tax to meet the target and then calculate the social costs in terms of GDP loss. In some of the studies, tax revenues are recycled into the economy through reducing other taxes. This should reduce the social costs, since most taxes do not correct externalities (such as pollution), and consequently induce a social loss due to distortions in resource allocation.

The social costs of abatement measures will also depend on the level of international coordination. Unilateral measures are likely to carry a higher cost than climate measures that are part of a coordinated climate policy involving many countries, particularly main trade partners. The reason for this is a possible loss of competitiveness for companies if unilateral measures, such as carbon taxes, lead to higher production costs than for their competitors. The sectoral economic structure of a country plays an important role in this relation. For Norway, as a major exporter of oil and gas, export revenues are a vital part of the national budget. If an international carbon tax on fossil fuels is introduced, these revenues will be reduced. Even if oil and gas revenues are reduced, the value of Norway's gas reserves could increase compared to the oil reserves if the tax is based on carbon content. In that case, the carbon tax of gas will be lower than the carbon tax on oil, and gas will become a more attractive energy source.

Two abatement cost studies are chosen as examples, one from the United States and one from Norway. In their study of the United States, Jorgenson and Wilcoxen (1993) show that stabilizing emissions at the 1990 level leads to a 0.55% reduction in GDP by the year 2020. The loss could increase to 2% of GDP for a more ambitious abatement policy. If the tax yield is employed to reduce other taxes (such as real capital taxes), there could be a net benefit for the economy. Moum et al. (1991) present some Norwegian results based on the KLØKT calculations carried out for the Inter-Ministerial Climate Group and based on the MODAG and MSG macro models developed by Statistics Norway. The carbon tax yield is substituted for other tax yields. The GDP reduction in the year 2000 is from 0.5 to 1.0% for a stabilization of CO_2 emissions at 1989 level compared to the reference path (which shows the development in the absence of any climate policy measures). The GDP reduction is 1% for unilateral Norwegian measures, but 0.5% in the case of an internationally coordinated climate policy. However, because of the reduction in the oil wealth, the reduction of consumption is largest in the case of an internationally coordinated climate policy.

5. Modelling Optimal Global Climate Policy

Few attempts have been made to model an optimal global climate policy. Nordhaus (1991 and 1993) has analysed potential climate impacts and damages and the cost of measures to abate greenhouse gas emissions. The model of the first study is further developed to DICE (Dynamic Integrated Climate-Economy Model) in the most recent study, which is a long-term dynamic optimization model. Most data are based on the

United States, but Nordhaus tries to generalize the results to other countries. He finds that the abatement cost for the first 10% of emissions is very low, but then the marginal cost for further abatement increases rapidly. Model results show that the optimal policy implies a moderate scale of measures to abate emissions, from 2 to 10% reduction initially as compared to the reference path. Peck and Teisberg (1992) show that the shape of the damage function – which gives the social cost of climate change as a function of atmospheric temperature increase – is decisive for the optimal climate policy. The results are similar to Nordhaus's results, since only moderate measures are optimal in the case of a linear damage function, where the marginal damage cost is constant when atmospheric temperature increases. In the case of a quadratic damage function – where the marginal damage cost increases with atmospheric temperature increase – more extensive measures will be optimal. In both cases the optimal choice is to delay measures until the next century. Cline (1992 and 1994) has undertaken a cost-benefit analysis, but unlike Nordhaus and Peck and Teisberg he finds that a more extensive climate policy is optimal. Such a policy would involve stabilizing annual greenhouse gas emissions at a level one-third lower than today; this would equal a 71% reduction compared to the reference path in 2050 and 90% in 2200. The deviation between the Nordhaus and Cline studies is caused by a higher discount rate and a linear damage function in Nordhaus's study, compared to Cline's non-linear damage function. In addition, Nordhaus has assumed no uncertainty and no risk aversion; he has chosen other parameter values and made other assumptions on the reference path than Cline. One policy implication of the deviation between these analyses is the importance of further research to improve our understanding of the climate system in general and damage functions in particular.

6. Short-term Climate Policy

A realistic climate policy for the next decade must be based on the FCCC. The FCCC is by nature a framework convention of few specific commitments for the Parties. Each Party to the Convention can formulate its own emission abatement target, while groups of countries, such as the EU, can coordinate their climate policies if they wish. The first Conference of the Parties CoP to the FCCC held in Berlin in March/April 1995. In its review of Article 4 of the FCCC on commitments the CoP, "... concluded that these are not adequate, and agreed to begin a process to enable it to take appropriate action for the period beyond 2000, including the strengthening of the commitments of Annex I Parties in Article

4.2 (a) and (b), through the adoption of a protocol or another legal instrument". Because of the weak commitments of the present FCCC, reluctant countries have many possibilities to keep a low profile on climate policy measures. Developing countries have even fewer commitments than industrialized countries, nor are they likely to get such commitments in the near future. According to the FCCC, industrialized countries are to cover the additional costs to developing countries which follow from introducing global climate and environmental considerations in their development plans. The Global Environmental Facility (GEF), established by the World Bank, UNEP, and UNDP, and funded by industrialized countries, is mandated to transfer the required financial resources to developing countries. Even if commitments for greenhouse gas abatement are somewhat strengthened for industrialized countries, comprehensive commitments and abatement programmes seem unlikely in the next few years. Likewise, extensive systems for carbon taxes or tradable quotas (and the inclusion of other greenhouse gases) are not likely either.[11] What is somewhat more probable is that a group of industrialized countries, such as the EU, may introduce tax or tradable quota systems (see Chapter 12). If all countries participated in tax or tradable quota systems, global cost minimization would be achievable, in theory.

Since such global regimes are less likely, however, we should rather focus on second-best options – where single countries introduce measures, or a bilateral cooperation is established, for example in terms of joint implementation arrangements under the FCCC,[12] or through coordinated policy measures within a group of similar countries. In the following, some important climate policy measures for single countries and groups of countries will be reviewed. These are taxes, tradable quotas, joint implementation, and efficiency standards. Emissions constraints are not discussed here, since these are considered to be the least cost-effective option.

6.1. Taxes

Numerous studies have focused on the design of an optimal carbon tax system, analysing the tax rates for fossil fuels reckoned in terms of the carbon content of each fuel. Many studies also discuss taxation of production or consumption. Other studies take up the handling of various market imperfections, such as monopoly, and asymmetric information. A review of national taxes, harmonized national taxes, and international

---

11 Tradable quotas is discussed in section 6.2.
12 Confer Art. 4.2.a of the FCCC.

carbon tax systems can be found in Hoel (1992). If a country or group of countries implement a coordinated climate policy, for example through a carbon tax, there can be a "leakage" that reduces the global emissions abatement effect. The market price of oil can be reduced due to lower demand for oil in participating countries, giving consumers in non-participating countries an incentive to increase their oil consumption. Kverndokk (1993) shows that in such a situation the group of participating countries could gain from paying non-participating countries to refrain from increasing their oil consumption. Another option would be to combine the climate policy of participating countries with a reduction of oil supply through buying oil reservoirs (Bohm 1993).

A few countries have already introduced greenhouse gas taxes. Norway, Denmark, Sweden, Finland and the Netherlands now levy a carbon tax on fossil fuels. Haugland (1993) has reviewed carbon taxes in these countries. The Norwegian tax was introduced in 1991. In addition Norway and many other countries have earlier introduced various energy taxes. The tax level varies, both the carbon share and the energy share of the total tax, and many economic sectors and energy commodities are exempted from the tax. Thus there is no clear relation between CO_2 emissions associated with energy commodities and economic activities, and the observed carbon and energy tax.

6.2. Tradable Quotas

Greenhouse-gas emission quotas that can be bought and sold on the market can implement a cost-effective distribution of emissions in the region where the quotas apply. The government (or governments) can decide the total emission quota and divide this into a number of identical quotas. These emission quotas may be shared between the emitters in the region (for example enterprises) in accordance with some principle, or they can be sold to the highest bidders. Since the quotas can be traded on a market, enterprises with low abatement costs can sell their quotas to enterprises with high abatement costs and earn a profit, while the latter group of enterprises will be better off buying quotas than going about the costly procedures needed for abating their own emissions. Thus a specific price for the tradable quotas is generated by the market, and cost-effectiveness can be achieved. The price of the quotas should then equal the marginal abatement cost, which is another prerequisite for an efficient solution. The quota market can cover a region of a country, be national, regional (e.g. the Scandinavian countries), or global. So far there are no examples of tradable quotas for CO_2 or other greenhouse gases, but the United States has established a market for SO_2 emissions.

Many technical and political problems must be solved before an effective international system of tradable emission quotas can be established. One political problem is to agree on the principle for initial distribution of quotas between countries, as well as agreeing on the possibility of redistributing the quotas after some time (Bohm and Larsen 1994).

6.3. Joint Implementation under the FCCC

According to the FCCC, climate policy measures can be implemented jointly by the Parties.[13] The idea is that the climate effect of greenhouse emissions is independent of the geographical location of the emissions, while the abatement cost varies substantially between countries owing to structural differences in economies and different resource bases. As long as a global tax or tradable quota system is less realistic, joint implementation arrangements can contribute to a greater cost-effectiveness than would be possible if all participating countries were facing equal percentage reductions of their greenhouse gas emissions. Joint implementation can be seen as an intermediate step towards an international tradable quota system. One example of joint implementation would be Norwegian investments in Poland to increase the energy-efficiency of a power plant in Poland if this is cheaper than domestic emission reductions in Norway per unit of CO_2 abated. Through Norwegian investments the new and efficient energy technology could be introduced in the power plant many years before Poland could make the same type of investment. Such investments should reduce the total of Polish and Norwegian emissions in this period. Emission abatement in Norway is relatively expensive because of its hydropower-based electricity sector and Continental Shelf emissions from oil and gas production. The investing country (Norway) receives emissions credits that are allowed to count against its national commitments, and is thus able to save some abatement costs, while the host country (Poland) gains through the transfer of new and efficient technologies, energy saving and reduced air pollution.

Private enterprises can have incentives for participating in joint implementation projects in other countries if the necessary framework in terms of environmental regulations and economic conditions can be established. Examples of such regulations are carbon taxes and license requirements for greenhouse gas emissions. The enterprises can be given credit for their joint implementation investments, e.g. through tax

---

13 After the first CoP in Berlin, such arrangements in the pilot phase (lasting until the year 2000) are referred to as "Activities Implemented Jointly".

exemptions. The mechanism for such transfer will have to be established by the national government in accordance with the FCCC. A precondition for bilateral joint implementation arrangements is that the investing country is given emission credits that can count against national commitments to reduce greenhouse gas emissions. Furthermore, the abatement effect must be measurable and verifiable (Hanisch et al. 1993b). A different option is to implement climate projects in other countries through funding of GEF activities; but in this case the investors cannot claim any emission abatement credits against national commitments.

The CoP in Berlin agreed upon a pilot phase for joint implementation that lasts until the year 2000. Both industrialized and developing Parties to the FCCC can take part in joint implementation projects during the pilot phase, but there will be no crediting against national commitments. Many issues and potential problems related to joint implementation will have to be analysed and solved to make the mechanism operational.[14] One of these issues is the definition of "baseline" in the host country for a joint implementation project. The baseline describes the reference situation in the host country in the absence of any joint implementation project, against which the abatement effect of the project must be measured. In principle only the additional costs of the more climate-friendly joint implementation project as compared to the best alternative (and less climate-friendly) project should be covered by the investor. Thus, local environmental and other national benefits for the host country should be accounted for. However, to give the host country incentives for participating in joint implementation projects, the cost-saving benefit for the investor country should be shared between the investor and host. Some incentive problems will remain due to asymmetric information. It is in the interests of the host country to exaggerate project investment costs and underestimate national environmental benefits, so as to increase the transfer from the investing country. In addition, uncertainty associated with the abatement effect leads to uncertainty as to the costs of reaching some specific abatement target.

Some developing countries and NGOs have expressed concern that joint implementation could lead to a new environmental colonialism. As with all other international arrangements, there will be a distribution of costs and benefits between participating countries, and this will depend on the rules. However, there is a potential for developing the joint implementation mechanism to benefit both investor countries, which are

---

14 Confer Torvanger et al. (1994) and Selrod et al. (1995) for a review and discussion of such issues.

likely to be OECD countries, and host countries, which could be developing countries or countries currently in transition to a market economy (i.e. East European countries and Russia). The benefit in terms of reduced cost for reaching a specific global greenhouse gas emissions target, should be shared among all participating countries. It would not be in the interest of a host country to accept joint implementation projects that were not to its benefit when all costs and benefits are accounted for.

6.4. Efficiency Standards

In some cases the most effective measure for reducing greenhouse emissions is to introduce technical efficiency standards. This could be the case if monitoring of emissions of pollutants to air is expensive, making the administrative costs substantial, and if taxing of these emissions is the alternative to efficiency standards. Efficiency standards prevail in many countries, since they seem more acceptable to politicians and enterprises than tax schemes (Fischer 1980; Bohm and Russell 1985). Technical efficiency standards should apply to all new purchases of equipment, and could be made relevant for maintenance and the technical standard of older equipment as well. Examples of such standards are energy efficiency and coolant requirements for refrigerators, and engine efficiency requirements for car manufacturers.

7. Long-Term Climate Policy

Future protocols or agreements can contain quotas for greenhouse-gas emissions or constraints for each participating country, a further development of the Joint Implementation mechanism, or a tradable quota or international greenhouse tax regime. The total cost of reaching a specific emission-abatement target should be lowest with the tradable quota and tax system, while a quota or constraint system pays more attention to the distribution effects between countries. These effects depend on the criteria for distribution of quotas and emission constraints among countries.

A more comprehensive joint implementation system may be established, e.g. involving a "market place",[15] where hosts present joint implementation project candidates, and potential investor countries can choose among them (Hanisch et al. 1993b). A "credits bank" is another option, where investors deposit funds and receive emission credits based on the average abatement yield of joint implementation projects carried out by the bank in host countries.

---

15 Confer the concept of a clearing house.

An extensive literature exists on the design of optimal tax systems, and the difficulties of inducing countries to participate due to different expected climate change impacts and costs, different abatement costs, and the general "free-rider" problem (all countries basically benefit from reduced climate change but prefer to pay as little as possible of the bill). Hoel (1991) shows that most if not all countries can benefit from a tax or tradable quota system, given a suitable redistribution of tax revenue or initial distribution of emission quotas.

An alternative to a comprehensive international tax system is to establish an international or global market for emission quotas of CO_2 and other greenhouse gases. Through the sale and purchase of emission quotas, the system can lead to greater international cost minimization. Rose and Stevens (1993) have analysed efficiency and welfare distribution consequences of an international tradable quota system. The gains from trade are large compared to fixed emission quotas for all countries. They find that the welfare distribution between countries depends on the choice of ethical principle for the initial distribution of quotas, whereas trade reduces these differences significantly, making the choice of ethical principle less important for the final welfare distribution. Manne and Richels (1991) have analysed the trade in emission quotas between five geopolitical regions. They find that the trade volume will be limited, since no region imports or exports more than 5% of the global volume of emission quotas.

8. Conclusion

The global costs of reaching a global emissions abatement target are minimized if policy options are implemented according to increasing cost-effect quotients independent of national borders. An optimal climate policy must be defined at the global level. It will be characterized by abating greenhouse gas emissions and investing in adaptation measures until the cost of the next policy option and investment option equals the benefit in terms of reduced damage from climate change.

Emissions of greenhouse gases can be abated if households and enterprises are given incentives to reduce activities that are greenhouse-gas intensive, for example through carbon taxes or efficiency standards. The main options are increased energy efficiency and fuel-switching, in addition to carbon sequestration through forestation. An efficient policy should evaluate the total environmental impact of climate-policy measures, including both the effect on greenhouse gases and other pollutants.

Climate-policy measures can be interpreted as insurance against rapid climate change that could have major impacts on the economy and environment, and could thus save large future costs. By nature, climate policy is long-term, so cost-benefit analyses will need to be supplemented by ethical considerations as to what constitutes a fair welfare distribution between generations.

Studies of climate change impact show cost estimates of 1 to 2% of GDP by the middle of the next century, but the estimates are open to substantial uncertainty. Impacts will be unevenly distributed between countries; developing countries with a vulnerable natural environment and economy are likely to face the highest costs. The social costs of abatement measures depend on the type of measure, the resource base and economic structure of a country, and the level of international coordination. Studies have indicated that the cost involved in stabilizing CO_2 emissions at the 1990 level could amount to 0.5–1% of GDP. Findings from a few studies on optimal abatement level show large deviations, from 2 to 33% reduction of present emissions of CO_2 being optimal.

The main options available for climate-policy decision-makers are taxes, tradable quotas, joint implementation under the FCCC, and efficiency standards, in addition to emission constraints. From a cost-effectiveness perspective, taxes and tradable quotas are the most attractive, and to some extent joint implementation, if the first two options are not realistic at the global level. The joint implementation mechanism could develop into a global tradable quota system.

The distribution of costs and benefits associated with climate policies is likely to be the most demanding issue for the international political scene (see Chapter 6). Also highly relevant will be the long-term distribution between generations. In the short term the climate policy costs should be distributed between countries so that as many countries as possible participate, and at the highest possible level of cost-effectiveness. In the long term, this is an issue of distribution between generations. The question is what costs we can allow ourselves to transfer to future generations, and what level of risk our generation engages in on behalf of future generations. To some extent, future generations can be compensated for expected risk increases through an increased savings rate and the transfer of wealth from our generation. The consequences of the climate-policy choices of our generation today will be met first and foremost by our descendants.

Part II
Critical Actors

Chapter 8

China and Climate Change

Christiane Beuermann[1]

1. Introduction

China's economic growth rate of more than 10% per year over several years causes industries and policy-makers in industrialized countries to focus on this market that *The Economist* labelled "A billion consumers" (1993). While many other countries face minimal growth or even stagnation, China is forced to curb its boom to avoid an overheating of the economy. Beneath all this economic enthusiasm the related environmental risks seem to be forgotten, dispelled or underestimated despite high pollution levels in many regions and heavy local impacts. The issue of climate change has been given much rhetoric. However, in international negotiations, China has so far rejected specific abatement commitments.

This chapter provides an account of the current Chinese climate policy. Starting with a description of the Chinese greenhouse-gas emission inventory and the assessed impacts of climate change on China, the main part of the analysis focuses on the willingness and the ability of China to develop and implement abatement strategies and measures. In this part, the Chinese position towards the adoption of targets and timetables and the implementation of policies and measures to reduce greenhouse-gas emissions will be reviewed. Through this review we will also record the Chinese perception of climate policy and consider the question of whether there is a difference between the Chinese Government's international position and actual national response.

Regarding the international climate change negotiations, this chapter

1 The author would like to acknowledge the valuable comments provided by Reinhard Loske (Wuppertal Institute) and Stefan Speck (Keele University), and, in particular, by the editor Gunnar Fermann of the Norwegian University of Science and Technology. Responsibility for remaining weaknesses, factual or interpretive, remains, of course, with the author.

focuses on the arguments which have been used to promote and defend China's interests and policy positions. The analysis concentrates on the Chinese ability to manage the issue of climate change and to implement an abatement policy. This leads to an examination of the role of Chinese science in the identification of climate change as an issue of political priority, and the integration of scientific research into political decision-making processes. A review of the development of basic environmental legislation and institutions and their application and adjustment to the issue of climate change follows. Finally, in order to evaluate the Chinese commitment, an assessment of currently implemented measures and proposed economic goals is presented.

2. China's Responsibility for and Vulnerability to Climate Change

Perceptions of responsibility and vulnerability are crucial in making climate change policies in any country. In the following section, the Chinese contribution to climate change is measured in terms of anthropogenic greenhouse gas emissions, and the most probable impacts of climate change on China are assessed.

2.1. Chinese Greenhouse Gas Emissions

As pointed out in chapter 6, responsibility is a multi-faceted and ambiguous concept. In this assessment, the Chinese responsibility is measured in terms of both total current and *per capita* greenhouse gas emissions. The anthropogenic sources of the carbon dioxide (CO_2) emissions are described in detail; for other greenhouse gases, the sources and figures of *total* current emissions are provided.

2.1.1. Carbon Dioxide

CO_2 emissions are determined by a number of factors, such as carbon intensity of fuels, energy intensity, GDP, population, to name a few. Since 1950, energy supply in China has grown rapidly, though there are still major deficiencies and shortcomings compared to the growth of energy demand. According to Marland et al. (1994), CO_2 emissions from fossil fuel consumption rose from 21.7 MtC in 1950 to 660.7 in 1990. In an international perspective, this corresponds to an increase from 1.3% of the world's total CO_2 emissions from fossil fuel consumption in 1950, to 10.7% 40 years later – a tremendous growth in both absolute and relative terms. Starting from a tenth position in 1950, China became the third largest emitter of CO_2 from fossil fuel combustion after the US (22% of the total) and the former Soviet Union (16%).

Ninety-five percent of China's commercial energy is produced from fossil fuels with a clear dominance of domestic coal. In 1990, the use of coal accounted for 86% of the total CO_2 emissions, petroleum for 13% and natural gas (without gas flaring) for slightly more than 1%. A sectoral differentiation shows that almost 70% of the CO_2 emissions are due to the industrial sector and power generation (Siddiqi et al. 1994). Accounting for less than 5% of the total, the share of the transport and agriculture sector combined is as yet fairly insignificant. Furthermore, in 1990, 2% of the total Chinese CO_2 emissions (25.5 MtC) were due to cement manufacturing.

As pointed out by Siddiqi et al. (1994), biomass is the dominant source of energy (80%) in the rural areas of China. In official statistics, the consumption of biomass in rural areas is reported to amount to 200 Mtoe (Loske 1993). CO_2 emissions are roughly estimated to equal the amount of carbon drawn from biomass (forests and crops). As regards CO_2 emissions from land-use changes, Smil (1994) concludes that estimates for deforestation and combustion of biomass are highly uncertain on both the global and national level. Nevertheless, Smil assumes that 5 to 10% of the global CO_2 emissions resulting from land-use changes are emitted in China.

For the energy sector, Siddiqi et al. (1994) assess that in a business as usual scenario, without any climate protection measures, CO_2 emissions will increase from 626 MtC in 1990 to 907 MtC in the year 2000; 1,354 MtC in 2020; and to 1,918 MtC in 2050. According to this scenario, China will become the largest emitter of CO_2 by the year 2020. In a climate protection scenario, an increase of CO_2 emissions to 888 MtC in 2000; 1,179 MtC in 2020; and 1,467 in the year 2050 are estimated. As an illustration of the huge abatement potential in China, the possible reduction of CO_2 for the year 2020 equals the current (1990) energy-related emissions of CO_2 of the UK. Moreover, the potential reduction for the year 2050 is larger than the total 1990 emissions of France and Japan together. This is not the least what makes China a critical actor in the international politics of climate change.

However, in a *per capita* perspective, China emitted only 2 tonnes CO_2 *per capita* in 1990, while the global average was 4 tonnes of CO_2. With 20 tonnes of CO_2 emissions *per capita*, the US has the highest *per capita* CO_2 emissions in the world. The former Soviet Union and unified Germany both account for 12 tonnes CO_2 emissions *per capita*, while Japan emits 9 tonnes.

2.1.2. Other Greenhouse Gases

With an estimated amount of 10–14 Mt methane (CH_4) in 1990, flooded rice fields are the main source of CH_4 emissions in China.[2] Other sources are related to coal mining CH_4 losses in the process of extracting oil and natural gas fields, emissions resulting from fossil fuel combustion, emissions from ruminant livestock, the burning of biomass as well as the decomposition of animal and municipal wastes or from sewage. However, problems remain in estimating the quantities of CH_4 emissions: A comparison of different data (Siddiqi 1994; WRI 1994; UNEP 1993; Smil 1994) shows that estimates of the total CH_4 emissions of China vary from 21 to 52 Mt per year – thus complicating an assignment of responsibility. Nevertheless, UNEP (1993) estimates that China was the largest emitter of CH_4 in 1990, accounting for about 15% of the world's total emissions, followed by India (12%), the former Soviet Union (9.5%), the United States (9%) and Brazil (3.5%).

As to nitrous oxide (N_2O), this greenhouse gas is mostly emitted from combustion processes (fossil fuels and biomass). Given the fact that the quantities emitted are highly dependent on factors such as the temperature of burning, estimates should be considered approximates. Other sources are the use of fertilizers and to a lesser extent emissions from soils and the production of synthetics (Siddiqi et al. 1994). UNEP (1993) calculates China's total N_2O emissions to be 2.24 Mt N in the year 1990.

The total Chinese consumption of chlorofluorocarbons (CFCs) in 1990 was 0.044 Mt; 44.4% for refrigeration, 42.5% foams, and 2.1% cleaning (Siddiqi et al. 1994).

2.1.3. Evaluating China's Emission Responsibility

An assessment of China's emission responsibility is necessarily ambiguous. It depends very much on the perspective chosen: In contrast to most industrialized countries, responsibility differs considerably depending on whether total or *per capita* emissions are considered.[3] Because of its high

---

2 Siddiqi et al. (1994) refer to measurements made by Chinese scientists which showed that "... methane emissions from a single rice field have very large diurnal, seasonal, and annual variation, and that emissions from different rice fields vary significantly in average emission rates and patterns of fluctuation". The emission level of a rice field depends, *inter alia*, on the climatic zone (local meteorological conditions), geomorphic characteristics (soil type) and farming practices.

3 Industrialized countries are generally characterized by both high total and high per capita emissions, while developing countries are characterized by low total and low per capita emissions.

population, China is bound to be one of the world's largest emitters of CO_2. On a per capita basis, however, China emits only one-tenth of that of the US and only half of the world average. Hence, China is both one of the most and one of the least responsible countries. While it is "almost inevitable" (Smil 1994) that China will become the world's largest emitter of greenhouse gases by 2020, the country's per capita emissions will still be modest in an international context. However, owing to the enormous size of its population, China will nevertheless be a key country in any concerted international attempt to combat climate change.

2.2. Impacts of Climate Change

The assessment of the impacts of climate change is complicated and confronted by great uncertainties. While regional predictions of impacts are of special interest to countries in their defining of positions and strategies in international bargaining processes,[4] such information is, at present, available only to a limited extent.

The assessment of China's vulnerability is particularly complicated because of its huge territory. Being the third largest country in the world, China's total area is about 9.6 Mio km^2. China can be divided into three major climatic zones: temperate latitudes in the northeast (Manchuria), desert-like dry areas in the northwest, and humid subtropics in the south (Statistisches Bundesamt 1993). Hence, impacts of climate change on the Chinese territory will very much depend on the climatic region in question.[5]

Some overall predictions of probable impacts of climate change in China can be found in the reports of the IPCC working group II, that

---

4 Provided there might be some winners of climate change, a strong climate policy will be hindered by an alliance of winning countries and countries that would suffer large economic losses by greenhouse gas mitigation policies (e.g. OPEC countries). In contrast, vulnerability in combination with non-responsibility is a strong incentive for a country to go for an effective climate protection policy. The best example is the Alliance of Small Island States (AOSIS countries). Though this group does not have much direct political influence, it plays an important role as a pusher – or a watchdog – for demanding climate change policy in the international negotiations. While the vulnerability of the AOSIS is very obvious, an assessment of China's vulnerability is more complicated because of the enormous size of the country and its several climate zones.

5 The existing climate computer models (global circulation models – GCM) allow only statements on large scale changes of the climate, such as different developments between the southern and northern hemisphere, between the continents and the oceans and the northwards drift of arid areas in the summer. Even a large country like China is too small for detailed regional predictions of climate change. Likewise, the assessment of impacts is on a very general level.

deal with impact assessment (McMcTegert et al. 1990; Watson et al. 1996).[6] Here, the IPCC concludes and strongly emphasizes that China will suffer from climate change. In fact, China has been judged to be one of the countries most vulnerable to climate change, along with Bangladesh, Egypt, Thailand and Brazil (McMcTegert et al. 1990: 2–20). This general conclusion is based on an analysis of the direct effects of climate change, such as effects on hydrology, water resources and natural terrestrial ecosystems, as well as indirect socio-economic effects, including effects on agriculture, forestry and human settlement. Without a consequent climate protection policy there are indications that China can expect winter monsoons to get weaker and summer monsoons to get stronger. Moreover, rainfall receipt could increase in already humid areas and extend further westward and northward. This would significantly increase the risk of flooding in southern China, and the already pressing problem of soil erosion would become more critical.

The consequences of climate change for the agricultural sector and in turn the Chinese food supply are ambiguous: A temperature increase of 1°C together with a precipitation increase of 100 mm is estimated to result in an increase in wheat and maize yields by about 10% nationally, although there might be a modest decrease in northern and eastern China. A temperature increase of 1°C without an increase in precipitation, however, is estimated to lead to a decrease in maize yields by an average of 3% in the eastern and central regions (McMcTegert et al. 1990).

It is also expected that owing to effects of sea-level rise (innundation, erosion, coastal and riverine flooding), the coastal zones of China would suffer from large farmland losses, up to at least 10 Mt of grain in the Yangtze and Yellow River region alone. Saltwater intrusion in surface and groundwater and a higher risk of flooding could lead to the salination of farmland and the loss of water for irrigation. Another consequence of land loss, damage to housing and infrastructure, etc., would be a dramatic increase in the number of environmental refugees: Estimates show that a rise of one metre in sea level would seriously affect about 70 million people along the coasts of China

---

6 This work is based on the assumptions of a doubling of the concentrations of CO_2 in the atmosphere between 1990 and 2025–2050, an increase of the global mean temperature in the range 1.5–4.5°C with an unequal global distribution (smaller increase in tropical regions, larger increase in polar regions), a sea level rise of ca. 0.3–0.5 m by 2050 and ca. 1 m by 2100, and a 0.2–2.5°C rise in the temperature of the surface ocean layer. These assumptions are in line with the estimates of IPCC WG I (scientific assessment) for a business as usual scenario (Tegart et al. 1991: 1; Houghton et al. 1990).

(Watson et al. 1996). Especially affected would be the Pearl River deltaic plain (city of Guangxhou), the eastern half of the Yangtze River plain (Shanghai), the eastern half of the North China coastal plain (city of Tianjin), and the southern half of the Lower Liao River plain (city of Yingkou). It is also projected that a one metre rise in sea level would destroy most of the salterns and seawater breeding farms and reconstruction would require 5.000 km² of land (McMcTegert et al. 1990). To counteract these threats, coastal defence improvements would have to be made and protection costs would rise drastically owing to the adjustment of dikes, sluices, land drainage and pumping systems.

An assessment of the impacts of climate change on the Chinese forest is also ambiguous, making it difficult to draw conclusions for the timber industry. Some species seem to be minimally affected, while major commercial timbers will suffer from increased climatic stress, disruption from fire and higher summer temperatures. Productivity is generally expected to decrease.

These examples suffice to demonstrate how serious the impacts of climate change may become for China – threatening both the social and economic development and the political stability of the country.

3. Chinese Climate Policy in an International Perspective

3.1 Targets and Time-Tables

The Chinese Government has not adopted a greenhouse-gas emission target and consequently has neither adopted a time-frame within which such a target should be achieved. The Chinese position in the international climate negotiations was and still is that developing countries, and in particular China, must be excluded from abatement commitments. This position is, for understandable reasons, based on the developing countries' and China's claim to their right to economic development. The opposition to targets and time-tables is not restricted to the reduction or stabilization of greenhouse gas emissions but also includes increase targets as announced by some member-states of the European Union (EU).[7] A Chinese increase target

---

7 Consider that the EU adopted a stabilization target committing the EU to stabilize CO_2 emissions at 1990 levels by the year 2000. Within the EU some burden-sharing is accepted: Some countries adopted reduction targets which are necessary to overcompensate increase targets adopted by other member-states, e.g., Spain (+ 25%), Portugal (+30–40%), and Greece (+25%).

could be interpreted as a ceiling (see also Section 5) and become a precedent for future negotiations. Chinese officials state that the increase of both total and per capita emissions of greenhouse gases will not be further specified (Qu 1992; Zou 1992; Siddiqi 1994). So far, specifications that go further than stating the aim of "working hard to reduce the increase of greenhouse-gas emissions" (WMO 1992) have not been made. The strong opposition of the developing countries and, in particular, China towards commitments has a serious impact on the further development of the FCCC. Discussions on the introduction of additional commitments, for example, nearly led to a stillstand of the negotiations during the first conference of the Parties (CoP) to the FCCC in 1995 in Berlin (Oberthür and Hermann 1995). To understand the rejection of targets and time-tables, an analysis of the determinants and justification of the Chinese Government's position is required.

3.2. Determinants of the Chinese Position

While climate change as an issue in its own right seems to be of minor importance, two basic driving forces have strongly influenced the way China handles international climate negotiations: foreign political influence and national sovereignty.

China has traditionally put much emphasize on foreign relations (Perlack et al. 1993). The Chinese government increasingly sees "environment issues" as an important part of China's foreign policy. The issue of climate change is regarded as an opportunity to demonstrate and defend China's claim and self-perception as a major player in world politics and to be the leader of and speaker for developing countries. The Chinese goverment considers this a naturally given role since China is the largest country by population and has demonstrated this perception on several occasions. For example, in June 1991 the Government of China hosted the Ministerial Conference on Environment and Development. The Conference was to serve as a platform for developing countries in formulating and expressing their points of view in preparation for the United Nations Conference on Environment and Development (UNCED) to be held in Rio de Janeiro in June 1992. Later, Li Peng, the Premier of the State Council, repeatedly expressed the view that the conference was a forum where the Chinese Government took the lead among the developing countries (WMO 1992).

During the negotiations of the Intergovernmental Negotiating Committee (INC), preparing the UNCED, the positions of the developing countries were represented by its political grouping G-77,[8] although there were divergent interests. China is not formally a member of this group, but drafts were typically introduced to the INC on behalf of the "G-77 and China". According to Bodansky (1993), the uniting argument within the group was that a convention must not hinder their countries' ability to develop. Apart from that, great diversity of interest was represented within the group. As a result, three interest subgroups of particular importance emerged: A group of semi-industrialized countries including the large developing countries such as China, India and Brazil; the AOSIS; and the oil-producing states. Because of struggles for power during the negotiations the group of semi-industrialized countries broke down and member-states were allowed to submit their own proposals. The remaining part of the G-77 was then led by India and China (ibid.; Oberthür 1983). India and China were in a continuous struggle over who was to act as the speaker (i.e. the political leader) of the group "G-77 and China" (Fischer 1992: 76). In practice, this struggle resulted in changes between, and even within, sessions.

The Chinese rejection of emission commitments under an international climate change regime also has roots in China's strong emphasis on national sovereignty. Abatement commitments were seen as being in conflict with the Chinese Government's sovereign decisions to use their natural resources in promoting their own economic development and environment goals. Such an understanding was officially declared in the Beijing Declaration on Environment and Development (*Beijing Review* 1991).

3.3. Justification of the Chinese Position

China justified its policy of "no targets and time-tables" on three lines of argument: the vulnerability of China towards climate change; the responsibility of the industrialized countries; and the scant resources available for abatement efforts in China.

Vulnerability: The argument is that China is among the countries that will be most negatively affected by climate change. This argument is generally included in the preamble of official documents and statements to explain and justify why the issue of climate change is important for China and why China is engaged in the international climate change

---

8 The Group of 77 was formed in the early 1960s in preparation for the UN Conference on Trade and Development. The name G-77 reflects the original membership of 77 countries which since has increased to more than 130 countries.

negotiations (e.g. Qu 1990; WMO 1992; Chen 1995). From observations during the negotiations, it is highly questionable whether the proposed negative impacts of climate change on China really had any influence on the Chinese positions during the negotiations of the INC, given the existence of other more tangible economic interests (Fischer 1992). For tactical reasons, vulnerability was only a secondary line of argument compared to the main argument of responsibility.[9] China's repeated questioning of the findings and results of the IPCC assessment may be seen in the same light (Loske 1996).

Responsibility: Chinese officials have repeatedly argued that the Chinese responsibility for historic and *per capita* greenhouse gas emissions is very low compared to that of other countries, and particularly compared to industrialized countries (e.g. Qu 1990; WMO 1992; Zou 1992; Chen 1995): Developing countries in general, and China in particular, cannot take on commitments if industrialized countries have not implemented reduction commitments in accordance with their historic responsibility. This line of argument coincides with the polluter-pays principle, which is not only the basis of the Chinese theoretical understanding of environmental policy (Section 4.1.2), but has also been recommended by the OECD since the mid-1970s.

Financial capacity: Finally, Chinese officials strongly argue that substantial abatement measures are dependent on the transfer of new and additional funds from industrialized countries (*Beijing Rundschau* 1992: 7–10). These political demands succeeded to some degree and have been integrated in the FCCC in several articles covering technology transfer, new and additional funds, and the opportunity to implement climate protection measures jointly. The call for new financial means for climate-protection measures reflects, of course, the scarcity of capital in most developing countries, including China itself (Perlack et al. 1993). Climate abatement measures which require large capital investment are judged unrealistic since such measures compete with other urgently needed investment. This view is supported by Lin Gan (1993: 270), who concludes in a study on the Global Environmental Facility (GEF)[10] that without GEF support "it would be difficult to initiate action on global climate change in China because of financial constraints".

---

9 Pushing for commitments and action of other actors, the assignment of guilt puts more moral pressure on countries than pointing to own vulnerability. Using vulnerability as an argument would also imply the general acceptance of own greenhouse gas reduction commitments.

10 The GEF is a source for development assistance under the World Bank, UNDP and UNEP. After the establishment of GEF in 1990, China was among the first countries to submit proposals to it.

4. National Chinese Climate Policy: Strategies and Institutions

While there are no Chinese targets and time-tables for emissions of greenhouse gases, the contours of a Chinese climate-protection policy can still be discerned.

The process of setting the political agenda has been basically influenced by international discourses and events. Since the formulation of a Chinese climate-change response strategy has yet to be finalized, implementation is at a very early stage. Difficulties and tensions arise because measures for climate change compete with other strategic, above all, economic, goals.

4.1. Formulation of a Chinese Response Strategy

The formulation of a Chinese response strategy is influenced by different institutions and their attempt to adjust to the rising environmental awareness of the Chinese population, climate change-related science, and environmental legislation. Stating this, it is also clear that China's development of a national response strategy towards climate change is influenced greatly by the international negotiations China has been taken an active part in.

4.1.1. Dissemination and Political Integration of Information and Knowledge

Public awareness and concern are important preconditions for the establishment of an issue on the national policy agenda. In particular with regard to long-term environmental problems, media coverage and public opinion have a significant influence on the process of setting the agenda and integrating these issues into short-term oriented political programmes.[11] Environmental awareness in China, however, is first and foremost a centrally steered process. For a number of years, the Chinese government has attempted to influence public opinion in order to persuade individuals to change behaviour in a more environment-friendly direction. Environmental education has been established throughout the Chinese school system. Taking high officials' frequent references to environmental issues and the growing interest of the media as indicators, it

---

11 In Germany, the three environmental issues acid rain, ozone depletion and climate change, appeared on the political agenda almost parallel to increasing media coverage (Jäger et al. 1993). This phenomenon can be found in most industrialized countries (Social Learning Group 1996).

appears that environmental awareness has improved in China (Perlack et al. 1993; WRI 1994). However, this optimistic view should be modified by the fact that access to education is generally limited in China (Hu and Wang 1991). The great differences in environmental concern evidenced in rural areas and in cities are due to both educational and economic variations. In particular, environmental concern rises where local pollution incidents negatively affect people, whether it be coal-related air pollution or noise in prospering regions (Ross and Silk 1986).[12]

Information and knowledge about the issue of climate change is to some extent integrated into the decision-making processes by a partial involvement of the Chinese scientific institutions into political structures. In a study on Chinese research on global change, the US National Research Council (NRC 1992) concludes that the Chinese scientific infrastructure is substantial, and that the Chinese global-change programme has a significant potential. The Chinese capacity for collaborating with foreign institutions and participating in the international scientific community is judged to be superior to that of other developing countries. In fact, the Chinese scientific community was already involved in international research programmes at an early stage and collaboration with other countries, in particular the United States, is close. According to the NRC (1992: 40), one of the explanations for this interest in research on global change, has been the personal commitment and efforts of Ye Duzheng, who was a former special advisor to the Chinese Academy of Sciences (CAS) and chairman of the Chinese National Committee for the International Geosphere-Biosphere Programme (IGBP). Ye recognized the relevance of the global change issues for China at the beginning of the international discussions. Since then, the CAS has taken the lead in the organization of global change research. Other institutions relevant to climate-change research are the State Planning Commission (SPC), the State Education Commission (SEC), the State Meteorological Administration (SMA), the State Oceanographic Administration (SOA), and the National Natural Science Foundation of China (NNSFC), which provides most of the research funds.

Chinese scientists have continuously participated in international programmes and fora concerning climate change. For instance, their

---

12 Except for these single citizen actions, China lacks an infrastructure of NGOs which are important actors for raising environmental awareness in industrialized countries. Lin Gan (1995) points to the difficulty of private engagement in opposing official policies in China.

participation in the IGBP started in the early 1980s. At the Second World Climate Conference in Geneva in 1990, the conference chairman was Zou Jingmeng from the Chinese SMA, who at the time was the President of the World Meteorological Organization (WMO). Participation of Chinese scientists in the working groups of the IPCC has also been prominent.

In 1987, the Chinese National Climate Committee (CNCC) was established by the State Science and Technology Commission (SSTC) in order to coordinate Chinese research on climate change. In doing so and in response to the need of interdisciplinary research, the CNCC evaluates and determines Chinese climate-change research projects. In 1990, CNCC presented a National Climate Programme for the years 1991 to 2000. Corresponding to the WMO climate programmes, the Chinese programme comprises five subprogrammes on (i) climate data, (ii) climate research, (iii) tropical oceans and global atmosphere, (iv) climate application, and (v) climate impacts (NRC 1992: 34).

Since governmental funding is the only way of financing and is generally limited and distributed to programmes of domestic priority, Chinese scientists focus in particular on social and economic impacts of climate change. Research and international collaboration on other areas than impact assessment depends heavily on external funding from bilateral and multilateral sources.

4.1.2. Adjustment of Political Institutions

The beginning of China's environmental policy can be dated back to at least the late 1970s: In 1979, a provision for the protection of the environment was added to the Chinese constitution. The same year, the Environment Protection Law was adopted, and amended in 1989. Since the late 1970s, several environmental laws have been enacted. Of special interest is the "Three Simultaneous Regulation", which determines "... that pollutant treatment devices must be designed, constructed and operated simultaneously with the design of the main project itself" (NEPA 1992: 16). Moreover, permission for operation is only given if a report on environment impact assessment has been approved by the administration. This approach has been successful in so far as compliance is judged to be increasingly good.[13]

---

13 Compliance rates with this regulation were assessed to have increased from 39% in 1979 to 94% in 1986 (Vermeer 1990: 59).

According to Qu[14] (1990: 104), the formulation of environmental policy in the 1990s is based on three elements: (i) the coordination of economic development and environmental protection, (ii) the search for cost-effective environment protection paths, and (iii) the strengthening of state environmental management systems.

Although economic development is clearly the dominant policy goal, Qu stresses that the Chinese Government consciously avoids repeating the mistakes of most industrialized countries believing that only rich economies can afford environment protection measures. This is reflected by the three principles supplementing the environment protection laws: the precautionary principle, the "polluter pays principle", and the principle of efficient environmental management. However, the overall priority of economic development was officially stated by Premier Li Peng in his speech before UNCED In 1992; in cases where the goal of economic development conflicts with the goal of environmental protection, economic development has first priority.

Following the judicial formulation of a Chinese environmental protection policy, an institutional infrastructure was established (GEF 1992; NRC 1992): The State Environment Protection Commission (SEPC) was set up in 1984 by the State Council (Vermeer 1990: 52). This is the highest environmental decision-making body and meets quarterly. The heads of all relevant ministries and agencies are represented in SEPC. Its tasks are to provide policy direction, coordinate between agencies, and review policies and strategic issues.

The Chinese National Environment Protection Agency (NEPA) was established as the implementing body of the SEPC and still functions as its secretariat. Since 1988, it has been an independent agency operating under the State Council. The NEPA is responsible for all kinds of environmental policy and focuses both on regional and local environmental issues. The NEPA maintains its own research institutes.

Based on the existing institutional infrastructure and in response to the emergence of the issue of climate change on the international agenda, the infrastructure was adjusted owing to the need for inter-ministerial and inter-agency coordination. As a result, the National Climate Change Coordination Group (NCCCG) was established in 1990. In addition to the NEPA and the SSTC, five government ministries are represented in the NCCCG. The group is responsible for the overall policy formulation on a variety of climate-change issues, and its tasks are similar to those of the IPPC. The NCCCG is divided into four working groups

---

14 Qu Geping is the Vice Chairman of the Chinese State Council Environment Protection Commission and the Director of the Environmental Protection Bureau.

on: (i) scientific matters (chaired by the SMA), (ii) impact assessment of global change (chaired by the NEPA), (iii) formulation of response strategies (chaired by the SSTC), and (iv) international negotiations on relevant conventions (chaired by the Ministry of Foreign Affairs) (GEF 1992).

As chair of the WG III, the SSTC also commissioned a study on a "National Response Strategy for Global Climate Change" with funds from the Asian Development Bank (see following section). Together with the Ministry for Foreign Affairs, SSTC has participated in international meetings on greenhouse gas issues (GEF 1992).

4.1.3. Development of a Chinese Climate-Protection Strategy

Not only was the development of a Chinese global environmental policy "to a large extent conditioned by the incentives and pressures from the international community" (Lin Gan 1993: 269), the formulation of the Chinese domestic climate-protection strategy itself was also influenced by the developments in international negotiations for climate change. Chen (1995) has pointed out the importance of the FCCC for the definition of domestic priorities in strategic decision-making processes and the prompt response of the Chinese government to the FCCC. As a matter of fact, China was the tenth country to ratify the FCCC. Aiming to comply with the obligations of the FCCC and with a statement made by Premier Li Peng, who attended the UNCED, that China would implement resolutions adopted at the conference, several programmes have been set up. Furthermore, the Chinese government announced that China would develop a National Agenda 21.[15] In July 1992 the Environmental Protection Committee of the State Council (EPCSC) decided that the State Planning Commission (SPC) and the SSTC should take the lead in organizing all appropriate ministries, departments and NGOs to work together to formulate China's Agenda 21, a "White Paper on China's Population, Environment and Development". Under the guidance of the Environment Protection Committee, a leading group was set up in August 1992 co-chaired by a deputy minister of the SSTC and a deputy minister of the SPC. At the same time the "Adminstrative Center for China's Agenda 21" was established, a working group composed of 52 ministries and agencies, and more than 300 experts. It was set up jointly by the SPC and the SSTC. The centre's functions are to: (i) participate in the

---

15 The following description of the Chinese Agenda 21 process was taken from China Network 1/1995 and the Preamble of China's Agenda 21.

formulation and implementation of China's Agenda 21 and take charge of its daily management, (ii) organize the formulation, coordination and implementation of priority programmes for Agenda 21, (iii) provide consultation services on sustainable development for government agencies, NGOs and the general public, (iv) participate in the international network for sustainable development, and (v) provide services for personnel training and international exchange, etc. (*China Network* 1/1995).

A first draft of China's Agenda 21 was completed in April 1993. It defined 80 areas for sustainable development strategies, policies, and frameworks for action related to population, economy, resources and the environment in China. The final version of China's Agenda 21 was discussed and approved at the Executive Meeting of the People's Republic of China on 25 March 1994. Subsequently, priority programmes have been worked out to support its implementation (*China Network* 1/1995). However, as the content of the Chinese Agenda 21 is vague, and since it is not clear what part of it will be implemented and what the concrete measures will be like, its actual impact remains to be seen (Kiefer 1995).

Another study on a Chinese climate-change response strategy was funded by the Office of Environment of the Asian Development Bank. The purpose of the project was to identify options that could be important elements of a response strategy and to examine economic, technological, social and environmental implications of these options. In doing so, the report highlights the following strategic options: (i) increasing the efficiency of energy utilization in the end-use sector, (ii) increasing the efficiency of the conversion process and reducing energy losses in the production process, and (iii) promoting the substitution of high-carbon by low-carbon fossil fuels and by greenhouse gas-free energy sources.

The report was prepared for the "Department of Science and Technology for Social Development" of the Chinese SSTC, which is also the implementing body for the recommended strategy. The project was justified with reference to the Chinese commitments as a signatory of the FCCC (Siddiqi et al. 1994). It is therefore likely that an amended version of this report will become the official report of the Chinese Government to the secretariat of the FCCC, as requested in Art. 12. As other developing country signatories, China is obliged to submit such a report within three years of the FCCC having entered into force, i.e. in March 1997.

4.2. Future Implementation of Abatement Measures Pending on Energy Demand and Population Growth

Even without climate change China is faced with a multitude of problems.

Of paramount importance for both the economic development of China and the issue of climate change are the increase in energy demand and population growth. Chinese policies for addressing and solving these two challenges will have a significant impact on both China's future greenhouse gas emissions as well as its climate change policy-making.

4.2.1. The Challenge of Increasing Energy Demand

The Chinese energy sector clearly reflects the rapid changes and developments in the Chinese society and economy. Between 1970 and 1990, energy consumption increased by 208%. Between 1980 and 1990, electricity use doubled, and between 1952 and 1990 commercial energy production increased twenty-fold (WRI 1994). In 1993, the total production of primary energy amounted to 1.12 billion tonnes coal equivalent (Zha Keming 1995), making China the world's third largest energy consumer following the United States and the former Soviet Union. From a *per capita* perspective, however, the situation is quite different. On average, a Chinese household uses less than 0.03% of the energy consumed by a US household (WRI 1994) and *per capita* commercial energy use in 1990 was only one-third of the world average (Zhong Xiang Zhang 1994). Consumption is characterized by a dominance of coal accounting for 76% of total consumption, followed by oil (17%), hydropower (5%) and natural gas (2%) (Wu and Wei 1991).[16] Large regional differences exist, however, particularly between rural and urban areas. Generally, rural areas are characterized by a heavy reliance on non-commercial energy use, such as biomass. Energy consumption in urban areas is much higher, because of the better provision of appliances. Furthermore, energy consumption is dominated by the industrial sector (67% of final energy consumption in 1990), followed by the residential sector (17%), services and agriculture (both 5%), and transport (4.5%) (Zhong Xiang Zhang 1994).

Despite economic growth rates of more than 10% in recent years, development of the Chinese economy is judged to have slowed down, limited by its energy supply capacity which is restricted by supply shortages and by energy inefficiency. Supply shortages are mainly due to limited transportation capacity because of the great distances between the main areas of energy production and consumption (Todd and Zhang

16 China is energy self-sufficient. Therefore, figures for primary energy production and consumption are almost the same. According to Zhong Xiang Zhang (1994), China's energy imports and exports accounted for 1.3% of the total national commercial energy consumption and 5.7% of the total national commercial energy production in 1990.

1994). The existing infrastructure (mainly railways and ports) is not sufficient to serve even the present needs. This state of affairs is further aggravated by the fact that energy efficiency in China is among the lowest in the world. This is due to several factors: the unusually high share of energy-intensive industrial production of the economy; the high share of energy-intensive manufacturing processes; the reliance on coal (e.g. coal-fired boilers are less efficient than oil and natural gas fired boilers); generally low standards of technology; the still unreasonably low energy prices (despite some adjustments); and the undervaluation of the Chinese GDP (Todd and Zhang 1994; Wu and Wei 1991).

While the problem of low energy efficiency was recognized by the Chinese Government at an early stage, counter-measures have proven to be of modest success. Energy efficiency improved by 5.6% between 1980 and 1988. Zhong Xiang Zhang (1994) interprets this, together with higher economic growth rates, as the beginning of a decoupling of economic growth and energy consumption. Indeed, there is a huge potential for energy conservation in China (Wu and Wei 1991; Haugland and Roland 1990). The motive for exploiting this potential, however, is not the implementation of abatement measures, but rather bridging the widening gap between energy supply and continuous economic growth (Huang 1991). The clear dominance of economic growth is demonstrated by the official goal of the Chinese Government to double the 1989 Chinese GDP by the year 2000.[17] To achieve this goal, an energy production target of +50% in that period has been set.[18] This target comprises: (i) Primary energy − 1.4 billion tons coal equivalent (corresponding to production targets of 1.4 billion tons raw coal, 0.2 billion tons petroleum, 30 billion cubic metres natural gas) and (ii) Electricity − 1200 billion kWh (including 240 billon kWh hydropower and 30 billion kWh nuclear power) (Qu 1992: 501)

In addition, by the year 2000 a 50% improvement of energy efficiency is to be achieved by technological progress and policy (Qu 1992). How the energy target should be achieved is not quite clear, as there are only vague and sometimes contradictory statements. Some officials point to the readjustment of the energy mix with an increase of the share of "clean" energy and secondary sources of energy like electricity. Others state that the dominance of the use of domestic coal is not questioned

---

17 Future economic growth rates are estimated to be in the range of 8–9% anually to the year 2000 (Lin Gan 1995) and 4.2–4.5% from the year 2000 to 2010 (Wu and Wei 1991).

18 Resulting from the socialist central planning tradition, the success of economic development and future targets are measured in physical quantities.

and will remain for at least the next 20 years (Zou 1992). Despite these inconsistencies, there is a consensus on the technological character of the strategy: Coal-specific measures aim to improve the efficiency of the use of coal in thermal power plants, small industrial boilers, etc. Renewable sources of energy such as windpower and hydropower and nuclear power are officially accepted and proposed to be further developed. Oil and gas reserves should be exploited despite long transport routes.

The mitigation of greenhouse gas emissions is not playing a significant role in these growth-oriented strategies, which are aimed at energy security rather than at greenhouse mitigation. It seems that the issue of climate change and other environmental considerations (e.g. reduction of SO_2) play only a minor role as they are used when it is possible to add an environmental justification to proposed measures and for improving international and national acceptance of policies and measures. For some of the proposed measures, climate change is used to induce a shift of risk perception stressing beneficial side-effects of the projects. For example, the risk of climate change is used to justify the building of nuclear power plants (safety issue) and implementation of the large hydropower project "Three Gorges Dam" against which strong opposition exists (WRI 1994). However, in focusing on the climate change issue, the policies stated and measures implemented do not go beyond no-regrets measures – that is to say, measures that have beneficial effects other than emission reduction. A comprehensive evaluation of the climate relevance of the announced strategies will only be possible when more of the proposed measures are implemented and show an effect.

4.2.2. The Challenge of Population Growth

Population growth is a paramount factor in most of the current problems of Chinese society and, particularly, for the climate change issue. All things being equal, the total emissions of greenhouse gases increase with the population growth rate.

In 1990, the population of China was 1.145 billion, 21% of the world's total population. The population growth rate declined from 2.61% in 1969 to 1.16% in 1979, and increased to 1.44% in 1990. The decline was due to the strong family planning measures, the so-called "one child policy", while the increase by 1990 was due to the increasing number of girls in fertile age (Siddiqi et al. 1994). Estimates of future population numbers vary depending on the assumptions of the future Chinese family-planning policies. Smil (1993) concludes that a total of 1.25 to 1.30 billion by the year 2000 seems to be most likely.

Reviewing factors influencing population growth indicates that economic factors are most important: In urban regions, the Chinese "one child policy" is more accepted and birth rates are generally much lower than in rural areas where strong reservations against these polices exist. Since urbanization is accelerating and the stringent family polices will be maintained, population growth is expected to decrease.[19] There is also a strong link between population growth and education. Hence, it would seem that population growth is influenced by the same factors that are important for the development of environmental awareness. Being interrelated problems, they should be addressed as such with the same set of measures.

5. Concluding Remarks: Future Prospects for an Effective Chinese Climate Policy

From an ecological point of view, the challenge of coping with the global problem of climate change allows only one answer: An internationally coordinated policy with commitments of participating countries to limit global greenhouse gas emissions.[20] In theory, this should be in China's interest since it is one of the countries most vulnerable to climate change. In almost every official statement the Chinese government stresses the importance of its climate change policy and assures its willingness to participate in international policy regimes as far as possible.

Nevertheless, China has been consistently eager to avoid any commitment to emission reduction targets and time-tables. Such commitments are judged not only to be unfair in view of the historical responsibility of the industrialized countries, but are, more importantly, also considered to contradict economic development based on increasing energy consumption. Both arguments are used to justify a significant increase in the future Chinese greenhouse gas emissions, especially of CO_2 (Siddiqi et al. 1994: 374). Thus, the more realistic challenge and objective of future Chinese climate change policy is to balance economic growth and emission abatement, resulting in a minimization of the expected emission increase.

---

19 Because of ethical implications, the Chinese family planning policy is subject to much criticism. For a more detailed discussion of this issue, see Smil (1993), who emphasizes the opportunities for other policies to reduce the wish for children, for example through education measures.

20 This is to some extent reflected in the FCCC, which in Art. 2 states that "the ultimate objective of this convention ... is to achieve ... stabilization of greenhouse gas concentrations in the atmosphere at a level that would prevent dangerous anthropogenic interference with the climate system".

Although China has not been favouring a strong and committing international climate change regime, international negotiations have induced some remarkable adjustments of Chinese institutions. An institutional infrastructure (legislation, scientific, organizational structure) has been established and several strategies for climate change protection have been carried out with support of bi- and multilateral funds. Nevertheless, climate change has failed to become a first priority issue, a fact cemented by the low public environmental awareness in China. It appears that additional funding has not significantly increased the national capacity to deal with climate change (Lin Gan 1995).

The general consensus is that Chinese society is facing more pressing national, regional and local economic and environmental problems than climate change. The solution to these pressing issues must be granted priority over that of climate change. As the analysis of the energy sector shows, however, policies addressing *other* problems may result in some limitation of greenhouse gas emissions. This is the case for energy efficiency measures implemented to fill the growing gap between energy-supply shortages and increasing energy demand. However, it is too early to judge the actual impact of such response strategies on Chinese climate change policies. For the time being, however, there is no indication of concise climate protection policies and measures; climate protection in terms of emission reduction or retardation of emission increases is simply a positive, beneficial side-effect of other economically-driven policies (e.g. energy efficiency measures). Moreover, emission reduction measures seem not to go beyond those that can be justified on other (economic) grounds as well (no regrets measures).

Even if climate change considerations were to be internalized into other policy objectives, the positive effects would at least partially be cancelled out by the growth of the Chinese population. Hypothetically assuming that due to strong climate protection policies Chinese *per capita* emissions of CO_2 could be stabilized or would only slowly increase to the present global *per capita* average (4t), the total CO_2 emissions by 2010 to 2025 would nevertheless increase considerably and still make China the world leader in greenhouse gas emissions. Therefore, the industrialized countries should have strong incentives to integrate China into an international system of targets and time-tables by acting as fore-runners, implementing strong policies themselves and providing additional funds in order to demonstrate that the issue of climate change is really a first priority issue for the global community. Otherwise, Chinese economically-driven interests justified in terms of minor responsibility for historic greenhouse gas emissions are likely to further dominate the negotiations.

The question of how such an integration of China into an international system of targets and time-tables could be achieved must incorporate the question of equity, responsibility and justice, and safeguard the right to economic development of the developing countries. In this context, Loske (1993) suggests a differentiation of political commitments between industrialized and developing countries. He maintains that a stabilization of CO_2 per capita emissions at a globally acceptable level can only be achieved if developing countries accept a certain ceiling for a moderate increase and the industrialized countries accept a clear reduction in per capita emissions of CO_2.[21] Two problems arise, however: One is that of establishing a globally acceptable level. The other, once again, is the challenge of population growth with the potential of cancelling out any mitigation achievements. Owing to the latter problem, total per capita emissions would have to be decreased over time if the globally acceptable level is to be maintained. In order to fulfil such a Chinese commitment, based on the already adjusted institutional infrastructure, the recently announced climate protection strategies could be a starting point for developing a more sustainable and advanced Chinese abatement strategy.

---

21 This view is echoed in Agarwal and Narain (1991), who speak of survival emissions for the developing countries and luxury emissions for the industrialized world.

Chapter 9

Brazil and Climate Change

Sjur Kasa[1]

1. Introduction

During the late 1980s, Brazil stood out as the environmental villain of an emerging global ecological order. The position of the Sarney government (1985–1990) on the issue of climate change was among the most defensive and hostile in the G-77 group. While generally denying any responsibility for global environmental problems, the country was forced by a coalition of the United States, the European Community, multilateral agencies, and international NGOs to adopt half-hearted reforms addressing the problem of climate change. Then, between 1990 and 1992, under the Collor government, Brazil's attitude shifted in a much more cooperative direction. At the same time, efforts to cope with the greenhouse problem were implemented with greater vigour. This phase culminated with the United Nations Conference on Environment and Development (UNCED) in Rio de Janeiro in 1992, and the impeachment of President Collor in September 1992. Under President Itamar Franco, came a reversal in Brazil's environmental policies, back to the defensive positions and the half-hearted efforts of the Sarney era.

In this chapter, I argue that these changes were influenced by a combination of the special characteristics of the sources of greenhouse gas emissions in Brazil, and the state and civil-society forces connected with these sources. The major share of these emissions comes from

1 I am grateful to Luiz Pedone, Don Sawyer and the Norwegian Embassy in Brazil for fieldwork assistance. Thanks to the Norwegian Ministry of Foreign Affairs for fieldwork financing.

deforestation in the vast region of Brazilian Amazonia.[2] But the political structures that linked the various interests created by this emission profile with federal decision-making changed considerably with the progress of democratization in Brazil. In sections 3–5, I explore the configurations of these forces, emphasizing the contrasts between the Sarney and Collor governments. A main point is that the process of democratization in Brazil implied a more progressive policy on the climate change issue as the influence of the military and their allies gradually weakened. First, however, we should survey some characteristics of the distribution of Brazil's greenhouse gas emissions as regards various anthropogenic sources and geographical regions.

2. Greenhouse Gas Emissions From Brazil: Conflicting Perceptions

At one level, the issue of climate change is a problem of the natural sciences, characterized by the struggle to reduce uncertainty and refine the methods of data analysis. At another level, perceptions of the issue are influenced by the strategies of several groups of actors, each presenting conflicting versions of an uncertain reality.

Worries about climate change in Brazil predate the scientific advances in atmospheric sciences from the late 1970s.[3] Such worries were especially focused on regional climate change as a consequence of deforestation in the Amazon region. Potter et al. (1975) simulated the climatic consequences of general deforestation in the zone 5°N to 5°S (of which Amazonia took the lion's share). Main outcomes from this model were temperature drops in this zone, along with changes in rainfall. Molion (1976) and Salati and Ribeiro (1979) warned that reduced rainfall might

---

2 The term "Amazonia" as used in most texts usually has one of two meanings. One is the "North Region" as defined by the Brazilian census agency IBGE, including the states of Pará, Amazonas, Rondônia, Acre, and Amapá in addition to the federal territory of Roraima. The definition "Legal Amazonia", that is used by the Brazilian planning agency for Amazonia, SUDAM, after 1979 also includes Mato Grosso, Tocantins and large parts of Maranhão. Legal Amazonia makes up more than 50% of the territory of Brazil.

3 Even though connections between human activity and global climate change as a consequence of increased anthropogenic emissions of carbon dioxide (CO_2) have been an object of scientific speculation since the 18th century, the launching of the World Climate Programme by World Meteorological Organization in 1979 was the first large-scale effort to estimate the seriousness of this problem (McCormick 1988: 190) – see Chapters 2–4 in the present volume.

be the outcome of continued deforestation in Amazonia – a possibility also taken up by political commentators (Bourne 1978). Speculations about the implications of deforestation for global climate change through the reduction of biomass and increased atmospheric emissions of CO_2 also emerged in this period (Bolin 1977; Adams et al. 1977; Wong 1978). For Amazonia, the issue was raised in connection with the enormous amounts of biomass represented by the Amazon rainforest.[4]

Though the net contribution of rainforest removal versus emissions from energy was clouded with uncertainty during the 1980s (Fearnside 1985: 80–81), most estimates concluded that the burning of fossil fuels had far greater effect than the removal of biomass from tropical forests (Dobson et al. 1989; The Economist, 11 February 1989: 19).

Parts of the reason for the strong focus on Brazil and Amazonia in relation to the global climate, despite the apparently moderate impact of tropical deforestation, were the new discoveries of the extent of deforestation in the Amazon region in 1988. Early that year, a research team from the Brazilian Forestry Development Institute (IBDF) and the Brazilian Space Agency (INPE) issued a report indicating that 325,000 fires had been detected in Brazilian Amazonia by the NOAA-9 meteorological satellite (Setzer et al. 1988). From the total number of fires detected, they concluded that in 1987 deforestation had demolished 80,000 km[2] of virgin forest, an area corresponding to the size of Austria. In turn, these estimates provided the basis for other reports concluding that Amazonian deforestation was gaining an enormous, exponential momentum (Mahar 1989; Myers 1989), and that Brazil was becoming a key contributor to global greenhouse gas emissions. The prestigious World Resources Institute (WRI) in Washington concluded in their 1990 survey of the global environment that deforestation in Brazil in 1987 accounted for greenhouse gas emissions greater than the use of fossil fuels in the United States (WRI 1990: 346). These assumptions brought Brazil to the top of total and per capita emissions of CO_2 among the larger nations of the capitalist world and focused enormous global attention on Brazil and its contribution to climate change. Amazonia was the front cover story of the 18 September 1989 issue of Time magazine.[5] State leaders like French

---

4 Salati and Ribeiro (1979) estimated the Amazon region to represent about 20 percent of the planet's organic carbon reservoir.

5 The front cover, with a skull of flames devouring a forest of innocent birds, panthers and beautiful snakes, confirms the popular image of Amazonia as an unpopulated, tropical paradise threatened by civilization. The poor urban populations of the large Amazonian cities as well as the rural poor at the agricultural frontier are not included in this image.

President Mitterand and German Chancellor Kohl, as well as rock star Sting, directed their attention to Amazonia.

In 1988 and 1989, the global discourse that linked Amazonia and climate change was in fact at odds with realities. Along with the emergence of Brazil in the role of villain, critical examination of previous deforestation estimates in Amazonia revealed methodological problems in the interpretations of remote-sensing images. The satellite images which provided the basis of the Brazilian Space Agency's estimates were based on spots of fire detected by infrared sensors. Such sensors are easily saturated by small spots of fires, which means that the areas covered by fire could easily be exaggerated (Fearnside 1990b: 214–215). In addition, Mahar's (1989) deforestation estimates were flawed by empirically unfounded exponential extrapolations of 1980 observations, grossly overestimating the deforestation in several Amazonian regions (Fearnside 1990a: 459–460). By the late 1980s, these facts were increasingly recognized by scientists, but Brazil was unable to exploit these discoveries to argue convincingly for more moderate estimates of the country's role in global warming. In the first place, the Sarney government's response to the global outcry that came in the wake of the new estimates was politically unwise. The INPE was ordered to make a crash estimate of deforestation refuting especially Mahar's widely cited study, which had gained credibility from being published as a World Bank paper. However, the study commissioned by the Sarney administration received considerable criticism. First, it used a questionable methodology to minimize deforestation and greenhouse gas emissions from Amazonian deforestation. The report concluded that total deforestation in the Amazon region was less than half of the 12% indicated by Mahar (*O Globo*, 6 April 1989). But it came to this conclusion by subtracting the "ancient" deforestation that took place before 1970. Furthermore, a second edition published on 2 May 1989 calculated deforestation as a share of the total area of Legal Amazonia, not originally *forested* Amazonia. As Legal Amazonia also includes large savannas, the numerator and denominator refer to different areas, thereby grossly underestimating total deforestation. Second, there are firm indications that the scientific process behind the figures was deliberately "rigged" to still international criticism of Brazil. Technicians at INPE complained that the report had been prepared too rapidly, and that the scientific staff had been excluded from the final data analysis (Anderson 1990: 20; *Folha de São Paulo*, 8 May 1989: C-3). This provoked heavy criticism from the national and global scientific community (*Folha de São Paulo*, 8 May 1989: C-3) and undermined Brazilian government credibility.

The clumsiness of the Sarney regime's handling of deforestation estimates thus tended to divert attention from the more moderate estimates that arrived somewhat later. In these estimates, Brazil's contribution to the greenhouse effect was drastically decreased, along with huge reductions of deforestation rates. In the 1992 version of World Resources, CO_2 emissions from Brazil were more than halved compared to the 1990 estimate. Deforestation was still the overwhelming cause of greenhouse gas emissions from Brazil, making up about 77% of total emissions, but no longer did it so grossly outpace total emissions from the other major "deforesting nation", Indonesia, or the *per capita* emissions of other countries with large natural forests (WRI 1992: 348–349, Table 24.2; Fearnside 1992). Furthermore, reports based on improved methods for measuring deforestation concluded that the years 1987–1988 had been exceptional in terms of deforestation. Special economic factors had contributed to a peak in biomass burning in the Amazon region just when global attention to the problem of climate change was emerging.[6]

A more general problem was connected with Brazil's position as a developing country: this was its inability to influence the perceptions that had placed the disappearing of the tropical forest at the top of the international agenda. For the people of Brazil, local pollution problems are much more of a threat than the spectre of climate change, perhaps with the lack of basic sanitation for the majority of the population as the most important problem (Viola 1992: 4). The biased representation of human needs represented by the ascendancy of climate change to the top of the global agenda is also reflected in the currents of international environmental financing. It has become far easier for Brazilian NGOs and researchers to obtain funding for forest-related projects than for projects related to improving basic sanitation.[7]

The outcome of the conflict over realities, however, was that Brazilian Amazonia came to occupy a very high position on the global ecological agenda. The region's contribution to global warming was a major reason for this. In turn, this led to strong international pressures on Brazil for more environmentally benign policies. As we shall see, the development of Brazil's policies on climate change took place in the intersection between these pressures and domestic politics.

---

6 More recent estimates conclude that annual average deforestation during the years 1978–1988 was 22,000 km². Between 1989 and 1990, the deforestation rate declined to 13,800 km²/year, and has since fallen further to just above 10,000 km²/year (Fearnside 1992:5).

7 Own interviews with Brazilian NGOs October 1992 and November 1993. For a more general discussion of this topic, see Agarwal and Narain (1991), and Beckerman (1992).

3. The Sarney Government and Climate Change

Deforestation in the Amazon region is very much influenced by development policies initiated in the early days of the military regime (1964–1985). In 1966, the military introduced a regional development policy for the Amazon region characterized by infrastructure extension and huge subsidies and tax breaks for industries, cattle ranching, and mining (Hecht and Cockburn 1989; Branford and Glock 1985; Pompermayer 1984). This policy was pursued with only minor changes through the 1970s and early 1980s. The main political motive was territorial security: the vast and unpopulated Amazon region was perceived as vulnerable to attacks from hostile foreign powers, or as a potential nest for communist subversion (Allen 1992).

In addition to huge public investments in roads and hydroelectric plants, the military development policy inspired a wave of private investments in cattle ranching, mining, and industry. While investments in industry were concentrated in the Manaus free-trade zone, those in cattle ranching and mining were more dispersed, having a stronger ecological impact.[8] Major initial investors were large companies from São Paulo, Brazil's south-eastern industrial powerhouse, and, to a smaller extent, transnational companies based in the United States and Europe. In addition to generous incentives to private investors, public investments in road-building tended to encourage private investments, as land values soared dramatically when road access was assured. In combination with strong general tax incentives for agricultural investments and land allocation rules that provided land title in accordance with the size of cleared forest area, these policies contributed to massive forest clearing in the Amazon region (Binswanger 1991). Moreover, from the 1970s, the prospects of huge speculative profits increasingly attracted investors from the Amazon region itself, independent of the subsidy programmes. Though local and regional in origin, most of these investors

---

8 The extension of cattle ranching appears closely linked to deforestation in the Amazon region. May and Reis (1993: 13–15) correlated planted pastures to Amazonian deforestation and found a very strong spatial covariation between concentrations of planted pasture and deforestation. However, as they based their calculations on IBGE data covering only agropastoral establishments, these data do not give information about non-agricultural causes of deforestation other than the construction of hydroelectric dams, which represents a fairly modest share of total deforestation. But generally, there is long-standing and broad consensus on the very strong effects of pasture expansion on deforestation, and that slowing down such expansion is a main prerequisite for a decrease in deforestation rates (Serrão and Toledo 1990: 196).

were also large-scale ranchers, manifesting the tendency towards land concentration in the Amazon region.[9] This makes assumptions about a close covariation between poverty and greenhouse gas emissions in Brazil far more problematic than commonly assumed.

In the mid-1980s, this trend in regional development climaxed in two very large projects. The first was the Great Carajás project, which included broad-based development of industries and agriculture in the eastern Amazon region. This project was organized around the infrastructure constructed in connection with the establishment of the extremely large Carajás iron mine.[10] In addition to the extension of large-scale ranching and agriculture, the Great Carajás project included a proposal for the construction of a series of pig-iron smelters utilizing virgin rain forest as its source of energy (Hall 1989). The net impact of the project, which was fuelled by special subsidies, was predicted to be extremely devastating in terms of deforestation (Fearnside 1986). The second project was the POLONORESTE immigration project, which involved the construction of a 1,500 km paved road connecting Cuiabá on the south-western fringe of the Amazon region with the capital of the western Amazonian state of Rondônia, Porto Velho. In connection with this road, 39 centres for planned agricultural immigration were established to attract thousands of immigrants to an area covering the whole state of Rondônia and the northern section of the state of Mato Grosso – an area equal to about 75% that of France (Rich 1988: 9). The outcome was a totally uncontrolled process of immigration into the region. From 1980 to the peak year of 1986, the annual number of immigrants entering Rondônia increased from 28,320 to 165,899 (Martine 1990: 30). Between 1978 and 1989, the deforested portion of Rondônia increased from 1.78% of the area of the state to 13.34%, implying the removal of about 27,000 km^2 of forest (INPE 1992). Both the POLONOROESTE project and the

---

9 The increase of land-use in Amazonia until the late 1980s was first and foremost the outcome of the expansion of planted pastures (May and Reis 1993: 13, table 4). There has also been a clear tendency for rural establishment to be concentrated in larger units. In the North Region and Mato Grosso in 1985, large agropastoral establishments (area greater than 500 hectares) covered 73.6% of the North Region and Mato Grosso, while small (area below 100 ha) and medium sized (area between 500 and 100 ha) establishments by and large divided the remaining area between them (May and Reis 1993:19). However, the predominance of planted pastures on medium-sized establishments and its substantial share of small agropastoral establishments suggest that livestock is an important source of income among all strata of Amazonian producers.

10 In addition to the establishment of the mine south of the city of Belém, this included the giant Tucuruí hydroelectric dam, the Ponta da Madeira deep-sea port near the city of São Luis, and a 900 km railroad linking the mine with the port.

Carajás central mining project – the cornerstone of the Greater Carajás Project – received World Bank financing.

These megaprojects continued uninterrupted during the first years of the Sarney government. In addition to its alliance with the military (Flynn 1993; Zirker and Henberg 1994), the conservative Sarney government had important political allies among federal deputies, senators and governors from the Amazon region.[11] As the Brazilian electoral system favours votes from remote regions like Amazonia, these allies enjoyed an influence out of proportion to the demographical and economic importance of the region.[12] This influence contributed to improve the basis of right-wing allies of the military forces, guaranteeing a very strong coalition in favour of continuing the commercial "development" of the Amazon region (Hurrell 1992; Zirker and Henberg 1994: 266–267). In 1985, before the relevance of Amazonia for global environmental problems was fully recognized by the international community, a comprehensive strategy for increased environmental protection in the Amazon region had been proposed as a federal bill to the Brazilian Congress in 1985, after discussions dating back to 1979. Apparently, the mentioned coalition was already operative in the national assembly. The bill was never voted upon by Congress, probably because of the fierce resistance of business interests in the region (Vieira 1993: 108).

As the link between deforestation and global environmental problems became well-known, and the surprisingly high estimates of Amazonian deforestation exploded in media all over the world, the Sarney government was confronted with mounting international pressure. The emergence of such pressure was facilitated by transnational channels connected to an ongoing campaign of international NGO protests against World Bank financed projects in developing countries. These channels emerged during the 1980s, when US environmental NGOs started informing the US Congress about the destructive effects of a series of projects in rainforest areas financed both by the World Bank itself and other

---

11 Sarney was a former top figure of the military's party ARENA, but created his own right-wing party, PFL (The Liberal Party) during the process of democratization. He became president because the first non-military president to be elected by the 1985 electoral college, Tancredo Neves, died of a heart attack the same year. While Neves was a highly respected democrat who had been elected by an electoral college which sensationally chose to act against the advice of the military, Sarney had served as a military alibi in this "package solution".

12 This was a deliberate strategy to increase right-wing influence by the retreating military regime in the early 1980s. The poor and peripheral regions of Brazil are characterized by the dominance of right-wing rural oligarchies who maintain their power through electoral clientelism.

regional development banks.[13] Twenty-four Senate hearings in which the policies of the Multilateral Development Banks were main topics, were held from 1983 to 1986 (Rich, undated: 20). Both the POLONOROESTE and the Carajás mining project were in focus in these hearings. This campaign was supported by a strategy of networking between US NGOs and NGOs in both Western Europe and developing countries. In Brazil, a broad network of activists including prominent individuals like José Lutzenberger,[14] NGOs organized around Amazonian rubber tappers,[15] like Instituto de Estudos Amazonicos and Conselho Nacional de Seringueiros, and NGOs organized around Amazonia's indigenous populations like União das Nações Indigenas (UNI).[16] As World Bank and Interamerican Development Bank projects were revised to include firmer environmental obligations mainly because of the activity of US directors on the board, pressure against the Bank was translated against the Brazilian government from the Bank. From 1985, this included temporal discontinuations of funding for the POLONOROESTE programme, demands for detailed environmental provisions in energy programmes, temporal discontinuations of funding for further road extension in the Western Amazon region and demand for improved environmental plans. A major blow to the Brazilian government was the indefinite discontinuation of negotiations on a crucial USD 500 mill. loan for a series of energy projects in early 1989. This energy programme included the construction of a series of very large hydroelectric dams in Amazonia which threatened

---

13 Greenpeace International, Friends of the Earth, Natural Resources Defense Council, Environmental Defense Fund, the Sierra Club, Environmental Policy Institute and Cultural Survival were important NGOs participating in the campaign. The campaign was coordinated by Bruce Rich (Natural Resources Defense Council), Barbara Bramble (National Wildlife Federation) and Brent Blackwelder (Environmental Policy Institute) (Arnt and Schwartzman 1992: 112).

14 Lutzenberger was leader of the first modern ecological movement in Brazil, AGA-PAN, founded in Porto Alegre (capital of the state of Rio Grande do Sul) in 1971. This required great personal courage, as it was done in the most repressive period of the military regime (Viola 1988: 214).

15 The rubber tappers are independent collectors of latex from the dense stands of rubber trees of Western Amazonia. They oppose the expansion of cattle ranching, especially in the state of Acre.

16 According to Albert (1992: 36), Amazonia is home to about 60% of the 236,000 Brazilian Indians; 98% of the total area of Indian lands (794,000 km²) is located within the region. Large areas of Indian lands in the Amazon are heavily forested and among the most interesting sections of Amazonia because of their richness in plant and animal species (World Bank 1994: 283).

to flood vast areas of pristine forest (Arnt and Schwartzman 1992: 115–116).[17]

Pressure from NGOs also activated bilateral pressure from Brazil's main trading partners and investors. US politicians like Senator Al Gore tried to link environmental reforms in Amazonia to a more liberal debt regime and the inflow of fresh investments in the Brazilian economy (*Correio Braziliense*, 18 January 1989: 12; Hurrell 1992: 406).

The response of the Sarney government to this pressure may be divided into two periods. The period until the second half of 1988 was characterized by the government's uncompromising refusal to recognize or address the environmental problems created by its policies in the Amazon region. Then a period of cautious reform came with the introduction of some planning study groups for the region, set up in October 1988, culminating in a series of reforms in 1989 and early 1990. Though some of these reforms implied significant breaks with former policies, there was still no change in the government's foreign policy positions on climate change and the environment, nor in policies related to Amazonia's indigenous populations.

3.1. The Sarney Government Before 1988: Intransigency and Increasing Pressure

Until the second half of 1988, the Sarney government refused altogether to accept that its policies in Brazilian Amazonia had any relevance for the global environment. This refusal was based on two main perceptions of Amazonia's position in the international system.

First, the government tended to see international considerations over the greenhouse effect and Amazonia as an international plot to challenge Brazil's sovereignty over the region. Such considerations were based on the military's assumption that the region was threatened by foreign powers. Seemingly far-fetched in the international context of the 1980s, this position was in fact grounded on a longer historical view. Around the turn of the century, Amazonia was the centre of the soaring world trade in rubber. At that time, a joint plan by the United States and Bolivia to break Brazilian control over transportation by declaring the Amazon River an international area provoked high tension in the region, ulti-

---

17 The message about the closure of negotiations on the power sector loan became publicly known in Brazil the same day as the Sarney government was insulted in public by a UNEP leader for its defensive environmental policies. In the newspaper *O Globo*, 31 March 1989: 7, these events are reported on the same newspaper page, providing a striking impression of the intense pressure on the Brazilian government.

mately leading to Brazil's successful takover of the Bolivian region of Acre in 1903 (Hecht and Cockburn 1989: 66–72). Also during World War II, worries over Amazonia's sovereignty came to the surface when Brazil declared war against Germany in 1942, the United States and Brazil joining forces to prevent German infiltration of the region. This cooperation involved the construction of airfields and the launching of a Brazilian expedition into the Amazon region (Cowell 1990: 17-22).

Second, there was also a feeling that the environmental campaign of the 1980s was launched by the industrial nations to prevent Brazil from becoming a world power thanks to the natural resources of Amazonia (Miyamoto 1989; Hurrell 1992: 405). This position was connected particularly with the North–South perceptions of the traditionally strong and influential Brazilian Ministry of Foreign Affairs (also called the Itamaraty). The Itamaraty viewed several international issues through the lenses of a broad North–South conflict over global resources – a perception which included the exploitation of Amazonia's natural resources (Miayamoto 1989).

These perceptions motivated a highly defensive policy on the Amazon up until the second half of 1988. The scheme of fiscal incentives for cattle ranching, which foreign and national critics saw as unprofitable and a main motive for deforestation, continued unchanged despite high costs in a situation of debt-imposed fiscal austerity. The forestry policy for the Amazon region outlined in the 1965 forestry law with later amendments remained on paper, with the agency responsible for its implementation, the Brazilian Forest Institute (IBDF), paralysed by decreasing funding (Prado 1986: 13–14) and a continued lack of political backing.

Also policies relevant to preservation of forest areas for the indigenous populations of Amazonia were formulated with an eye to commercial interests. The National Indian Foundation (FUNAI), responsible for the Indian tribes of the Amazon region, was widely known as a corrupt bastion for military and mining interests (Albert 1992).

The foreign policy of Brazil on the issue of the Amazon region and its effects on the global climate remained extremely defensive and anti-cooperative throughout the Sarney presidency. The government's attitude to Amazonia's contribution to climate change reflected the importance of the region for the military. For example, President Sarney withdrew his participation in the 1989 Global Conference on the Protection of the Atmosphere in The Hague after deliberations with his military ministers (Zirker and Henberg 1994: 266). Furthermore, the Brazilian delegation to the meeting protested against any references to Amazonia in the final communique and against President Mitterand's proposal to invest the UN with authority to intervene in cases in which global environmental interests were threatenened (Hurrell 1992: 406). Proposals suggesting debt-for-nature swaps as a combined remedy for Brazil's debt-burden and the country's contribution to global environmental problems were forcefully denounced by the government as attempts at foreign intervention. The refusal of the World Bank to continue negotiations on the energy sector loan in 1989 was met by enraged protests by Brazil, deviating from the rules of diplomatic conduct in such cases (*O Estado de São Paulo*, 7 March 1989: 47). The military suspected foreign NGOs of being spearheads of an imperialist takeover of the region (*O Estado de São Paulo*, 9 February 1989: 13). Though mostly focused on foreign influences, the classic army "think-tank", the Superior War College (ESG) also identified Brazilian artists, intellectuals, the Church and transnational companies as a possible "fifth column" behind the international campaign against Brazil (*Folha de São Paulo*, 29 May 1990: A-4).

3.2. The Sarney Government from the Second Half of 1988: Cautious Action

As of the second half of 1988, the Brazilian government started to respond to external pressure. It was clear that Brazilian trade, the debt regime and a revitalization of foreign investments in the country were dependent on improvements being made in Brazil's record in Amazonia.

In October 1988, President José Sarney held his first speech concerning the environment, in which broad policy changes related to Amazonia were introduced (Hurrell 1992: 409). A new plan for environmental protection in Amazonia was decreed on 12 October 1988, "The programme for the defence of the complex of ecosystems in Legal Amazonia". It was marketed under the label *"Nossa Natureza"* (Portuguese for "Our Nature") and its aims were to address Amazonian problems broadly: to restrain predatory actions, discipline the exploitation of the region, regenerate the region's ecosystems, and protect indigenous populations and the rubber tappers of the region (Decree No. 96.944/88). Concerns for the global climate emerged as environmental policy motives for the first time, as is clearly stated in the report from the executive committee of Nossa Natureza in February 1989 (SADEN 1989: 26, my translation from Portuguese):

> The destruction of tropical rain forests, in particular the Amazon forest, attracts the interest of industrialized countries, not just because the forests constitute the ecosystems with the most biological species, *but above all because of their possible influences on global climate* (my italics).

The core of the first phase of the programme was the establishment of six interministerial working groups on topics related to environmental problems in Amazonia.[18] These groups were coordinated by an executive interministerial committee headed by a general from the National Defense Secretariat (SADEN).

This measure was introduced together with a decree that also prohibited the approval of fresh fiscal incentives for ranching and agriculture in Legal Amazonia and the Atlantic rainforest for a period of 90 days, later extended indefinitely. The decree also banned the disbursement of official agricultural credits for new ranching projects in these areas. In light of Sarney's earlier reluctance to change this set of incentives, and the protests it provoked from important organizations of ranchers in the region, this move signalled a highly significant retreat.[19]

In February 1989, the final report from the working groups was delivered to President Sarney. During the following months, several laws, decrees and administrative acts were approved. A package of actions by the government was published on 10 April 1989, of which only the most important will be mentioned here. One decree prohibited the disbursement of official rural credit and fresh fiscal incentives for agricultural and ranching enterprises in forest areas in Legal Amazonia until a definitive zoning of the area had been prepared. A second decree prohibited the use of fire for clearing new land. A third decree made it obligatory for companies that consume timber as a raw material, such as sawmills, metal industries and cellulose industries, to present integrated plans for meeting their demands with planted raw material to the newly created environmental body IBAMA.

IBAMA, the Brazilian Institute for the Environment and Renewable Resources, was also created at that time, by merging a set of already existing small and neglected federal bodies responsible for resource exploitation and the environment.[20] These were the Special Secretariat for the Environment (SEMA), the Brazilian Forest Institute (IBDF) and two other agencies responsible for fishing and rubber production. For

---

18 The groups addressed the following topics: forest cover in Amazonia, chemical substances and inadequate processes in mining, structure of the administrative system for environmental protection, environmental education, research and protection of the environment, indigenous communities and populations involved in extraction activities.

19 For protests against Sarney's cancellation of fiscal incentives, see for example the article: "Emprésarios criticam as restrições impostas a projetos agropecuários", *Folha de São Paulo*, 12 October, 1988: C-2. Confirmed in my interview with Fernando Mesquita, president of IBAMA under Sarney, November 1993.

20 For a further analysis of the neglect of these bodies, see Guimarães (1991) and Foresta (1991).

the first time, an executive body dealing both with urban pollution and the use of natural resources was established. The new body was an autarchy under the Ministry of the Interior, a legal status that enabled the body to add income from the distribution of fines and other activities to its government funding. It was also decentralized, with the majority of its employees in special state branches. IBAMA was put under the command of Sarney's press secretary, Fernando Mesquita, who was also a close political ally of Sarney from the state of Maranhão.

The main accomplishment of IBAMA in 1989 was to implement a surveillance and inspection programme for the Amazon region – the Emergency Program for Legal Amazonia (PEAL). PEAL was co-financed by transferring 30% of World Bank funding for the POLONOROESTE programme in Rondônia/Mato Grosso to IBAMA; 70 inspection groups (50 of them mobile) comprising about 1,000 persons from the Federal Police, the Road Ministry, the Marine Ministry, state environmental secretariats and the Forest Police patrolled the Amazon region. The programme was assisted by observations from the meteorological satellite NOAA (which is especially suitable for fire detection) channelled through the Brazilian Space Agency, and by observations from airline pilots. Nine helicopters also assisted the surveillance activities, increasing the mobility of the inspection groups (Arnt and Schwartzman 1992: 287). As of November 1989, 838 penalties had been applied in the Legal Amazon (IBAMA 1989: 3, undated). Rondônia was the state with the highest number of penalties, 382 (ibid.).[21] The PEAL programme provoked strong protests among ranchers and their organizations in the Amazon region. Complaints from governors and federal deputies from the Amazon region representing farmers and commercial interests indicate that the

---

21 Also other key regions received increased attention. In 30 days of monitoring and inspection, the 6 teams that accomplished the so-called "Operation Carajás" started legal proceedings against 3 pig iron producers (which had not carried out their required reforestation plans), 22 sawmills (which had acquired timber illegally) and 1,110 vehicles (for not having licences for timber transport and for carrying species whose exploitation was prohibited) (Arnt and Schwartzman 1992: 287). According to Hall (1991: 298), in July 1989, the SIMARA pig-iron smelter in the city of Marabá was fined the equivalent of USD 0.5 million by IBAMA for buying illegally cut timber, and was forced to suspend its operations (see also Margolis 1992: 127). The larger COSI-PAR smelter (owned by the Minas Gerais steel company Itaminas), also located in Marabá, was fined the equivalent of USD 25,000 for similar offences, but continued its operations (Hall 1991: 298).

new surveillance programme constituted a major, qualitative change of government attitude to law enforcement.[22]

Other Amazonia-related areas of reform under Sarney were of a more contradictory nature. Though a series of new national forests, extractive reserves for rubber tappers and reserves for indigenous populations were created, these decisions do not reveal the same willingness to discipline private investors as the monitoring and surveillance programme. The establishment of a series of national forests between 1986 and 1990 should not be considered as indicating that environmental concerns were to have priority over commercial interests. National forests in general permit some extraction of natural resources, like minerals, and are usually badly policed. In addition, the establishment of national forests in areas belonging to Indians was according to the 1988 Constitution and was used as a deliberate instrument to secure access for small-scale placer mining as well as other commercial interests to indigenous reserves (Albert 1992). This was most pronounced in decisions regarding the huge Yanomani reservation in Roraima. Sarney's response to the gold rush of placer miners into the Yanomani reserve, poisoning the forest ecosystem with mercury pollution and mistreating its native inhabitants, was a 1989 decree that split the reserve up into 19 pieces and encouraged further immigration (Albert 1992).[23]

3.3. Sarney's Policies: Evaluation and Explanation

Though important changes were introduced to curb greenhouse gas emissions from 1988, the Sarney government's policies were seen as half-hearted by the international community. A key reason for this was the lack of reform related to some of the focal points of external pressure. In

---

22 Though these actions involved a clear step-up of law enforcement in Amazonia, legal barriers curbed the effects of government action. According to Fernando Mesquita, IBAMA managed to collect only about 2% of the fines meted out to trespassers of the forest law (*Jornal do Brasil*, 11 October 1990: 12). This happened because the routines for collection were excessively bureaucratic. The offenders first turned to IBAMA for reassessment, and after that had their cases tried by the court. Up to a year could pass before the fine was paid by the offender (Arnt and Schwartzman 1992: 289). Attempts to solve this problem were made during the last months of the Sarney government. Laws and provisional measures gave IBAMA the authority to collect fines without waiting for court decisions from February 1990. However, bureaucratic and judicial problems have continued to hamper IBAMA's capacity in this respect (interviews with IBAMA presidents 1990–1993).

23 The fragmentation of the Yanomami territory was formulated in Interministerial Directive 160 of September 1988 and Interministerial Directive 250 of November 1988.

contrast to institutional and legal reform related to the forest code and the environmental bureaucracy, neither foreign policy positions nor policies related to indigenous populations were changed. The Nossa Natureza environmental plan was wrapped up in a nationalist hard-line rhetoric that tended to overshadow the substantial progress which this new policy represented.[24] In addition to continued international outcries regarding the situation of Amazonia's indigenous populations and the unwillingness of the Sarney government to reform its policies here, the lack of foreign policy reform precluded any substantial decrease of pressure on Brazil.[25]

The policy shift under Sarney may be analysed from two angles. First, we may ask why reforms emerged at all. Second, we may ask why the strong reluctance to push on with reforms in certain areas persisted.

The partial shift away from intransigent hard-line nationalism starting in 1988 can be explained almost completely by external pressure. Though there was an important environmental opposition which had emerged after the end of the military dictatorship in 1985, this movement was clearly not strong enough to put the problem of climate change on the Brazilian political agenda, or to exert effective pressure on the government. In the context of economic crisis and hyper-inflation after 1986, it was very difficult for Brazilian NGOs to raise the issue of climate change to a significant position on the political agenda. In the wake of the national debate on Amazonia, the media showed increasing interest in environmental issues in general, but the environment and Amazonia took on only moderate importance in the presidential campaigns in 1989, and seemed to assume only secondary importance for the electorate (Hurrell 1992: 413). In addition, global problems like climate change were victims of the heterogeneity of the Brazilian NGO movement. While some new, middle-class-based NGOs like FUNATURA[26]

---

24 The following statement from IBAMA's first president, Fernando Mesquita, during my visit to Amazonia in February 1989, is typical of the Sarney government's position (my translation from Portuguese): "... the Brazilians do not permit the improper interference of strangers in the preservation of the ecology of Amazonia and consider this as an offense to our sovereignty" (*Correio Braziliense*, 10 February 1989: 9).

25 Unfortunate circumstances also played a role. In December 1988, Chico Mendes, the world famous leader of the rubber tapper movement in Acre, was brutally murdered by ranchers. This event provoked an international outcry that associated the murder with the Sarney government and the military's policies in the region.

26 Of the two organizations, FUNATURA has the most extensive focus on Amazonia as it has done research for sub-sections of the PMACI (The Plan for the Protection of the Environment and Indigenous Communities) for the pavement of the Porto Velho-Rio Branco highway financed by the Interamerican Development Bank (Arnt and Schwartzman 1992: 303). FUNATURA also arranged a series of seminars on Amazonia in 1988-1989, bringing together NGOs and scientists to make suggestions to guidelines for environmental management in the region (FUNATURA 1990: 2).

and SOS Mata Atlantica (both created in 1986) were focused on global environmental problems, other environmental NGOs had a socio-environmental view based on the needs of the urban poor, and political loyalties to parties on the left. On occasions, these NGOs supported parts of the anti-imperialist rhetoric of the Sarney government, for example by denouncing debt-for-nature swaps.[27] Such cleavages made it difficult to establish a united environmental stance (Hurrell 1992: 413; Viola 1988).

However, the environmental movement did have real impact on one important aspect of the environmental policy-making process under Sarney. A "green bloc" of federal deputies from a broad spectrum of political parties was successful in getting several articles on environmental protection included in the new 1988 Constitution. One of these articles (art. 225, 4), mentions Amazonia as a "national patrimony" that requires special federal attention, along with four other major Brazilian biomes (Pantanal, the Atlantic rainforest, the coastal zone and the Serra do Mar mountain range). Other articles (231–232) also demand a protective attitude towards indigenous populations and confirm the property rights of these populations to their territories. These are clauses highly relevant to the preservation of Amazonia, as the overwhelming majority of Indian lands are situated in this region.

But it is difficult to argue that such constitutional clauses on the environment were decisive for the adoption of the "Nossa Natureza" plan. Fernando Mesquita, president of IBAMA from its creation until the end of the Sarney administration and a close associate of Sarney, emphasizes that neither national NGOs nor the constitutional requirements had any significant impact on this decision. Foreign pressure was the dominating motive of the government.[28] This is also illustrated by comparison with the case of indigenous populations. Constitutional requirements for the protection of the lands of indigenous populations were much clearer than the requirements for protection of the Amazon. However, while the constitutional articles on indigenous populations were sabotaged in the name of national security, as described in greater detail below, the more general article on Amazonia was followed up to a surprising extent.

The environmental bodies of SEMA and IBDF were a second possible candidate as a force behind the change under Sarney. They were, however, possibly an even weaker national impetus behind environmental reforms than the environmental movement. SEMA was established in

---

27 For this conflict, see the interview with FUNATURA's president Maria Tereza Jorge Padua in *Jornal do Brasil*, 30 March 1992: 3.

28 Own interview with Fernando Mesquita, December 1993.

1973 to provide an environmental image for Brazil and to satisfy needs for an environmental agency as a required counterpart for international financing. Both bodies were extremely peripheral in the Brazilian bureaucracy (Guimarães 1991: 143–211; Viola 1992: 9). The head of SEMA, Messias Franco, resigned in 1988 because the Minister for Social Affairs, Prisco Viana, refused to sign a decision prohibiting the use of mercury in mining in Amazonia (*Folha de São Paulo*, 20 September 1988: 2). More generally, Franco justified his letter of resignation by pointing to the lack of political will to deal with the environment in a serious way (Guimarães 1991: 200, *O Globo*, 25 September 1988: 18).[29] The IBDF was in a state of disarray and economic crisis during the late 1980s, hardly able to maintain its internal regime, let alone exert influence on the government (Foresta 1991: 163–188).

Now, given that the change of policies related to climate change under Sarney is most adequately explained by external pressure, we may ask: why were reforms so unevenly distributed over various areas? While parts of the policy related to environmental monitoring and incentive changes were rather progressive, foreign policies and the policies related to Indians very much resembled policies under the military dictatorship. Most of the answer to this question is to be found in the persisting influence of the military. The military enjoyed a very high degree of influence over the policy-making process under the Sarney government, especially in relation to Amazonia (Allen 1992; Flynn 1993; Zirker and Henberg 1994).

Both the Nossa Natureza plan and other regional planning in the Amazon region were in the hands of the military. The main military actor behind the shaping of this policy was General Rubens Bayma Denys from the Secretariat for Defence and Strategic Affairs (SADEN/PR). Bayma Denys, the Secretary-General of SADEN/PR, headed the executive committee responsible for the Nossa Natureza plan. Five other members of this body also participated in the executive committee (Albert 1992: 50). SADEN/PR was represented in all the interministerial planning groups. Bayma Denys and SADEN/PR were already involved in the planning of the Calha Norte project, which aimed to increase military and civilian presence in border areas in the Amazon region. The body was generally perceived as a staunch supporter of the military's grand vision

---

29 For details of Franco's more general criticism of the environmental policies of the Sarney government and the Minister of the Interior, João Alves Filho, see "Messias Franco reafirma criticos a Ministro em simposio" (*O Globo*, 25 September 1988: 18). Two years before, Paulo Nogueira Neto had resigned as head of SEMA for the same reasons as Franco (Guimarães 1991: 200).

of a conquest of Amazonia justified by the need to decrease border vulnerability (Allen 1992).

The persisting influence of the military's vision was especially significant for policies related to indigenous areas. Albert (1992) has shown how SADEN was able to take full control over this policy area and implement a *de facto* fragmentation and opening of the lands of the Yanomani tribe in the state of Roraima for gold miners and for the prospecting activities of the larger mining companies. These acts were guided by the strategically motivated fear that the indigenous populations of border regions could pose a security threat through subversion and alliances with foreign powers. A later statement (1990) from the military commander of Amazonia that "the presence of the *garimpeiro* (gold miner) has strategic importance for the occupation of the territory" clearly reveals the motive for this policy (*Isto é*, 4 April 1990).

The foreign policy of the Sarney government was shaped through cooperation with the military, most notably SADEN, and Itamaraty. For Amazonia, this cooperation was especially clear, as official reactions to the pressure on Brazil in this period touched upon both the perceived strategic vulnerability of Amazonia and the region's importance for growth and development. As noted by Miyamoto (1989), cooperation in the 1980s between the Ministry of Foreign Affairs and the army regarding a complex of diplomacy and military policies intended to decrease the vulnerability of the border regions of the Amazon confirms this picture of foreign policy-making. Miyamoto analyses both the military Calha Norte border project and the revival of the Brazil-initiated Amazon pact (1978) in the 1980s as an unitary political initiative, shaped by the cooperation between Itamaraty and sections of the armed forces. This interpretation is supported by the fact that a proposal for revitalizing the Amazon Pact was included in the report from an interministerial group responsible for outlining the Calha Norte project in further detail (Allen 1992: 74). The Amazon Pact is a treaty between Brazil and its neighbouring countries in the Amazon region, basically intended to pre-empt territorial disputes and rivalry. Under President Sarney, it was revived both as an instrument under the Calha Norte project and as a tool for building a coalition of Amazon states against ecological criticism from the industrialized countries (Santilli 1989). In May 1989, an Amazon Pact summit was held in Manaus, on the initiative of Brazil. All members of the Pact supported Brazil's nationalist position as restated in a declaration which included a rejection of strings and conditionality on multilateral financing motivated by considerations for global environmental problems (Hurrell 1992: 407; Cleary 1991a: 25).

Zirker and Henberg (1994: 264–265) discuss these findings in light of

Alfred Stepan's (1988) observations of the military and politics in Brazil. Stepan describes the Brazilian political regime after 1985 being only marginally democratic because of the existence of far-reaching military prerogatives, providing privileged military access to decision-making in the government, the legislature, and state companies. In light of the definition of democracy put forward by Rueschemeyer et al. (1992: 43–44), where the accountability of the state to an elected parliament is one of three key indicators of a democracy, such doubts as to the democratic nature of the Brazilian political regime under Sarney may be justified.[30] However, as we have noted, other segments of the state also enjoyed privileges over environmental decision-making in this period. The Itamaraty shaped policies related to Amazonia and climate change, largely on the basis of their own initiative and their cooperation with the military. Moreover, the resistance to change in other centres of the Brazilian bureaucracy should not be underestimated. The military's geopolitical vision of Amazonia may have penetrated the Brazilian bureaucracy in a more general way, as many technocrats and government advisers have attended courses in "politics and strategy" held by the military think-tank ESG (*Folha de São Paulo*, 25 July 1991: A-12).

These observations on the continued influence of the military over climate change policies may also suggest that the 1989 presidential elections – the first such democratic elections to be held since 1960 – might provide an impetus for change. And this is exactly what happened.

4. The Collor Presidency 1990–1992: Reforms and Waning State Power

Brazil's first democratically elected president in 20 years, Fernando Collor de Mello, was inaugurated in March 1990. His presidential period, 1990–1992, was marked by vigorous Amazonia reforms motivated essentially by concerns for global environmental problems, with climate change at the fore.

The first document on the environmental policies of the Collor government was a 77-page document presented between the first and the second round of presidential elections in 1989 (Collor, undated). It was written by Helio Setti, an activist from the environmental movement focused on the protection of Brazil's Atlantic rainforest. The document

---

30 Rueschemeyer et al. (1992: 43) define the other two key indicators as regular, free and fair elections of representatives with universal and equal suffrage, and the freedoms of expression and association as well as the protection of such individual rights against arbitrary state action.

provided the main background for the environmental reforms of the Collor presidency (1990–1992) also in relation to Amazonia, and was written with more than one eye to global environmental problems. Priority areas were identified as effective zoning efforts in the region, aiming to identify areas for protection and economic exploitation, a system of permanent monitoring of the most vulnerable regions, the creation of new conservation units and better inspection services in these areas, development of less destructive agricultural systems, strengthened state and municipal environmental administration, cancelling the natural forest-based iron production facilities in the Great Carajás programme, increasing the amount of extractive reserves and accelerating the demarcation of indigenous areas with a special priority on the Yanomani reserve (Collor undated: 17–22). However, these propositions were not implemented through coherent "policy packages" to the same extent as under the Sarney government. Rather, reforms came in uneven clusters.

The first step taken by the Collor government was to establish a secretariat for the environment, SEMAM. SEMAM's status as a secretariat directly connected to the president of the republic meant that Brazil for the first time got an environmental agency with a voice at government level. IBAMA was transferred from the Ministry of the Interior, becoming the executive body of the new secretariat. The new secretary of the environment was José Lutzenberger – deep-ecologist and renowned veteran of the Brazilian environmental movement.

The system for incentives to encourage deforestation was further revised in the 1990–1992 period. The suspension of fiscal incentives for ranching in the forest areas of Legal Amazonia was continued until a new law for the application of regional fiscal incentives was adopted in January 1991. This new law, together with subsequent decrees, made the approval of fiscal incentives for crop and livestock activities dependent on the previous consent of the Secretariat of Strategic Affairs (SAE, the successor to SADEN), the National Institute of Colonization and Agrarian Reform (INCRA), IBAMA, SEMAM and the National Indian Foundation (FUNAI). It also gave IBAMA the authority to scrutinize the environmental aspects of all projects and if necessary cancel them. This implied a noteworthy extension of power for IBAMA, SEMAM, and FUNAI.

The Great Carajás area programme was also revised further under the new government. At government initiative, new directives were established that significantly altered the legal basis of the programme. These directives required that the steel and iron companies in the Great Carajás project comply with a new set of regulations in order to be eligible for regional fiscal incentives. The new regulations implied the obligation for companies to fuel their factories with energy sources other than native forests and to use energy-conserving technologies in production.

Together with the general environmental criteria for approval of SUDAM fiscal incentives for projects, these detailed provisions meant a considerable change of incentives for ecologically destructive projects connected with wood-consuming industries and agriculture. In addition to these incentive changes, a 1990 decree also charged IBAMA with the obligation of giving priority to the Great Carajás area in their surveillance and monitoring operations.

Surveillance and monitoring efforts were further increased in the 1990–1992 period. In June 1990, the Collor administration introduced a programme to combat deforestation and fires in Amazonia, popularly called "Operation Amazonia" (*Gazeta Mercantil*, 5 June 1990: 15). It involved a more sophisticated linkage between inspection and satellite monitoring. In cooperation with the Brazilian Space Agency, a new system was successfully introduced which could link satellite fire observations with IBAMA's regional offices within hours (Cleary 1991a: 39). Field personnel and journalists reported an increase of IBAMA inspection in Amazonia in the dry season of 1990 compared to the previous year (Margolis 1992: 308).[31] More fines were meted out and routines for their collection were considerably improved.[32] Politicians from Amazonia speaking on behalf of economic interests that were punished by IBAMA in 1990 and 1991 complained about these changes.[33] According to IBAMA sources, the body was able to detect and punish trespassers of deforestation prohibitions over an area of 96,305 hectares in the dry season up to 30 October 1991.[34] This was just below 9% of estimated deforested area that year, according to the Brazilian Space Agency (INPE), a

---

31 According to Dr. Luis Alberto Vieira Dias at the Brazilian Space Agency INPE, "Operation Amazonia" used an improved methodology. In PEAL, weather satellites (NOAA) for fire detection were used, while "Operation Amazonia" used more detailed images from the land monitoring satellite LANDSAT. Personal communication March 1995.

32 Interview with former IBAMA leader Tania Munhoz in November 1992. On the improvement of the collection of fines, see *Jornal do Brasil*, 11 October 1990: 12.

33 Jarbas Passarinho, senator from Pará, concluded in interview that there had been a change and that the implementation of rules became stiffer from 1990. See also the discussion later on in this chapter regarding the opposition against "Operation Amazonia". Dr. Luis Alberto Vieira Dias at INPE emphasizes that even though funding for "Operation Amazonia" was far from sufficient, the programme represented a distinct improvement compared to previous years: "... to make ten helicopters available for this programme was an unheard effort in Brazil. Previously we had none!"

34 Sources at IBAMA.

figure which strengthens the assumption that "Operation Amazonia" had a substantial demonstration effect in many areas. Several large companies were heavily fined.[35] However, although monitoring was given high priority, it is also clear that funding for the environmental bureaucracy declined substantially under Collor. In 1992, this hampered monitoring and surveillance operations, as funding began to dry up.[36] The Collor government's general record with reference to protected areas other than indigenous reserves must be characterized as poor. Though the definition of new conservation units continued, hefty budget cuts for IBAMA left the already underfunded national parks in a state of disarray.[37] At this level, policy changes under Collor were more rhetorical than substantial (Viola 1992: 14).

Another area of reform under Collor was the increased attention to the demands and rights of Amazonia's indigenous populations. This stands in clear contrast to policies under Sarney. Though the first year of the Collor government was characterized by only symbolic policies, like destroying the airstrips of gold miners in the Yanomani reserve in Roraima without any further action to improve the legal status of the reserve, things changed quite rapidly after Collor sacked FUNAI's president in June 1991 and put the respected career Indianist, Sidney Possuelo, in this position. Possuelo launched a more determined policy of definition and demarcation of areas for indigenous populations, backed up by President Collor and the Minister of Justice, Jarbas Passarinho.[38] In addition to the famous decision of November 1991 to recognize and demarcate the whole Yanomani reservation in Roraima as belonging exclusively to the indigenous populations, a long series of other reservations were defined in the 1991–1992 period. Even though demarcation of the areas has proceeded slowly, the contrast with the Sarney government

---

35 Examples of this are the company Mineração Taboca, a subsidiary of the Brazilian mining giant Paranapanema, and the Jerdau logging company. Mineração Taboca was fined about USD 1.1 million for its operations in the cassiterite mine at Pitinga in the state of Amazonas (Jornal do Brasil, 11 October 1990: 12). The company had been infamous for years because of its unpunished ecological disturbances of the Waimiri Atroara Indian reserve close to the mine (Fearnside 1990b: 210).

36 See Correio Braziliense 24 May 1992: 18. The state superintendent of IBAMA in Pará, Reginaldo Anaisse, complained that funding for air surveillance had decreased to a very low level in 1992 (own interview November 1992).

37 Between 1989 and 1992, federal expenditures on conservation units seem to have been reduced by more than 75% (World Bank 1994: 280).

38 Own interview with Possuelo, July 1994. Possuelo's cooperation with Collor was remarkably intensive, given the peripheral position of FUNAI in the bureaucracy. At least once a month he had personal meetings with Collor, when questions related to FUNAI and indigenous populations were discussed.

is substantial and clearly recognized by Brazilian NGOs and other observers.

Foreign policy positions on the environment in general, and on climate change, were also dramatically revised. Collor's position here stood in marked contrast to that of his predeessors. In his inaugural speech, only the economic liberalization programme received priority over environmental questions (Cleary 1991: 117). This change of positions is demonstrated by the following statement by Ambassador Marcos Castrioto de Azambuja, Ministry of Foreign Affairs, in December 1990 (Brazil 1991: 9):

> It is precisely because these are also concerns of the Brazilian society and of Brazil's Government that we understand that the international community has a right to be concerned by the violation of human rights and by the damage done to the environment wherever they may occur. In dealing with these and other issues, Brazil no longer [my italics] resorts to allegations of sovereignty to deflect criticism. Instead, it shoulders its responsibilities, conscious that its actions have repercussions for the whole planet.

This shift also meant a new willingness to admit the necessity of actions to address and counteract the problem of climate change. A specific forest protocol under the Framework Convention on Climate Change (FCCC) was also supported to address the problem of deforestation (in boreal, temperate, subtropical and tropical forests) – a marked change of position from the Sarney government. The Brazilian government now also officially supported joint implementation of the FCCC, providing states with the opportunity to obtain credits under the Convention by investing in measures against greenhouse gas emissions also in other countries.

Preparations for UNCED, which the Collor government had worked hard to locate to Brazil, also involved unique opportunities for influence for environmental NGOs, who participated in the groups that prepared Brazil's positions.

However, we should also note that the foreign policy change was partially neutralized by the negotiators from the Itamaraty. Both the interim Secretary for the Environment, José Goldemberg, and Brazilian environmental NGOs were dissatisfied with the operationalization of the Brazilian position in the negotiations on climate and biodiversity.[39] This

---

39 See the article "ONGs pressionam por posição oficial", *Jornal do Brasil*, 14 November 1991: 7. Goldemberg, who participated in the groups that defined Brazil's international positions in preparations for UNCED and was actively involved in shaping Brazil's official positions, describes the ministry's role in the negotiations like this:

brings us from mere descriptions of policy change under the Collor government to closer analysis.

4.1. Collor's Policies: Evaluation and Explanation

The main motive for the more cooperative position on the environment and climate change under Collor is to be found in his economic programme. Collor's core political project was to modernize Brazil through extensive privatization, trade liberalization, and increased cooperation with the industrialized countries. Both in his inaugural speech as a president, and in the introduction of the environmental planning document, he links the change of positions on the environment explicitly to an improvement in Brazil's relationship with the industrialized world and the economic reform programme (Collor, undated: 4, my translation from Portuguese):

> I am conscious that if Brazil does not confront the environmental question internally and externally, this will make our programmes for economic development and international financing difficult.

Most commentators conclude that the main explanation for the *change* of environmental policies is to be found in this connection between the environment (especially Amazonia) and the new economic programme, and not in internal pressure from NGOs, whose influence on Collor and his presidential campaign was quite marginal (Hurrell 1992; Viola 1993). This interpretation is also explicitly supported by key members of the government.[40] In addition, the prospects of the Rio-Conference (UNCED) increased this motivation, owing to the greater international attention to environmental policies in Brazil. However, to see this as external pressure of the same kind as the pressure on the Sarney government is misleading, as the decision to locate the conference in Brazil

---

"Itamaraty was always defensive and at one point they were ordered to follow instructions on the Climate Convention and take less confrontational attitudes to the industrialized countries" (Personal interview with Ana Maria Fonseca (FBCN), who confirmed Goldemberg's statement, and complained about Itamaraty's defensive attitude on patents in the biodiversity negotiations). The generally negative impression of Itamaraty's ability to respond to the positions defined in the groups was also confirmed by Cesar Victor do Esperito Santo, adjunct leader of the environmental NGO FUNATURA, in an interview in November 1993. See Viola (1993: 18) for a slightly more sympathetic evaluation of the Brazilian position.

40 Own interviews with ex-ministers José Goldemberg and Jarbas Passarinho, November/December 1992.

was wholeheartedly supported by the Collor government itself. Thus, UNCED was rather the hallmark of a less defensive environmental policy than a new source of external pressure.

Alternative explanations pointing to the strength of environmental NGOs or the environmental bureaucracy as explanations of the Collor reforms are more difficult to defend. Some of the IBAMA presidents, most notably Edouardo de Souza Martins (1991–1992), who was also adjunct secretary of SEMAM, had an extensive informal dialogue with the NGO movement.[41] Of course, also the choice of Maria Tereza Jorge Padua, leader of the environmental NGO FUNATURA, as president of IBAMA for a few months in 1992 demonstrated a willingness to involve NGO representatives in decision-making. But none of these NGO representatives were involved in Collor's initial decision to change Brazilian policies related to Amazonia. The political resources controlled by environmental NGOs were still too modest to enable these organizations to press successfully for major changes.

Neither was the environmental bureaucracy influential enough to bring about policy change. Though it was strengthened through the establishment of SEMAM, this was a symptom, not a cause of the policy change. IBAMA experienced serious organizational difficulties, and was not very active as an environmental lobby.[42]

There were also strong economic forces that hampered reform efforts. In 1990, Collor introduced a Draconic austerity plan to curb inflation and deal more effectively with the huge deficit in the public sector. As part of these policies, funding for public bodies like IBAMA was considerably decreased (Carvalho 1991). These cuts did not damage policies like the changes in fiscal incentives and foreign policy, but they reduced the capacity of IBAMA, FUNAI, and SEMAM to carry out their day-to-day work, not to mention the long-term construction of a viable public environmental sector. Though the drastic economic measures chosen by Collor were exaggerated and also ineffective in solving the country's major economic problems (inflation rates continued to skyrocket), they were perhaps also symptoms of the very narrow range of options open to developing countries charged with the necessity of environmental reform and fiscal austerity imposed by the hostile economic climate of the 1980s (Altvater 1987).

Continued protests from the military and their civilian allies also contributed to obstruct reforms. Though the military was in a much more

---

41 Own interviews with Edouardo de Souza Martins, November 1993, and Ana Maria Fonseca (FBCN) November/December 1993.
42 Various interviews at IBAMA, October 1992 and December 1993.

peripheral position under Collor than under Sarney, it still commanded substantial political resources. In the Collor government, the military ministers were main opponents of reforming policies in Amazonia.[43] But military opposition was not limited to intra-governmental protests. The army and regional politicians orchestrated a campaign against what they called the "internationalization of Amazonia". These accusations were promoted by regional politicians like Gilberto Mestrinho, the Governor of Amazonas, and by military bureaucrats like the Commander of the General Staff of Amazonia, Thaumaturgo Sotero Vaz. ADESG, a diplomatic affiliate of the war college ESG, joined in the campaign against environmental reforms with their support to the campaign against the "internationalization of Amazonia" in 1991 (*Folha de São Paulo*, 25 July 1991: A-12). So did the military institutions Escola de Comando e Estado Maior do Exército (ECEME) and Centro Brasileiro de Estudos Estratégicos (CEBRES), by hosting a seminar on the external threat to Amazonia in October 1991 in which high-ranking officers as well as regional politicians participated (*O Globo*, 14 October 1991: 7).

The campaign against the "internationalization" of Amazonia was rhetorically centred on foreign policy and the demarcation of indigenous territories, but was of course also inspired by resistance from commercial interests. These interests pressed for favours like the cancellation of fines, licences for environmentally dubious projects and the appointment of easily controlled political clients in positions as IBAMA state superintendents (Kasa 1994).

Amazonian politicians were involved in the dismissal of Tania Munhoz, president of IBAMA in 1991,[44] and the firing of Lutzenberger as Secretary for the Environment in 1992.[45]

A special congressional investigation (CPI) on the "internationalization" of Amazonia in 1991 was led by Atila Lins, deputy from Amazonas. The CPI was widely supported among Amazonian politicians. Among the proposals of the CPI were the reduction of indigenous territories, limitations on IBAMA's surveillance and inspection activities in the region, the creation of new investment funds, and the revitalization of the Calha Norte military border protection project (*Jornal do Brasil*, 29 November 1991: 4). During the investigation, the special committee also

---

43 Own interview with former Secretary of the Environment, José Goldemberg in November 1992.

44 *Folha de São Paulo*, 3 October 1991: A-10, own interview with Goldemberg in November 1992.

45 *Correio Braziliense*, 23 March 1992: 5, interviews with Feldman and an anonymous SEMAM adviser.

demanded the dismissal of Munhoz and Lutzenberger (*Folha de São Paulo*, 10 September 1991: A-14). The military was heavily involved in the CPI as ghostwriters of the first draft of the report from the investigation for the Amazon congressional group (Bernardo and Bastos 1993: 18-19).

The importance of this combination of external pressure and a liberalist economic programme in explaining the Collor reforms was demonstrated by the government that followed Collor's impeachment in September 1993.[46] After UNCED, a certain fatigue seems to have overcome the global environmental movement, also concerning climate change. This decline of attention drew attention away from Brazil and the country's role in global environmental problems. At the same time, Itamar Franco, Collor's vice-president, emerged as a highly conventional and weak president who worked closely with the military and the Ministry of Foreign Affairs. The appointment of Coutinho Jorge, senator from the Amazon state of Pará, to the new position as Minister for the Environment in 1992 confirms this. Though Coutinho Jorge did not represent a full return to the destructive policies of the former military regime, his administration was marked by a considerable decrease in initiatives related to Amazonia. Together, these impulses produced a breakdown of Brazil's wave of reforms in Amazonia.[47] This has also been the case for policies on climate change. At various conferences on climate change, like the March 1995 Conference of the Parties in Berlin or the UNDP conference on joint implementation in Rio in December 1994, Brazil once again adopted non-cooperative and confrontational positions.[48] The reversal of environmental policies also seems to have hurt the capacity of the environmental bureaucracy. Following the takeover of Franco, political parties regained control over employment policies in some of the most important regional branches

---

46 Collor was impeached by the Congress after the disclosure of an enormous corruption scandal; see Flynn (1993).

47 The following illustrates the negative role of the Itamaraty. According to Sidney Possuelo at FUNAI, the Itamaraty refused to receive grants of USD 18 million from Germany to facilitate the demarcation of indigenous territories in February 1994. (Personal communication with Possuelo, August 1994). In 1993, FUNATURA top staff complained about the negative attitude of Itamaraty to the ITTO 2000 rules on sustainable forestry, which were perceived as threats to Brazil's sovereignty. (Personal meeting with FUNATURA staff in November 1993).

48 Own interviews with CICERO observers present at both conferences. There are still proponents of international cooperation in Brazil: At the UNDP conference, former Secretary of the Environment José Goldemberg was seen taking a nap in full public to demonstrate his dissatisfaction with the representative of the Itamaraty who was delivering his speech.

of IBAMA, like the Amazon state of Pará. State superintendents were once again selected on the basis of connections to political parties (most notably the Brazilian Democratic Party, PMDB), making it very difficult to implement any forest policy motivated by concerns for climate change.[49]

5. Conclusion

The Brazilian military had demonstrated a strong interest in continuing a development policy for the Amazon region which would also imply massive emissions of greenhouse gases. This meant that a weakening of military prerogatives also opened opportunities for revision of the hard-line nationalist position adopted by the Sarney government. Collor used the power given to him by an electoral majority to exploit this opportunity. However, for Collor, a revision of Brazil's policies and positions on climate change was mainly perceived as preconditions for his economic programme. Under Collor, environmental reforms became as much a subsection of foreign policies and economic policies as a policy in itself. This implied both continuity and change related to the Sarney government. *Continuity* because the environment was still seen as an element within a much broader set of policies, and *change* because it advanced from being a subfield of security policy to a precondition for a major economic programme. In part, this reflects the nature of environmental policies, cutting across most sectors of public activities (Guimarães 1991: 175–176). And partly it reflects changes in the global context: The Sarney government's perception of Amazonia's vulnerability was inspired by the military's perception of the ever-present danger of Communist infiltration and subversion typical of the US–Soviet rivalry, while Collor's perception of Amazonia was more linked to considerations of international trade and investments, which took on greater importance in the post-Cold War, capitalist, global political and economical order. But the lack of an independent environmental policy also reflects the weakness of Brazilian social movements working for the preservation of Amazonia and a revision on positions of climate change. Thus, on a topic like global environmental problems, and especially in relation to the vast and sparsely populated Amazon region, Brazil has yet to reach the stage in

---

49 This point is based on interviews with employees at the headquarters of IBAMA in Brasilia in December 1993, and needs further documentation before it can be presented as a fact.

which civil society can effectively challenge the state apparatus and change policy outcomes unaided by external pressure.[50]

---

50 In this respect, the NGO campaign to save another important Brazilian forest biome, the Atlantic rainforest, may be a good comparative case. The Atlantic rainforest is situated along the Brazilian coast from the north-east to the south. It represents a unique ecosystem with very high biological diversity, but has been reduced to only small patches of its original extent because of agricultural extension and logging. Especially in the economically advanced and populated southern and south-eastern states of Brazil, well-organized NGOs have enjoyed a high degree of influence over environmental decision-making both at state and federal level, and have been heavily involved with IBAMA at various levels regarding surveillance and law enforcement as well as definition and protection of conservation areas. But it should be noted that a more advanced civil society is not the only contrast between Amazonia and the Atlantic rainforest that may explain stronger and more successful political mobilization over the latter issue. Mobilization over the Atlantic rainforest is also facilitated by the absence of military interest in the region. (Interview with Ana Maria Fonseca, FBCN and staff member of the Atlantic Rainforest Network (Rede da Mata Atlântica), December 1993). The Atlantic Rainforest Network is a consortium of more than 70 NGOs and research institutions with strong activity on issues connected to the preservation of the Atlantic rainforest.

Chapter 10

Africa and Climate Change

Ewah Otu Eleri[1]

1. Introduction

The management of the global climate has become one of the most crucial issues confronting international society, and is likely to remain so in the foreseeable future. Both the potential physical occurrence of global warming and the measures suggested to remedy it will affect the regions of the world disproportionally. Likewise, the capacities of different regions to adjust domestically and contribute internationally to efforts in addressing climate change will vary.

To date, sub-Saharan Africa's contribution to the build-up of greenhouse gases has been low; the region therefore does not pose a major threat to the global climate. However, the occurrence of climate change might pose formidable challenges to already deteriorating conditions of human development in that region. Both national contributions to emissions of greenhouse gases and the presence of considerable forest cover as sinks vary within Africa just as widely as the vulnerability of various countries to the impacts of climate change.

The appearance of the climate issue on the international agenda in the latter half of the 1980s coincided with a remarkable period in the economic and political life of sub-Saharan Africa. This period was marked by a growing international indebtedness, increased dependence on international aid and political conditionality tied to both debt-relief and aid flow. The "African crisis" has been compounded by rounds of droughts as well as by intensified ethnic conflicts. In many countries, national survival and improvement in living conditions have become dominant issues that transcend international concerns.

1 Part of this chapter draws on previous work done at the Fridtjof Nansen Institute. Grants from The Norwegian Research Council under the SAMMEN programme are gratefully acknowledged.

It is as yet too early to draw any definite conclusions about the climate policies of various African countries. Uncertainties also prevail as to the potential impact of climate change and the effect of the climate change problem-solving regime on the actions of countries, sub-national actors and individuals. However, a preliminary assessment can be made on the basis of the present and potential future interaction between the region's prevailing political economy and the emerging demands for international action on global warming.

In Section two, evidence is presented on Africa's contribution to the total world greenhouse-gas emissions and the available reservoir of sinks. In Section three, the potential impacts of climate are reviewed, including physical vulnerability and costs of adjusting policies to mitigate climate change. In Section four, an assessment of policy responses and positions adopted by regional bodies, national institutions and among sub-national actors is presented. In Section five, a suggestion is given about the factors likely to shape climate policies in the region; and finally, in section six, future prospects for the development of climate policies in the region are discussed.

2. Africa's Contribution to Climate Change

Africa's present contribution to the total global anthropogenic emissions of greenhouse gases is still very low; on a per capita basis, its contribution is the lowest among all the regions of the world. Two-thirds of CO_2 emissions from Africa stem from land-use, with the remainder due to industrial activities and transport. Emissions of CO_2 from energy combustion represent barely 3% of the global total. Studies suggest that Africa emits between 1% and 7% of the world's total methane (CH_4) emissions and about 3% of chlorofluorocarbons (CFCs).[2]

The distribution of current greenhouse gas emissions varies widely in Africa. Nigeria and South Africa represent the largest sources of industrial greenhouse gas emissions. Nigeria is the world's largest flarer of natural gas. By 1990, the country was releasing about 21 billion m^3 of associated natural gas from its oil production (Homer 1991). This is Africa's largest single source of CH_4 emissions. With the country's substantial tropical forests, which serve as sinks and reservoirs for CO_2

---

2 There is considerable disparity in the assessment of the region's share of the total global emission of greenhouse gases. Davidson (1993) estimates emissions from Africa to be about 1%; Agarwal and Narain (1991) about 3%; while the World Resources Institute's (1991) figure is 7%. Subak et al. (1992) have estimated 6.9% for CO_2; 21% for CO, 9.5% for CH_4; 15.8% for N_2O; and 4.4% for CFCs.

emissions, Nigeria plays an important role in the overall regional carbon flow. Much of South Africa's emission stems from coal-mining and combustion for power generation This places the country globally among the top ten energy-related emitters of greenhouse gases – and Africa's largest.

The other group of countries with significant capacities to affect carbon flows from African sources on account of their profuse tropical forest cover includes, Côte d'Ivoire, Zaire, Sudan, Madagascar, Malawi and Cameroon. Eighteen per cent of the world's tropical forest is in Africa (WRI 1991). These resources have come under pressure from land-use, especially agriculture and logging for timber, bush burning and charcoal production. Three to five million hectares of tropical forest is lost every year in Africa. This is particularly precarious in some countries: for instance, Madagascar has lost an estimated 90% of its original forest, reducing not only sinks for CO_2 emissions, but, even more seriously, its rich biological diversity (Seragelding 1990).

Much of Africa's potential increase in greenhouse gas emissions will depend on the direction of economic development and population growth rates. This will strongly influence energy requirements and the pressures brought to bear on land resources.

2.1. Potential Growth in Energy Use

Sub-Saharan Africa is well endowed with modern energy resources. Oil is extracted in Gabon, Côte d'Ivoire, Nigeria, Cameroon, Congo and Namibia. Proven reserves in the sub-Saharan region amount to 20.5 billion barrels, equivalent to 2.3% of the world total. It has 3.5% of world proven natural gas reserves, located in Angola, Namibia, Nigeria and Gabon. At the current rate of consumption, natural gas reserves can supply the sub-region for over 120 years. As a region, Africa can also boast of 200,000 megawatts of untapped hydropower, enabling the current installed capacity of power facilities to be boosted twenty-fold (WB 1989). Substantial coal reserves exist in South Africa, Botswana, Swaziland, Zimbabwe, Nigeria and smaller amounts in other countries – together representing about 7% of world reserves. Reserves of uranium are found in Namibia, Zaire, Niger, Gabon and South Africa. Of even greater significance is the huge potential in renewable resources such as biomass and solar energy.

Despite the abundance of these resources, many countries of the region are engulfed in an energy crisis. Firstly, a large percentage of foreign exchange revenue is spent on the procurement of energy, particularly petroleum. Furthermore, hydropower infrastructures are in poor shape and perform on an average capacity of less than 40%. Finally, biomass – the chief source of fuel for the great majority of the region's inhabitants – is scarce, as population increases have driven agriculture and the search for fuelwood into marginal lands.

Energy policy objectives among the countries of Sub-Saharan Africa centre on the following: (i) growth in the productivity and quanta of energy use at all levels of activity; (ii) shift of the mix of energy resources from primary fuels such as biomass towards modern secondary forms such as electricity; and (iii) efficiency in the transformation, distribution and utilization of all energy resources, irrespective of type (Wereko-Brobby and Nkum 1991; ECA 1991a; Davidson and Karakezi 1992).

Table 1: Total Primary Energy Supply in sub-Saharan Africa (Mtoe) (WB 1989)

Energy source (projected)	1960	1986	2020
Coal	3.5	4	10
Petroleum	5.6	24	140
Natural gas	0	3	30
Hydroelectric power	0.5	3	20
Subtotal (commercial fuels)	9.6	34	200
Wood fuels	-	66	200
Total	-	100	400

The key challenge to Africa's energy sector in the immediate future seems to be the expansion of the energy supply base in an economically efficient, socially equitable and environmentally sustainable manner (Eleri 1996). Growth in the supply of commercial energy, according to the World Bank, is likely to increase six-fold by the year 2020 to support a projected growth rate of 5% (WB 1989). With no change in the mix of energy sources, this will imply an enormous increase in Africa's emission levels. Such an optimistic energy-use scenario can wipe out some of the most ambitious CO_2 reduction targets in OECD countries, and will set Africa's energy development in direct collision course with the abatement intentions of the 1992 Framework Convention on Climate Change (FCCC).

2.2. Potential Emission Increases from Population Growth and Land-Use

Trends in the demography of the region may also exert substantial impact on emission levels. With Africa's 3.0% average annual growth in population, the continent is likely to double its present population to an estimated one billion people over the next two decades. By 2025, Africa's population may be close to 1.6 billion. The urban population is likely to quadruple. Overall, more than 40% of the new inhabitants in the region will be under 14 years of age (OAU/UNEP 1991). Such a demographic scenario will dramatically heighten the pressure on land resources and energy demand.

The contribution of population growth to greenhouse emission depends greatly on the intensity of the use of technology in production as well as consumption patterns: the slower the change in technology and consumption patterns, the more population increase accounts for much of a region's greenhouse emission. One study estimates that increase in population accounted for 68% of the increase in sub-Saharan Africa's CO_2 output between 1980 and 1988 (UNFPA 1992: 28).

Population increases are inextricably linked to the intensity of land-use and the encroachment of agriculture into forest land. An estimated 80% of Africa's people subsist on agriculture. Both agriculture and forestry are advancing into marginal lands. Africa is losing between 3 and 5 million hectares of tropical forest each year through deforestation – an area greater in size than the country of Togo and larger than several European countries. Should this rate of loss continue, tropical forests in Africa will be gone within 60 years. The continuation of this trend will have serious implications for the provision of sinks for CO_2 as well as the protection of some of the world's richest diversities in flora and fauna (Serageldin 1990).

3. Impacts of Climate Change

Both the physical occurrence of global warming and the regime formulated to address it will have severe implications for human survival and development in Africa. The following section reviews the potential physical consequences of climate change and the impacts of the FCCC on aspects of development in Africa.

3.1. Some Potential Physical Impacts

Many parts of the region are already experiencing severe environmental stress from climate variability. The intensity of recurring drought and desert encroachment, particularly in Southern Africa and in the Sahel, are key parameters shaping development and survival in these parts of Africa. Moreover, resources are meagre and economies and political institutions in many parts of the continent are so weak that their ability to cope with further stress is limited.

Several studies that seek to provide better knowledge on potential impacts of global warming in Africa are currently being carried out. Some of these have been commissioned by the United Nations Environment Programme (UNEP), the Global Environmental Facility (GEF) and bilateral donors. The IPCC has recently released preliminary guidelines for assessing the potential impacts of climate change. However, a considerable degree of uncertainty will remain as to the impacts on Africa. A few qualitative speculations are emerging. By and large, increases in greenhouse-gas concentrations in the atmosphere lead to increases in temperature, changes in weather patterns and sea-evel rise. Temperature increases could increase evaporation, and potential evapotranspiration, leading to increased drought. The 1992 IPCC Supplementary Report suggests that it might well become hotter in the tropics as well as in the northern and southern parts of the region. This would seriously affect agricultural patterns and consequently the vulnerable food-security situation. Views on possible changes in rainfall patterns are generally quite negative: rainfall shortages, reduced soil moisture, crop failures, human migrations, lack of appropriate coping mechanisms, among other impacts (Kelly and Hulme 1992; Glantz 1992). Furthermore, Africa will be vulnerable to sea-level rise occurring as a result of the thermal expansion of seawater and the melting of land-based glaciers. Loss of wetlands, increased rates of beach erosion, flooding and increased salinity of ground water will result (Ibe et al. 1991). Finally, Africa's small island states – such as parts of Equatorial Guinea, Sao Tome and Principe and Comoros – could be particularly threatened. Low-lying coastal regions, especially in West Africa and in Egypt, will be threatened as well.

3.2. Impacts of the Climate-Change Regime

The impacts of climate change are not only directly physical; measures to address and adapt to them are also costly and might distort development in the countries implementing them. During the various rounds of negotiations, leaders of Third World countries have remained apprehensive of international action to mitigate global warming, mostly for fear that these measures will impose serious restrictions on development processes within their societies.

In broad outline, the principles embedded in the FCCC stipulate that industrialized countries have agreed to bear much of the burden in managing climate change and its consequences. Developing countries · on the other hand have secured an understanding that economic development, rather than combating global environmental problems, is their primary concern. The industrialized countries were thus obliged to provide the financial and technological requirements to aid industrializing countries in meeting their commitments. The obligations and rights of African countries are the same as those of all Parties to the Convention, particularly for developing-country parties.

The climate-change regime exerts both direct and indirect impacts on Africa. Among the most direct impacts, along with other parties, countries in Africa are obliged to prepare national reports on their inventories of greenhouse-gas emissions. Industrializing countries would be obliged to submit such reports no later than March 1997. Country inventories on sources of emissions and sinks have been completed for Burkina Faso, Cameroon, Gambia, Nigeria, Senegal, Seychelles, Tanzania, Uganda, Zambia, Zimbabwe, among others, while climate-change impact studies have been completed for Burkina Faso, Kenya, Mauritius and Nigeria. The United States Country Studies Programme is presently financing on-going studies on both mitigation and adaptation strategies in several African countries (USCSP 1995). Furthermore, industrializing countries have committed themselves to outline national strategies to combat climate change. Studies outlining greenhouse-gas mitigation options have been commissioned in several countries.

At the First Conference of Parties (CoP) in Berlin, March 1995, GEF was endorsed as the financial mechanism for the Convention for a period of four years. This came as a negotiated middle ground between the suggestion of industrialized countries to retain the GEF as the permanent mechanism for resource transfer, and the scepticism of developing countries regarding their influence over the institution.

There are no specific references in either the FCCC or from the Berlin Mandate on the mechanism for the transfer of technology as stipulated by the Convention. Fears abound in Africa that technological transfer might tend to benefit larger and more technologically advanced developing countries, as these have the capability of tapping and employing patented and non-patented technologies (Juma 1993).

The agreement reached on joint implementation constitutes a direct potential source of impact on national finances of countries in Africa. Joint implementation refers to mechanisms by which a country with relatively high costs of emission reduction can invest in emission-reducing measures in a country with lower reduction cost and be credited, in whole or in part, for emission reductions in its own climate-gas accounts (Nordic Council of Ministers 1995: 8).[3] The overriding aim is cost-effectiveness and flexibility in measures to mitigate climate change. However,

---

3 See Chapter 7 for an elaboraton on the principle of joint implementation.

joint implementation may potentially have serious political and economic impacts (Ojwang et al. 1995; Selrod et al. 1995).

Developing countries fear that by joint implementation attention might shift from emission reduction in industrialized countries to developing ones. The mechanism could also weaken the incentive of the North to accelerate the transition to cleaner technologies. Many industrialized countries are already unable to meet their modest targets for emission stabilization. For instance, Norway, far from stabilizing CO_2 emissions at the 1989 level by the year 2000, is expected to increase emissions by 16%. Joint implementation might also have important political ramifications by virtue of its potential to involve the North increasingly in decisions concerning major investments in the South. For *conditionality-weary* African countries, this is a source of major concern.

Despite this scepticism, joint implementation may offer interesting future opportunities for African countries. With the declining investment in Africa's energy sector, opportunities may arise for additional resources to be transferred from the North to cover these shortfalls. However, there is bound to be stiff competition for these resources from more advanced industrializing countries. If some donors should resort to redirecting part of their development assistance budget to such projects, commercial considerations will increasingly override political aspirations to help the poorest countries (Eleri 1994).

Indirect impacts of measures to abate climate change might be expected to accrue from adjustments in domestic policies by industrializing countries and transnational corporations. This might influence investments, trade and aid. The trend of increasing energy and carbon taxes among OECD countries, partially in response to climate change, has already raised significant concerns among oil-exporting countries, many of these West African countries. Should energy taxes be extended to coal use, Southern African exports of coal will come under pressure.

It is also likely that the industrialized countries' need to meet emission targets will necessitate a greening of multi-national energy corporations, thus influencing the investment agenda for new energy projects. While these developments might accelerate the transfer of cleaner technologies, there might be a negative impact on political and social choice within these countries.

Integrating global environmental protection into development aid is under discussion in both bilateral and multilateral development agencies. For instance, the World Bank has been contemplating the concept of "global overlays", providing incentives within aid packages for projects that benefit the global environment. Bilateral development assistance agencies are discussing how to incorporate global environmental concerns into their work. In related fields – such as human rights, gender equality, democratization and good governance – attaching strings to aid continues to give rise to concern among recipient countries and NGOs (Clayton 1994).

Africa might not be more threatened by climate change than most other regions of the world; nevertheless, the continent remains in a special situation due to its relative dearth of the capabilities needed to adapt to drastic changes. The economic decline of the region over the past decade has reduced its eagerness to launch major catastrophe prevention schemes. Existing physical infrastructure and national institutions have become weakened under the austerity programmes imposed by structural adjustment programmes. The region is also vulnerable to international pressures that may not promote its efforts towards recovery and development. The current economic situation – receding political sovereignty due to indebtedness and aid dependence – continues to reduce the scope for the pursuit of the national interest, and might moreover aggravate the vulnerability of African states and societies to the impacts of the international climate-change regime.

4. Climate Change Policy Positions and Activities

Global environmental problems – including climate change – have neither gained priority over development nor over other more immediate environmental concerns among governments, the private sector or individuals across sub-Saharan Africa. This is not to imply that African countries care less about the environment: but today their priorities are to meet more pressing economic and environmental problems facing their peoples. This partly accounts for the apparent lack of enthusiasm for formulating national policies on climate change. Moreover, since current obligations placed on this group of countries have not been stringent enough to seriously threaten present national interests, there might not be sufficient incentives to induce them to work seriously on climate-change policy at the domestic level.

4.1. The African Common Position

Despite the divergence of interests, African countries have participated actively in the international dialogue and negotiation processes at both the regional and global levels. Moreover, views from the region have been expressed through international coalitions under the auspices of the Group of G-77, the Alliance of Small Island States (AOSIS), and the

Organization of Petroleum Exporting Countries (OPEC). Regional bodies such as the Economic Commission for Africa (ECA), the Organization of African Unity (OAU) and the Sudano-Sahelian Countries were also important actors in galvanizing regional perspectives prior to UNCED.

As part of Africa's regional preparation for UNCED, countries of the region, in close consultation with NGOs, adopted the African Common Position on Environment and Development (Common Position) at the Second African Regional Ministerial Conference for the UNCED in Abidjan, Côte d'Ivoire, November 1991. The various intergovernmental regional conferences were hosted by the ECA and OAU. Drawing on previous regional blueprints for economic and political development, the Common Position represented Africa's policy response to the challenges of a deteriorating global environment.[4] Africa's current economic predicament and the emerging consensus on an alternative development path for the region provided a foundation upon which to build Africa's expectations on the FCCC as well as other issues on the UNCED agenda.[5]

According to the Common Position, poverty is one of the major factors that has perpetuated the underdevelopment of resources, low levels of technological development – with the consequent low level of production in all sectors – as well as the exacerbation of environmental degradation. The difficult economic situation of Africa, particularly mass poverty and crushing foreign debt burden, collapse of commodity trade, inadequate transfer of appropriate and environmentally sound technology, the reverse flow of financial resources and backward scientific and technological capabilities, has led to severe constraints on the continent's development capacity. The Common Position concludes that this

---

4 Prior to the Abidjan Conference, the first Regional Conference on Environment and Sustainable Development in Africa was held in Kampala, June 1989. There were also the OAU Pan African Conference on Environment and Sustainable Development held in Bamako, January 1991 and the First African Regional Preparatory Conference for UNCED in Cairo, 11–16 July 1991. The declaration of the OAU Heads of State at the Abuja summit, July 1991, on the need to forge an African Common position was instrumental to negotiating an African Agenda in November 1991.

5 A vigorous debate has been raging between the World Bank and the ECA on the appropriate strategy for African recovery and development. Important in this debate has been an emerging consensus among African countries for strategies towards a self-reliant and self-sustaining development path. These common views have been expressed in the Monrovia Declaration (1979), the Lagos Plan of Action (1980), the African Priority Programme for Economic Recovery (1985), the United Nations Programme of Action for African Economic Recovery (1986), the African Alternative Framework for Structural Adjustment Programmes for Socio-Economic Recovery and Transformation (1990), and the African Charter for Popular Participation in Development (1990).

circumstance of very limited resources has been a major hindrance to Africa's capacity to participate effectively in global development and environmental efforts (ECA 1991a: 24).

According to the Common Position, the industrialized countries are principally responsible for the human activities that result in global warming. Hence, primary responsibility for combating the problem should be borne by them. African countries particularly emphasize the linkage between the global warming problem and the plague of drought and desert encroachment in the region – a relationship attended by considerable uncertainty. The Common Position particularly deplored the rate at which deforestation of tropical forests – because of energy needs, drought, desertification – is reducing the natural capacity to absorb CO_2 in Africa.

Central to the regional opinion on climate change was the issue of drought. According to the Common Position, drought "is a global problem requiring the effective contribution of the international community in afforestation, combating desertification, and general eco-system rehabilitation activities" (ECA 1991a: 24). At the fourth INC session, the African Group proposed a Green Plan designed to promote massive reforestation and integrated management of forest cover with a view to expanding Africa's absorptive capacity of CO_2 and other greenhouse gases (UN 1991). At UNCED, the African group pressed for and secured international acceptance for the formulation and signing of a United Nations Convention on Halting Desertification.

4.2. National Positions and Activities

While few African countries have adopted specific national policies on climate change, it might be possible to review their positions in relation to the emerging negotiation blocs among industrializing countries in the climate change negotiations. Three broad categories of negotiating blocs can be distinguished. The most pro-active negotiating blocs – urging for more binding international commitments to reducing greenhouse emissions – are the AOSIS group of 36 countries,[6] apprehensive that sea-level rises might wipe out their countries. The AOSIS can be termed as "pro-actives" for increased international action on global warming. Second, is the G-77 representing most of the developing world – arguing that their emissions would necessarily rise due to their efforts to meet the development needs of their people. This group of developing countries holds that industrialized countries are primarily responsible for present and historical

---

6 AOSIS is composed of 36 developing and industrialized countries.

emissions of greenhouse gases, and therefore have the primary responsibility of mitigating climate change. The G-77 has forcefully negotiated for increased financial and technological transfers to the developing world to enable it to offset incremental costs of adaptation and mitigation. The G-77 might be termed the "mediators" – not for their mediating positions between the extremes of Third World concerns, but for their role in negotiating for an equitable distribution of burdens among Parties in the regime. The final group of Third World negotiating blocs comprises oil exporters; most notably OPEC countries – fearful that reducing the role of petroleum in economic activities will jeopardise their economies, they have pushed for a slower pace in policy implementation, arguing that the present scientific uncertainty warrants caution in the implementation of drastic measures.[7] We may term this group "draggers" in international efforts to combat climate change (Nazer 1993).

Can Africa's positions on climate change be grouped into categories of proactives, mediators and draggers? At the regional level, states in Africa participate in various fora representing developing nations. Apart from peculiar issues touching upon Africa's development, such as drought and common regional perspectives on economic development, the ECA/OAU system, as reflected in the UNCED preparation process, has in general been supportive of the positions of the G-77, without prejudice to the proactives and draggers among its ranks.

Beyond regional bodies, the geographical location of countries and their energy endowment determined their membership in international organizations – helping to define their positions in the regime formation process. The island states of Mauritius, Madagascar, Comoros, Equatorial Guinea, Sao Tome and Principe have joined ranks with AOSIS in supporting accelerated international action on climate change. African OPEC members such as Algeria, Nigeria, Gabon and Libya have been part of an international lobby spearheaded by Saudi Arabia and Kuwait to slow down policy implementation on curbing emissions of greenhouse gases. On the other hand, several countries in Africa have been traditional supporters of the aims of the G-77. Algeria and Ghana, for example, have contributed substantial leadership in upholding Third World solidarity within the G-77, and its efforts directed towards an equitable redistribution of global economic welfare. The positions of the G-77 are reflected in African regional positions as well as in the national reports to UNCED from various countries.

---

7 For an elaboration on the international negotiating blocs on climate change see Ian Rowlands (1995). Some might choose to further differentiate the G-77, to distinguish the impacts of larger states such as Brazil, China and India (see e.g. Bergesen et al. 1995).

Sub-national actors, notably Africa's major energy corporations and various NGOs, have invested considerable interest in the issue of climate change. The two largest indigenous energy corporations in Africa – the Nigerian National Petroleum Corporation and South Africa's electricity utility, Eskom – have followed climate negotiation processes closely (Lennon 1993). While this has not resulted in the crystallization of clear policy preferences, these corporations have conceived participation in UNCED as a learning process on issues that will form their business environment. Africa's NGO community has also been effective in maintaining state attention to issues on global development and environmental protection. Several conferences, studies and training programmes have been initiated. Many official studies have been carried out by local NGOs in association with industrialized-country counterparts. The Climate and Africa Project under the auspices of the African Centre for Technology Studies and the Stockholm Environment Institute is a good illustration of this.

Beyond negotiating positions, several activities aimed at meeting the obligations of countries to the Convention are underway in the region. By November 1996, all members of the OAU (including Morocco) except Equatorial Guinea and Somalia had signed the FCCC. Thirty-two of these countries have either ratified or acceded to its terms. This clearly signals the intention of a majority of African states to participate in mitigation activities and to be eligible for resource transfers from this endeavour.

Several countries, especially in Southern Africa, have organized national post-UNCED conferences to discuss national strategies. Other follow-up activities might be grouped into two categories. The first relates to national climate studies in preparation for meeting their reporting obligations according to the Convention; the second relates to pilot projects initiated to mitigate emissions of greenhouse gases and the enhancement of sources of sinks.

In collaboration with the UN Collaborating Centre on Energy and Environment in Risør, studies on the Methodological Framework for National GHG Abatement Costing have been conducted on Senegal and Zimbabwe. Country inventories on sources of emissions and sinks have been completed for Burkina Faso, Cameroon, Gambia, Nigeria, Senegal, Seychelles, Tanzania, Uganda, Zambia, Zimbabwe, among others. Impact studies have been completed for Burkina Faso, Kenya, Mauritius and Nigeria. The United States Country Studies Programme is financing on-going studies on adaptation to climate change in several African countries. Some training programmes have also been initiated. Several climate change-related studies and training programmes are being planned with financing from the industrialized countries.

Several countries have sought GEF funding for pilot projects in greenhouse gas mitigation projects and carbon reservoir expansion projects, among them the Nigerian-associated natural gas utilization project and the Zimbabwean photovoltaic project. There are also attempts to attract GEF funding for the Tanzanian Songo Songo gas development project. In all these three instances, efforts are being made to involve the private energy sector. Several such projects are being planned in other countries as well.

5. Factors Shaping Present and Future Policy Positions

Several explanations have been offered for why political actors in international society choose certain positions on cooperation in solving common problems. For our purpose, three broad categories can be discerned: "national interest", "pluralist" and "knowledge" based explanations.

Traditional explanations focusing on *national interests*[8] are usually quite pessimistic about the prospects for international cooperation. According to these, the international system lacks government and is made up primarily of self-centred states – ever in pursuit of supreme national interests – speaking only the language of power. Positions on climate change will therefore necessarily reflect that country's degree of vulnerability, and the marginal costs of abatement and adaptation (Bergesen et al. 1995: 13), as well as the relative costs that others are willing to bear (Underdal 1992). Seen in this perspective, African countries may be putting their national priorities first – in balance with their relative power in the international society.

Pluralists generally offer more optimistic explanations of the prospects and desirability of international cooperation. They remind analysts that there are more actors than states, and more interests than national ones – and that these shape interactions across national boundaries. Although political leaders may be ambivalent to climate-change issues, their local and transnational NGOs, transnational corporations, media, individuals – and, in fact, state-owned energy utilities – may seek to influence the political process both nationally and internationally (Keohane and Nye 1977).

Knowledge-based explanations suggest that international cooperation will be possible only when there is information which "commands sufficient consensus at a given time among interested actors to serve as guide to

---

8 It is difficult to lump so many strands of a rich tradition into one basket; however, the realist perspective has been one of the most influential.

public policy designed to achieve some social goal". Adherents to this perspective hold that the introduction of new ideas, knowledge and beliefs by communities of experts and scientists can facilitate cooperation among states (Haas 1990: 52). Here, knowledge is power, and whoever controls knowledge has an edge in climate-change regime negotiations. For instance, concerns have been raised about Africa's limited participation in the work of the IPCC, INC sessions, and in research relating to climate change, all of which is deemed important in securing various regional objectives.[9] According to this perspective, African positions – or lack of them – can be explained by how much access to relevant information and scientific bodies were available to key decision-makers in the region.

5.1. Power and National Interests

What are the interests of African countries on the issue of climate change and in international climate politics? There is apparently a genuine concern among government officials across Africa that climate change, if it proves real, will further aggravate the deep crisis already facing the continent. However, at best this concern coexists but is essentially subordinate to the developmental and other environmental problems facing these countries. The foremost concerns of states in Africa today are economic recovery, national political sovereignty and stability, poverty alleviation and ecological restitution.

In a sense, Africa has become the Third World of the Third World. While countries like Nigeria and Ghana in the mid-1960s had *per capita* incomes in excess of South Korea and Indonesia, respectively, the quality of life has with the rest of the region dropped below 1960 levels (WB 1989). While the number of poor people in the world is expected to decline between 1985 to the year 2000, Africa remains the only region with a substantial growth in poverty (WB 1990). Of the 41 Least Developed Countries of the world, 28 are African. This represents an increase of seven African countries in this category since the 1981 United Nations Conference on the Least Developed Countries (UN-DPI, 1990). In the 1980s, sub-Saharan Africa's *per capita* GDP was on average falling at the rate of 4.2% a year.

A large proportion of the continent is stricken by severe environmental stress, especially drought, soil erosion and desertification. The 1992 drought in the southern region left 40 million people with severe food shortages. The precarious food balance in the Sahel and particularly in the Horn is already well known. In a 1991 study commissioned by the

---

9 These concerns have been raised by the WMO Secretary-General, G.O. Obasi (1991).

22-member Sudano-Sahelian Countries, climate change ranked as only 17th of the 20 most important priority concerns for their countries (CILSS/UNSO/IGADD 1991).

Thus it is safe to say that states in Africa have several pressing development and environmental problems that are perceived as far more threatening than climate change. Rather than worrying about climate change, leaders fear that the climate-change regime itself will impose both economic and political costs as another conditionality in addition to human rights, gender, democratization and good governance. On this background, arguments can be made that Africa's interest in climate-change politics is that of shielding the state and society from costly and externally-imposed policies. In this perspective, the call for Africa to contribute to global environmental efforts can be viewed as a distraction from more urgent regional challenges.

Africa's climate positions might depend on the balance of power between the state and external actors. State sovereignty is challenged in Africa; from below by ethnic tensions; from above by political concessions in return for economic aid from external actors. Internally, twelve African states are currently at war, two are in an early post-war phase and fourteen are experiencing or have a recent record of significantly high levels of political violence. More than half the countries in the region – 28 in all – have recently been afflicted by serious violent conflict (Sandberg and Smith 1994). In countries like Liberia, Somalia and Zaire, the identity of official representatives of governments is not always clear. On the external front, high indebtedness and dependence on aid has led to the concession of major policy-making powers to external actors (Plank 1993). In over 40 countries where structural adjustment programmes have been implemented, few significant economic policy initiatives are decided upon without clearance from bilateral and multilateral donors. Not only is economic decision-making increasingly dominated by external actors, policy implementation is also increasingly dominated by technical assistance and NGOs funded by external actors.

With the receding political sovereignty in most of sub-Saharan Africa, there is fear that the power and the interest of external actors might define future positions and activities on climate change. In most cases, it is bilateral and multilateral development agencies that have financed the preparatory activities preceeding UNCED, research on national inventories of greenhouse gases, adaptation and mitigation strategies.

National interests and the balance of power with external actors vary greatly in Africa, and might in the future explain differences in policies among countries in the region. Although this study lacks adequate

statistics on debt, aid and economic dependence on various categories of countries, it can be expected that both South Africa and Nigeria – states with relatively higher levels of economic independence – might have better chances of promoting their own national interests than the more economically dependent countries of the continent.

5.2. Beyond the Nation-state

The primacy of national interests in determining positions in climate change negotiations might not adequately explain the several activities relating to climate change that are already taking place in Africa. Despite the divergence of interests, important initiatives are being taken by regional bodies, NGOs, sub-state actors and the media.

In the early phase of the regime formulation stage, especially under the preparation for the UNCED, the ECA and the OAU were the foremost catalysts in attempts to formulate a unified regional premise for global environmental conventions. Several factors underscored the key position of the ECA as a spearhead of Africa's international economic and environmental diplomacy in the early 1990s. First, the Commission has since the mid-1970s been responsible for driving the process towards the formulation of a regional economic blueprint for development and cooperation. Secondly, its long-drawn altercation with the Bretton Woods institutions, particularly the World Bank, over the usefulness of structural adjustment programmes positioned the ECA as the most important regional agent in international economic relations. This meant that within the ECA a mechanism existed for collating regional responses on environment and development.

Parallel with the process of economic and political transformation in the 1980s has come a growth in the influence of the NGO movement. Through generous funding from the North, African environmental NGOs have engaged in elaborate regional workshops and conferences. The first major milestone in incorporating the perspectives of NGOs was laid in Tanzania with the First African Regional Conference on Environment and Sustainable Development in 1989 (OAU 1989). This process culminated in an NGO summit, ECO-92, held in Cairo, June 1991 in conjunction with the African Regional UNCED Preparatory Conference.

African NGOs have generally focused on the development component of the UNCED process, and have urged for active African participation in negotiations. They have provided sources of information, analysis, and sometimes training for policy-makers. Interaction between policy-makers and NGOs varies within and between countries. While some maintain relative distance from the country's mainstream political institutions and processes, others maintain close interaction with their national institutions. With few exceptions, the business community in Africa has shown little interest in the climate regime. Eskom of South Africa actively participated in the Earth Summit and became a direct signatory of the International Chamber of Commerce's Business Charter on Sustainable Development (Eskom 1994). The utility also participated in the first CoP in Berlin, held in March 1995. Within South Africa, its representatives are key figures in the national committee studying the climate change regime, its implications and possible response strategies. In Nigeria, the Nigerian National Petroleum Corporation headed the national committee in the preparation for climate negotiations.

5.3. Science and Policy

Local and international scientific networks have been important players in providing policy-relevant information and analysis on climate change. Recent years have witnessed a remarkable growth in the number of scientific communities working on the documentation of greenhouse gas inventories, mitigation strategies, adaptation studies and the analysis of both regional and national response options. Their work continue to serve as a reference and base for national and sub-national decision-making processes; and many of the scientists serve as advisers to their governments and to external agencies.

Researchers at the University of East Anglia in England have provided substantial inputs to natural science studies of climate-change impacts (Kelly and Hulme 1992). Collaborative research programmes such as the African Centre for Technology studies and Sweden's Stockholm Environment Institute have focused on policy issues. The United States Country Studies Program addresses mitigation and adaptation to climate change, while the European Union has provided funds for inventory analysis carried out in several African countries. Transnational institutions have engaged in research collaboration with African institutions with the aim of providing relevant knowledge and information for decision-making and implementation of the climate regime. A notable example is the information and analysis provided by the Climate Network Africa – an affiliate of other regional climate-change NGO networks.

6. Future Prospects

Much remains uncertain as to how climate change will affect Africa, and as to the probable responses by countries, sub-national actors and external actors to the issue. The relatively recent emergence of the climate change issue and the uncertainties facing key actors in Africa, have limited policy-making. Moreover, since climate change has not yet figured as a priority issue for these actors, few nuances in differences among countries have emerged.

Since the international politics on climate change is in an evolutionary process, national positions are likely to develop and change according to the issues on the agenda, and how they affect vital national interests. This would tend to determine which actors are likely to participate; which interests will be mobilized and how they shape national responses. It may be useful to suggest three stages in this evolutionary process: first is the pre-regime formation stage; second, the pilot project phase; and third, a full implementation stage – where international commitments will be expected to be domesticated and internal policy adjustments made.

The pre-regime formulation stage has been marked by a very weak issue tangibility; it has not always been clear how much countries were expected to adjust domestic policies. This was even the more so as many of the responsibilities for stabilization and reduction of emission rested on the shoulders of the industrialized countries. The stake and obligations of industrializing countries were, and indeed continue to be, ill-defined. As the threat to national interests was not apparent, state agencies were not adequately mobilized, which made room for actors such as regional organizations (ECA and OAU) and non-governmental organizations to fill the vacuum.

Since the first CoP in Berlin, March 1995, we have seen more and more tangible international decisions. Of importance to African countries is the initiation of the pilot phase of *joint implementation* and the adoption of the GEF as the key financial transfer mechanism. Depending on the volume of financial and technological transfer that industrialized countries are ready to make available, states in Africa are likely to step up activities in wooing investments in energy and land-use sectors. Backed by the growing wealth of information and knowledge about climate change, this phase is expected to increasingly involve state institutions and the private sector.

Depending on developments in the science of climate change, the frequency of serious natural disasters and the preparedness of industrialized countries to step up mitigation policies, several developments might be set in motion. One such development would be a growing demand for developing countries to set emission-reduction targets, though this pressure may rather be directed towards the Newly Industrializing Countries and emerging market economies. Such a scenario may affect such countries as South Africa and Nigeria in particular. The other probable development would be the integration of climate concerns into development assistance, and thus the subsequent emergence of climate conditionality on debt and aid. That scenario would probably ring the wake-up call for most sub-Saharan countries, and mark the incorporation of climate politics into mainstream African international economic relations. In such a scenario, bilateral and multilateral development institutions and officials of key economic ministries would be the most likely key actors. Experience from other issue-areas suggests that national interests might succumb to the mobilization of economic power by donors and creditors. Despite national hesitation in such a scenario, chances are that mitigation programmes would be formulated and implemented partly through the mobilization of technical assistance.

Chapter 11

Russia and Climate Change

Friedemann Müller

1. Introduction[1]

Russia, a country between the extremes of domestic destruction and superpower claims, plays a major role in the framework of global climate policy. As the country with the largest fossil energy reserves in the world, today Russia is in third position in energy consumption, after the United States and China. Prior to the post-Communist recession it contributed more than a tenth to world emissions of carbon dioxide (CO_2). In addition, other greenhouse gases, like methane (CH_4), are emitted to a degree which gives reason for concern. Natural gas production and leaks in the pipeline system release at least 8 bill. m^3 of CH_4 – probably much more. Russia, however, also has a vast potential for sinks, like the forests of the taiga that absorb the major greenhouse gas, CO_2.

The Russian political agenda, however, is focused on other more immediate environmental problems. The threat of long-term climate change is at present perceived to be an exotic or at least mainly academic problem. Nuclear devastation, air and water pollution are deemed much more serious than climate change. During the late 1980s, environmental problems played an important role in the emancipation process among the Soviet and Russian population supported by Gorbachev's glasnost. Indications of this new awareness are the dropping of the plans to divert Siberian rivers to revitalize Lake Aral in 1989, and the closure of the Semipalatinsk nuclear test site in 1991, both under public pressure. However, the economic decline after 1990 changed the ranking of priorities among the general public, as the problems of everyday life

---

1 While several Russian contributions have been made to the science of climate change, there is a remarkable lack of contributions dealing with Russian climate-change policy-making. Articles written in Russia typically become vague as soon as it comes to politics. Hence, the present chapter relies heavily on the author's personal interviews with Russian experts and decision-makers.

became more urgent. Since 1995, environmental issues have again gained in importance in public priority. However, among the many NGOs dealing with environmental issues, there is hardly any focus on climate-change problems. The driving force in Russian policy for dealing with climate change derives not from an environmental basis, but from science on the one hand and international political developments on the other. Neither of these can exert much force, considering all the problems Russia has to deal with during its transition period.

The following analysis seeks to explain the official Russian position and interests towards an international climate policy and what can be expected during the coming years. The chapter describes the current political and scientific constellation influencing the formulation of a political position in Russia, followed by a presentation of economic structures influencing Russia's climate change policy-making. The fourth section deals with Russia's position towards the international political process of the Framework Convention on Climate Change (FCCC). Finally, some probable future Russian positions in this field are inferred from the previous discussion.

To summarize the official Russian position: it is defensive when it could be offensive, and it tries to keep all options open instead of accepting any clear commitment. At the Berlin Conference, March 1995, Russia made great efforts to improve the position of the countries in transition by reducing their obligations and increasing their possibilities for financial and technical assistance within joint implementation programmes. On the other hand, Russia – as a country with one of the highest reduction rates in CO_2 emissions – did not go beyond the obligation of reducing its emissions to the 1990 level by the year 2000. Mainstream thinking in official Russian policy sees any reduction of energy consumption as being proportional to economic decline. The idea of decoupling energy consumption from economic growth, as happened in most Western countries after the oil shocks of the 1970s, has not yet gained acceptance among influential forces in Russian environmental policy. This will require the completion of the transition process towards a modern market economy – which only a minority clearly aims at in the current situation. For that reason, Russia is missing a big chance to be a driving force instead of a retarding one in a field where a new international reputation could be built up.

2. Climate Change and the Political Structure: Mapping the Actors

Although Russia is obviously a country in transition it has a political infrastructure not so different from Western countries in the field of ecology. Ever since Gorbachev's Soviet times, the Ministry of Ecology has been led by Danilov-Danilyan, a figure with an international reputation. Another internationally well-known figure is Professor Alexey Yablokov, the former adviser to President Yeltsin and now the Director of the Center for Russian Environmental Policy. The Duma, of course, has its own committee on the environment. These political bodies are supplied with information from respected scientific institutions. In the case of climate change, this is the Russian Federal Service for Hydrometeorology and Environmental Monitoring (RFSHE), which represents Russia in the Intergovernmental Negotiating Committee (INC) provided for in the FCCC. The RFSHE works closely together with the Institute of Global Climate and Ecology (IGCE) of the Russian Academy of Sciences. Its director is Yuri Izrael, an academic and climatologist with high international reputation and former vice chairman of the Intergovernmental Panel on Climate Change (IPCC). There seems to be rivalry between these two institutions on the one hand and the Ministry of Ecology on the other – at least, the INC responsibility was transferred to the RFSHE against the vote of the ministry. To balance differences in the competition for influence on the decision-making process an Interagency Commission on Climate Change (ICCC) was established in 1994. Members of the commission are representatives of the Russian Academy of Sciences, among which are both the IGCE and the RFSHE; also, members are representatives of the energy industry, the transportation sector, the agricultural and forest sector, the Foreign Ministry and the Ministry of Sciences.

On the parliamentary side it is still difficult to establish an influential position for the Commission of Ecology, mainly because of the unclear position of the Parliament in the power struggle between the President and the Government. In this context two things seem to indicate that the Parliament has a weak position:

One is the presidential decree on "The National Strategy of the Russian Federation for the Protection of the Environment and the Securing of a Sustainable Development" (Presidential Decree 1994), obviously a substitute for a law which might have been passed by the Parliament. It deals with the whole complex of environmental problems in Russia, including the most urgent hot-spots. In its Chapter 4 – on "participation in solutions of global ecological problems" – the "prevention of anthropogenic climate change" is only briefly listed among eight other points.

The other is the fact that the Parliament did not ratify the FCCC negotiated at the United Nations Conference on Environment and Development (UNCED) in 1992, although governmental institutions and the ICCC urged the Parliament to do so during the first half of 1994. The official reason was that there is no tradition of ratifying international agreements on the environment. However, this reference to tradition means Soviet tradition – which in many other fields the Duma has been eager to overcome. A more convincing interpretation might be that the Parliament is willing to accept an international committment only if a direct benefit for Russia is visible and can be explained to the public.

A major problem in this structure is that most institutions work as if there is no transition process underway. As in the past, the academic institutions try to be as apolitical as possible, so as not to sacrifice their standing as scientific bodies. Governmental institutions work mainly in terms of projects and balancing lobbyism. The real challenge of a profoundly changed situation, particularly in the field of energy consumption, seems to be inadequately dealt with or represented within this institutional framework. What is meant by this challenge is explained in the following section.

3. The Context of Economic Transition

According to the FCCC definition, Russia belongs to the "countries in transition to a market economy". Up to now, this transition has brought negative growth rates of the Gross Domestic Product (GDP) for six years in a row, and has led to an accumulated halving in the size of the Russian economy. The negative growth rate might be – after the annual two-digit rates began to decline – only 3.5% in 1995. The overall decline of the Russian economy is much deeper and broader than in other transition countries, like Poland or the Czech Republic. There are, nevertheless, similarities in these processes which follow the line of a deep recession: the more centralized and monopolized the economy, the deeper and broader is the decline curve.

Table 1 presents the virtual breakdown in industry as well as in energy consumption – the latter following the GDP decline with a time-lag of one to two years. The decline in energy consumption, however, was less than the GDP negative growth. The result is an increase in energy intensity, that is to say a decrease in energy efficiency, which was already extremely low compared with Western standards.

International comparison of GDP between countries with convertible currencies and those without, like Russia, are admittedly fairly inaccurate. GDP calculations on countries like Russia might differ by a factor of two or three, depending on which method – purchasing power or currency exchange rate comparisons – is used. If, however, the Russian Ministry of Ecology assumes that the energy intensity in Russia is twice

Table 1 Russian Federation: Changes in Comparison to Previous Year (in %)

	1990	1991	1992	1993	1994
GDP	−11.0	−13.0	−18.5	−14.0	−15.0
Industrial production	−0.1	−8.0	−18.8	−15.0	−20.9
Energy consumption	0.9	−5.2	−7.4	−8.6	—
Energy intensity	13.4	9.0	16.6	6.3	—

Source: *Plan Econ Report*, 19 December, 1993, p. 6 and 7 April, 1995, p. 13; *DIW Wochenbericht*, 21 October, 1993, p. 582; Data on energy consumption from Committee for Productive Forces and Natural Resources (KPES), Moscow.

or three times as high as in West European countries this certainly is a cautious calculation. Western estimates indicate from four to ten times the West European level.

It is particularly important for the global warming problem that there is a huge potential for improving energy efficiency in Russia. This potential can be divided into two stages: one is the revision of the increase in energy intensity since 1989 of about 50% and, second, the adjustment to a common Western level of energy efficiency. This implies that an absolute decline in energy consumption is possible even under the assumption of continued economic growth.

Table 2 shows that production of fossil energy was reduced by about a quarter during the first 5 years of the 1990s. The decline in oil and coal production was dramatic, while natural gas production in 1994 was only slightly less than at the end of the 1980s. As a consequence, the share of natural gas in available fossil energy grew from 39% to 52%. The accumulated decline in energy production between 1989 and 1994 was 26%, while the decline in industrial production was 50%.

There is an almost equal decline in energy consumption as in energy production: 20% accumulated decline between 1989 and 1993.[2] Like energy production, consumption also shows a shift from oil and coal towards natural gas. In 1989 the share of natural gas in energy consumption was 47%, in 1993 it went up to 52%, while the share of oil decreased during the same time from 30 to 27% and that of coal from 23 to 21%. Because of the lower carbon content per energy unit of natural gas in comparison to oil and coal, the decline in CO_2 emissions was

---

2 The Russian Statistical Office (Goskomstat) does not provide figures for energy consumption. The available figures have been compiled by the Committee for Productive Forces and Natural Resources (KEPS), Moscow, in cooperation with British scientists.

Table 2 Russian Federation: Production of Fossil Energy (in Mtoe).

	1989	1990	1991	1992	1993	1994
Oil	551	516	461	397	354	316
Natural gas	496	518	518	516	496	487
Coal	212	205	183	174	144	129
Total	1285	1236	1161	1087	994	932
Changes to previous year (%)		−1.8	−6.1	−6.4	−8.5	−6.2

Source: KEPS, Moscow: Deutsches Institut für Wirtschaftsforschung, *DIW Wochenbericht*, 25 May, 1995, p. 377.

higher during this period than the decline in energy consumption, which is more than 20%. If we take the year 1990 as the basis year in climate policy, the decrease in fossil energy consumption until 1993 was 22%. That means CO_2 reduction was approximately 25%.

Russia's energy consumption decreased by 25% between 1990 and 1995 (Schipper and Martinet 1993: 974). This was mainly due to the decrease in the industrial sphere, and in smaller part to the transportation sector. The potential for savings in households and municipal areas is unused up to now, because of the still extremely low energy price compared with world market prices. Under such circumstances, there is no incentive to engage in energy conservation measures. The transformation to a more efficient market system is not yet reflected in more efficient energy consumption. Energy pricing as well as the tax system provide scant incentives to energy-saving measures.

While the reduction in energy consumption until the mid-1990s is obvious, there are divergent estimates about future development in the case of economic growth, predicted from 1997 on. It is striking that Russian predictions assume that regaining economic growth means growing energy demand, whereas Western estimates are based more on assumptions that energy conservation will bring about slower growth in energy consumption than in GDP. In his long-term projection, Yuri Sinyak assumes a rise in CO_2 emission until the year 2050 of between 50% and 250%.

In comparison, the International Energy Agency (IEA) projects two scenarios. The first ("Capacity Constraints") assumes an increase in energy consumption between 1992 and 2010 from 1228 Mtoe to 1323 for the Former Soviet Union (FSU) as a whole. This corresponds to a yearly average growth in energy consumption of 0.4%. The figures for the alternative "Energy Saving" scenario show a decline to 1157 mtoe (yearly

average growth rate is –0.3%). However, in both cases the low or negative growth rate is related to the strong decline in the 1990s and a positive yearly average growth rate is expected for the first decade of the 21st
century. The "Capacity Constraint" scenario expects a 2.2% growth rate,
the "Energy Saving" scenario a 1.4% growth during these 10 years (IEA
1995: 322, 331).

The World Energy Council (WEC) estimates a similar growth rate in
energy consumption until the year 2020, in this case for the whole
region of the former Soviet Union and East Central Europe. It projects an
average economic growth rate of 2.4% between 1990 and 2020, with
the average annual growth in energy intensity between –1.2 and –2.7%
(WEC 1993: 48). Consequently, energy consumption would rise between
–0.3 and 1.2% annually.

Russian mainstream thinking is obviously still dominated by the former
Soviet terms of extensive growth: growth can be achieved only by growing input, not by relying on productivity improvements (intensive
growth). More input, at the same time, means more waste and burden to
the environment. The transformation towards thinking in terms of intensive growth is an urgent task for the coming years in Russia, and the outcome of this will heavily influence the development of Russian CO_2
emissions. As yet, this thinking has not influenced the shaping of a
Russian climate policy position. A new generation of policy-makers will
probably be required to transform those thoughts into a real driving force.

4. Russian Interests and Options in International Climate Policy-Making

The climate policy of the Russian Federation is greatly influenced by four
framework conditions. Firstly, the goal of industrial countries to restrict
CO_2 emissions to the level of 1990 can easily be reached in the case of
Russia, even if there were a considerable economic growth in the second
half of the 1990s and even if energy consumption was to grow in proportion. In addition, there is a huge potential for energy saving, which
gives Russia the chance to act in a progressive way also beyond the year
2000.

Secondly, the policy instruments recommended for the reduction of
greenhouse gas emissions – competitive energy prices, greenhouse gas
taxes, reduction of subsidies, determination of standards, obligations to
the construction and transportation industry to reduce the energy intensity, etc. – can only be realized in a long-term frame, because of the
insufficient level of administrative organization prevailing in Russia
during the transition process.

Thirdly, some government institutions and those experts dealing with the climate problem have now realized the necessity of an internationally coordinated climate policy. The population at large and the Parliament, however, do not as yet appreciate why such a policy should be in the Russian national interest. Nor can they see that Russia is in the midst of a modernizing process that gives the chance to combine economic growth with energy consumption reduction, for the sake not only of the environment but also for the efficiency of the economy and the wealth of the country; nor do they feel themselves responsible for a problem of global dimensions. They are simply not used to thinking in terms of common international responsibilities.

Finally, the government is principally interested in participating in shaping the international climate policy to counter international isolation. Russia's problem, however, is that it does not know to which group it belongs. There is an affinity to the G-77 position with its claim to burden the Western industrialized countries with responsibility for the problem and its solution. There is also an inclination to support the OPEC countries, because it is not in the interests of energy-exporting Russia that a demand reduction should put downward pressure on the world market price. Russia is somehow also locked into a common position with Western industrialized countries, since even after the CO_2 emission decline, its *per capita* emission is much more like that of Western countries than of average G-77 countries. The special treatment Russia receives in the FCCC by belonging to the countries in transition does not mean in the long term that it can wait and see whether its high *per capita* emission can be tolerated by the participating states.

Regarding the Russian climate change policy-making process before and after the First Conference of the Parties in Berlin, March 1995, the following points should be stressed:[3]

1. It was not possible to persuade the Duma to ratify the climate convention. However, the government undertook measures to minimize the damage and prepared an equivalent, so as to become a full member of the conference of signatory states.
2. Russia is not ready for a concrete commitment to greenhouse gas reduction. It is quite clear that the goal for the year 2000 – not to exceed the 1990 emission quantity – will be met. However, the

---

3 The following points rely to a large extent on a Ministry of Ecology paper by E. F. Utkin (1994), as well as on my own interviews with the Russian head of delegation to the INC, members of the delegation and the director of the Institute of Global Climate and Ecology, Moscow, Professor Izrael.

widespread assumption of an emission increase in combination with economic growth prevents any firm longer-term self-restriction. It is a widespread assumption in Russia that damage limitation of climate change will be less costly than climate change prevention.

3. There is an interest in a protocol to include all greenhouse gases not regulated in the Montreal Protocol. This is due to the fact that Russia has a huge potential to reduce CH_4 emissions in the context of natural gas production and transportation. Gazprom, the still monopolist in natural gas production and transport, asserts that not more than 1.3% of overall natural gas production is lost unburned to the atmosphere – that is, 8 bill. m^3 of natural gas. Other estimates are considerably higher. No one can verify these losses, however, without being granted access to all relevant stations by Gazprom. At present, it is unclear what emissions would count in the base year.

4. Russia underlines its participation in the group of countries in transition, and in this context mentions the difficulties of choosing optimal instruments to implement a climate policy considering the state of political, economic and social restructuring in the country. Russia has, however, also pointed out that the division of labour and its coordination in the Interagency Commision offers, in combination with the presidential and governmental decrees, workable conditions for an active climate policy.

5. Russia is involved in international energy policy as a major exporter of oil and natural gas. It tries to implement a balanced policy towards its competitors, mainly OPEC countries, by not undermining their policy too obviously. It also cooperates in selected fields like the Caspean Pipeline Consortium, in which Russia, Kazakhstan and Oman are members. The very fact that OPEC has major reservations towards a strict international climate policy does not mean for Russia that it is not in favour of the FCCC; it will, however, make clear its interests during the negotiations of the protocols in a way that can give priority to its economic goals.

6. In the framework of Russian environmental policy, climate change does not play an eminent role. However, even the experts at the RFSHE have pointed out that Russia does not feel it will be a potential winner of climate change due to its geographic location. There are fears that climate change would create a major dry zone that could do major harm to the vast forests. A retreat of the permafrost area would have catastrophic effects on the statics of a major part of construction in that region, like railways and electricity grids. Priority over a national climate policy has been given to: (i) the improvement of air, drinking water and agricultural land quality in general, as

well as the clean-up of major nuclear devastation areas; and (ii) the clean-up or major improvements in extremely polluted areas (major cities, Volga River, Lake Ladoga, etc.), in particular. These areas are much more stressed in ecological education than are the threats of climate change.

7. Active participation in international climate policy is mainly based on economic interests, as reflected in the following areas: (i) the utilization of the huge potential of energy conservation is necessary to prevent a situation in which Russia becomes a net importer of energy, not least because the revenues from energy exports are badly needed. (ii) Furthermore, Russia hopes that an efficient international climate policy will bring major investments and technology transfer. Russia is very much in favour of the joint implementation philosophy, assuming that Western companies will engage in projects for reducing Russian greenhouse gas emissions. (iii) Russia considers it necessary to be represented in all relevant bodies, like the INC and the secretariat in Bonn. It insists that all documents are to be translated into all six official United Nations languages, including Russian.

8. Russia sees itself as a country with an extraordinary potential for CO_2 sinks, particularly forests. The ability to absorb CO_2 is said to be similar to that of Brazil. According to one Russian study, the absorption of CO_2 by sinks exceeds the total natural and anthropogenous CO_2 emission on its territory. Russia, for this reason, is interested in establishing an overall inventory for greenhouse gas emissions and absorptions. It hopes that the information gained through such an inventory would support joint implementation projects on Russian territory.

9. The Russian government expresses its preference for state regulations and economic instruments rather than voluntary agreements and public consciousness-building measures, even if reality shows that regulations are not adequately enforced and economic instruments are rarely implemented. The government assumption is that there is no tradition of the latter type of instruments. This, of course, is correct. The transition process, however, would require that the basis for new and better traditions be built up.

10. As long as the climate change problem is disconnected from other political problems there will not be a real climate policy or lobby. The interests of the population and of green movements is focused on shorter-term problems that are more visible. This priority is likely to remain valid for a long period, and Russia's climate change policy will be established within this mainstream thinking. The scientific community is convinced that more urgent problems are air, water, land pollution, nuclear contamination and erosion of agricultural

land. Even the most prominent Russian climatologist, Yuri Izrael, assumes that one dollar put into the reduction of air pollution in cities saves many more lives in Russia than the same money put into greenhouse gas emission reduction. His proposal to the IPCC is to shift capacities to provide reliable calculations on these questions.

To conclude, Russia concentrates its efforts on influencing the international process by promoting (a) the establishment of an inventory to measure emissions and sinks of greenhouse gases, (b) projects to support technology transfer, and (c) reforestation programmes. Any definite reduction goal is unlikely to find Russian support in the near future.

5. Conclusion

Since the dissolution of the Soviet Union, Russian foreign policy has reflected the incalculabilities of its transition process. As in many other fields of Russian international engagement, the question "what are the Russian interests" leads to the question "who represents the Russian interests" and "how consistent are interests articulated on one issue with other Russian interests". On other functionally related issues (environment vs. energy), it is obvious that Russia finds itself in a state of instability, which renders planning and prediction extremely difficult. Hence, instead of guessing who and how strong the next Russian President will be, the question should rather be "what factors will determine the long-term climate policy of Russia?"

The strongest factor in this context is the existence of a huge energy-saving potential in Russia. The most important change will have to take place in the mainstream thinking among those who influence the decision-making processes in the framework of the overall transition process. If the *per capita* energy consumption in Russia in 1995 is almost equal to that of Germany and the prediction for Germany is a slight decline in energy consumption over the next 15 years (ESSO 1994), there is no reason to believe that an increase in Russian energy consumption is inevitable. Rather, it is a question of implementing adequate instruments for bringing Russia's energy consumption into a relation to GDP, which is at least not more than twice that of West European countries. This must be a no-regret measure which also lies in the economic interests of Russia, not least because savings in domestic energy consumption may mean higher revenues from the export of energy.

If this philosophy gains support in the political process, it can bring Russia into a different position in the international climate change discussions. Russia could take the offensive, in contrast to its defensive and almost unnoticed behaviour during the first Conference of the Parties. While the outside world's image of Russian foreign politics is influenced by civil wars within the Russian Federation, and Russian resistance to NATO extension, Russia has the potential for developing a more progressive climate change policy. This will all depend, however, on its own belief in the completion of its transition process towards a more efficient market economy.

Besides this challenge to participate in a future international coordination process, there is unlikely to be rapid change in the domestic list of priorites within the ecological sector. Air, water and soil pollution will remain urgent problems in the perception of the population and politicians for years to come. The burden of the past is so overwhelming that it will take many decades of extreme effort to reach a standard acceptable to the people of Russia. It is not difficult to predict that ecological movements will not be the force for promoting an active climate policy in Russia. If climate change is to receive higher priority in Russia's foreign policy, the issue will have to be promoted by those economists and politicians who see a chance to improve Russia's international reputation by making their country a driving force. If for no other reason, economic considerations make it inevitable that Russia will have to reduce its greenhouse gas emissions in the long term anyway.

Chapter 12

The Climate Change Policy of the European Union

Jay P. Wagner

1. Introduction

Climate change arguably represents one of the most serious challenges facing European energy and environmental policy, now and in years to come. Although the European Community (EC), now the European Union (EU),[1] took a prominent role on climate change when the problem was formally recognized as a threat to the global environment in the late 1980s, and adopted a decision in October 1990 to stabilize carbon dioxide (CO_2) emissions by the year 2000 at 1990 levels for the Community as a whole,[2] the EU's current strategy has run into serious problems. Despite repeated reaffirmations of the 1990 stabilization target and the commitments made under UN auspices – the EC signed the Framework Convention on Climate Change (FCCC) itself, alongside its member-states in June 1992 and ratified it in 1993 – it now seems doubtful whether the 1990 targets will be met. Indeed, according to the Commission's Second Evaluation of National Programmes on Carbon Dioxide and other Greenhouse Gas Emissions, published in March 1996, CO_2 emissions in the EU could increase by as much as 5% by the year 2000 from 1990 levels (IER 1996a: 252).

Should the EU fall short of achieving its emission-stabilization targets

---

1 Although, as a result of the Maastricht Treaty, the EC is now referred to as the EU, the Community remains the relevant legal entity. The EC ratified the Framework Convention on Climate Change – not the EU, which has no legal personality. For this reason, the EC is still often referred to. Use of the term EU is thus restricted to more general references.

2 Commission of the European Communities – Communication from the Commission to the Council: A Community Strategy to Limit Carbon Dioxide Emissions and to Improve Energy Efficiency, SEC(91) 1744 final, 14 October 1991.

and fulfilling the international commitments it has entered into, this would weaken the FCCC in the very area in which the EC itself fought hardest for stronger wording. It would also be a major embarrassment to the EU and would provide other industrialized as well as developing countries with a reason to avoid substantive action. Indeed, most analysts agree that if the EU fails to live up to its goal of stabilizing CO_2 emissions it will become very much more difficult to reach any effective, let alone equitable, solution globally. Conversely, success in implementing the stabilization goal could serve as a potential model internationally (Grubb et al. 1994: 6).

The EU, in other words, belongs to the rather exclusive but heterogeneous group of "critical actors", alongside the Federal Republic of Germany (FRG), Japan and the United States, capable of contributing substantially to either the solution or the worsening of the problem of climate change (Fermann 1994: 1–2). Considering the contribution of the EU to the problem of climate change – the EU, as the world's largest trading bloc, accounts for slightly less than 15% of the world's estimated CO_2 emissions – and its formidable technological and financial resources, the position of the EU on climate change is of crucial importance internationally. Indeed, most developing countries now regard action by industrialized countries such as that outlined in the EC's 1990 decision as a precondition for any consideration of substantive action or strengthening of the FCCC on their part (Grubb et al. 1994: 6).

With the entry into force of the FCCC on 21 March 1994 and the conclusion of the first Conference of Parties to the Convention (CoP) in Berlin in March 1995, attention has increasingly come to focus on the implementation of the Convention. Particular attention in this context is being given to the stance of key actors such as the EU. This concern is all the more timely as it emerges that the greenhouse-gas emission targets of major industrialized countries such as Germany and Japan are themselves in serious doubt (IEA 1994; EC 1994; Fermann 1994; Cavender and Jäger 1993).

Against this background, this chapter provides an overview of the evolution of the EU's climate-change policy, and assesses the prospects for EU climate-change policy in the future. Since efforts to curb greenhouse-gas emissions are multisectoral and costly, with major implications for economic policy and social habits, it is necessary to place climate change into the wider context of the EU's environmental policy.

Policy, however, is not all this chapter is concerned with. Equally important is the question, as seen in the title of this book, of the *politics* of climate change: the *process* by which goals on climate change are selected, ordered in terms of priority, and implemented. To this end, the chapter focuses on the political dimension of EU climate-change policy with particular attention to the determinants or driving forces underlying the EU's approach to climate change.

Policy, in its strictest definition, is a set of decisions taken by a political actor or group (e.g. the EU) concerning the selection of goals and the methods of attaining them relating to a specified situation (e.g. limiting greenhouse-gas emissions). A policy may be simple – consisting, for example, of a single decision – or it may be a complex set of contingency plans. It may be a decision to postpone decision ("wait and see"). It may be relatively concrete (e.g. a decision to increase energy taxes or raise a levy on the carbon content of fuel), or relatively abstract (e.g. a policy of voluntary industry agreements).

Strictly speaking, therefore, there is no such thing as *a* climate-change policy. Indeed, few issues interface with as many other issue-areas (energy, transport, land-use, etc.) as does the question of climate change. A policy to address climate change therefore is no more geared to one goal (e.g. limiting CO_2 emissions) than it is to another (e.g. increasing energy efficiency). Climate-change policy is thus merely one aspect, albeit a vital one, of the EU's wider energy and environmental policy.[3] For the purpose of this chapter, climate-change policy can be defined as the common policy decided on by the EU member-states, and the implementation by the individual countries, regardless of national environmental policy. To this end, climate-change policy is considered as an extension of the EU's environmental policy. The emphasis will be on energy-based CO_2 emissions, these accounting for the vast majority of the EU's combined emissions. Key questions addressed in this chapter include: (i) What instruments underlie EU policy and what is their legal status? (ii) What factors have shaped the EU's climate-change policy? What are the main obstacles to an effective EU climate-change policy? (iii) Is the EU a catalyst or a hindrance on the road to implementing the FCCC? (iv) What are the proposed or planned actions with regard to EU climate-change policy?

There are some things this chapter does not attempt to do. In the first place, it does not seek to promote any particular course of action or any

---

3 Formally, EC legislation only comes into existence once the Council of Ministers adopts a written text. Proposals do not become policy until they have been approved by the Council of Ministers, but they can give an indication of possible policy.

particular idea of environmental strategy.[4] Secondly, it does not seek to trace the intricacies of internal EU decision-making processes on climate change: instead, it deals primarily with the policy outcomes of internal EU decisions.

Ultimately, EU climate-change policy depends on the prospects of the emission-control strategies of all the member-states. Structural weaknesses at national level automatically affect the approach to the problem at EU level. While taking this into account, this chapter will focus on the EU as an actor in its own right.

2. The EU as an Actor in International Environmental Affairs

In the 1970s, Europeans frequently looked to the USA for inspiration when developing their environmental policies. Today, as the environmental agenda has become ever more internationalized, third parties increasingly find themselves monitoring the EU's actions in the environmental field and treating the EU as an important actor in its own right. The importance of the EU as a collective actor in international environmental affairs is perhaps most clearly underlined by its efforts to establish an effective global regime to control greenhouse-gas emissions.[5] Indeed, the EU signed and ratified the FCCC and has thus effectively committed itself to participate, even lead, in international programmes to reduce CO_2 emissions.

However, the role of the EU in international environmental affairs, as indeed in the wider field of foreign policy itself, is one of ambiguity. This ambiguity derives from the fact that the EU does not easily fit into established models of public policy according to which policy is split between domestic and foreign affairs (Haigh 1991: 165). While the EU is clearly not a federal state, it is not just another international organization like the OECD or UNEP, within which states collaborate without giving up important aspects of sovereignty. With the exception of the UN Security Council, it is one of the very few international organizations with the

---

4 There are few overall reviews of EU climate-change policy. For important contributions to the literature on this subject, see Grubb et al. (1994); Grubb (1995a); Bergesen and Haigh (1994); Rotmans et al. (1994); Skjærseth (1993); and Wynne (1993). Some consider the issue from an economic or legal standpoint, whereas others (e.g. Rotmans et al. 1994) treat climate change in terms of impacts and modelling of adaptive measures. Still others focus on specific policy recommendations (e.g. in relation to the carbon tax) (Vellinga and Grubb 1993; Grubb et al. 1994) or dwell on the effectiveness of international regimes (Skjærseth 1993).

5 For a discussion of the role played by the EC in this area, see Benedick (1991).

power to adopt decisions that directly bind member-states without further approval or ratification by their national parliaments. In the environmental field, for example, EC legislation has attained a degree of influence which makes it impossible to understand the policies of any member-state without reference to EC policy (ibid.: 164).

On the other hand, while it undoubtedly displays elements of supra-nationality, the EU has not achieved the quality of state or nationhood. For example, it has no head of state or a constitution designed for dealing with all eventualities. Most importantly, it does not have the power to raise taxes directly from citizens and lacks the institutions capable of enforcing compliance with its decisions. According to some analysts, therefore, the EU might be better understood as a union of states established originally among six, later 12 and now 15 member-states (Hovi 1991; Brewing 1994).

As indicated, EU environmental policy and, by extension, its climate-change policy have both an internal (intra-EU) and an external dimension. The process leading to the adoption of EC legislation, for example, is in some ways comparable to the making of an international treaty involving intergovernmental negotiations behind closed doors. On the other hand, once adopted, EC legislation becomes directly applicable in the member-states and has the same status as national legislation, quite unlike an international treaty (Haigh 1991: 165).

The EC also has competence to act externally. In fact, following a ruling by the European Court of Justice (ECJ) in 1971, where the EC has legislated this automatically confers the powers to act externally. Since 1971, the ECJ has repeatedly decided that competence for external affairs can be implied by the Treaty of Rome, as well as by the acts of the institutions performed under the Treaty (Haigh: 169).[6] Over time, therefore, as the powers of the EU have increased, and its internal policies have started to influence member-states more deeply, so the EU has been strengthened in its ability to act in its own right on the international stage. For this reason member-states carefully review EU policy, as the adoption of internal EC legislation in some ways leads to a loss of external competence on their part.

The external competence of the EU in environmental issues is perhaps best understood with reference to the general principles of competence underlying all Community and member-state action. According to the Treaty of Rome (Arts. 210, 228), the Community can enter into binding international agreements, including environmental agreements, and

---

6 Commission v. Council (ERTA) Case 22/70, ECR 19/1 263, quoted in (Haigh 1991).

thereby achieves what in international law is known as *international personality* (Hession 1995: 155).

Competence in this context refers to the scope of the Community to act, by negotiating, signing and ratifying international treaties, speaking at international conferences and adopting legally binding implementing measures. Indeed, the EC is now party to well over 30 international environmental conventions, in many cases alongside its member-states. Mixed agreements involving dual participation by the European Commission and member-states raise the question of competence between the Commission and the member-states. As long as there is unanimity within the EU, this need not be problematical. Where, however, there is no agreement (e.g. in the Paris Commission regarding land-based marine pollution), disputes can arise over the powers of member-states to adopt provisions differing from those negotiated and signed by the EC (Haigh 1991).[7]

The role of the EU in international environmental policy thus raises complex legal and political questions, principally as to competence (i.e. whether the EC itself has the right to be a party to an international convention) and the transfer of power from member-states to the Community when the EC joins an international convention. While disputes over general Community competence in the area of the environment were largely laid to rest with the adoption of the Single European Act in 1987, governments continue to cast a wary eye on EC participation, especially since EC participation in an international convention may diminish the ability of member-states to act on their own (Redgwell 1994: 131).

Community competence in environmental matters, on the other hand, is not exclusive. In accordance with Article 130T of the Treaty on European Union (Maastricht Treaty), member-states have a right to adopt more stringent measures. There are also, as will be seen, provisions for the adoption of lower standards than those provided for in EC legislation. Community competence is therefore limited by subsidiarity (Treaty of Rome, Article 3b). This has important consequences for environmental policy in general and climate-change policy in particular.

3. The International Significance of EU Climate-change Policy

Energy-related CO_2 emissions in the EU amounted to 3026 million tonnes in 1992, or 29.8% of total OECD emissions. Following the accession of

---

7 In the case of international conventions, the Commission does not have the same right to insist on implementation as it does in the case of directives and regulations.

Austria, Finland and Sweden in 1995, it is estimated that the EU's share of global emissions has increased slightly, from 14.02% in 1992 to about 15% (IEA 1994: 26). This leaves the EU in the dubious position of being the second biggest emitter of greenhouse gases after the United States, which accounts for about 25% of global emissions, followed by the former Soviet Union and Central and Eastern Europe, as well as Japan which ranks fifth in the world. On the other hand, the EU's CO_2 *per capita* ratio of 8.72 tons per person was significantly below the OECD average of 11.73 (WRI 1994: 201; *International Environment Reporter*, 3 April 1996: 252).

Beyond this direct measure of the EU's salience lie institutional aspects of its global role. First, the EU took an early lead in the climate change debate by declaring its 1990 emissions stabilization target. Second, the EU in many ways represents a microcosm of the climate-change problem internationally. Not only do the member-states vary considerably in terms of economic development, environmental conditions, institutional structure, policy culture, and perception of environmental problems, there is also a North–South dimension to EU politics, resembling the wider North–South divide at the international level (Grubb et al. 1994: 12; Wynne 1993). Third, the EU has formidable technological and financial resources and hosts many energy-related industries of global importance such as international oil and gas companies, equipment manufacturers, electricity companies and vehicle manufacturers (Grubb et al. 1994).

On the other hand, as Rotmans et al. (1994: 112) argue, if the EU were to take strong action against climate change while the rest of the world did nothing, the resulting CO_2 concentrations by the year 2000 would only be about 6% lower. Nevertheless, EU failure to meet its goals would not only be embarrassing; this would weaken the FCCC and would provide other industrialized as well as developing countries with a reason to avoid substantive action. Conversely, success in meeting its declared climate-change objectives (and fulfilling its obligations under the FCCC) would serve as an example internationally. In this sense too, the stance of the EU is of enormous symbolic importance.

4. EU Climate Change and Environmental Policy: Legal and Political Basis

4.1. The Political Framework

The political framework for EU environmental policy is provided principally by the Community's action programmes on the environment. To date, five such programmes have been adopted, the latest being the Fifth Environmental Action Programme (Towards Sustainability) (*Official Journal of the European Communities*, (1993) No. C138, 17 May). Each of these programmes represents, in a sense, the basic reference charter for Community environmental policy. The programmes generally cover a period of four to five years and define and describe all of the actions envisaged by the Community in the field of environmental protection and sustainable development (Johnson and Corcelle 1989: 3).

The Community's action programmes are adopted in the form of Council Resolutions, thereby indicating the political will of the member-states to take action on the issues specified in the programme, without implying any legal obligation to do so. They are primarily political declarations of intent and do not constitute a legal basis for Community environment measures (Krämer 1990: 2). Indeed, a review of past Community action programmes reveals that many of the actions provided for have in fact never been implemented (Johnson and Corcelle 1989: 11).

The Fifth Action Programme, which is to guide Community environmental policy into the next century, emphasizes the need to integrate environmental concerns into the EU's other major policies such as energy, industry and transport. It also focuses on key economic sectors such as energy, and calls for a reduction in consumption of non-renewable energy sources. Moreover, it foresees the need to broaden the range of policy instruments to achieve the necessary reductions in greenhouse-gas emissions.

A further important political basis of EU environmental and climate-change policy is the tendency of the European Council of Heads of State and Government, which is the highest political authority of the EU, to consider environmental issues. In 1978, for example, in the wake of the Amoco Cadiz oil disaster off the coast of Brittany, the Council decided on a specific Community action on marine pollution. Similarly, in 1983, the Council underlined the urgent need to speed up and reinforce actions to combat acid rain. Climate change has also been addressed by the Council.

The first mention of the climate-change issue was made on the occasion of the Rhodes meeting of 2 and 3 December 1988, when, on the initiative of the Greek EC Presidency, the European Council adopted a Declaration on the Environment which underlined the desire of the Community to play a leading role in the action needed to protect the world's environment, particularly as regards such global problems as depletion of the ozone layer and the greenhouse effect (Johnson and Corcelle 1989: 21). Echoing this sentiment, the European Council in

Dublin in June 1990 called for the early adoption of targets and strategies to limit greenhouse-gas emissions. Since then, climate change has repeatedly been taken up by the Council, particularly in the run-up to and since the 1992 United Nations Conference on Environment and Development (UNCED).

4.2. The Legal Framework

The legal basis for Community action to protect the environment is provided by the 1957 Treaty of Rome, as amended by the 1987 Single European Act and the Treaty on European Union of 1992 (Maastricht Treaty). The Maastricht Treaty replaced and expanded, inter alia, the environmental provisions of the Single European Act (Article 130R-S) with a new Article 130R-T. Of particular importance are the procedural changes which provide that environmental legislation is to be passed by qualified majority voting, subject to important exceptions. The Treaty amendments also reinforce the authority of the EU to deal with environmental matters, and the role of environmental issues as an important consideration in formulating other EU policies.

Under Article 130R(1), the aims of EU policy, as set out in Article 130R of the Single European Act, are retained; namely, to preserve, protect and improve the quality of the environment; to protect human health; and to ensure a prudent and rational utilization of natural resources. In addition, a new objective is introduced: that of "promoting measures at international level to deal with regional or worldwide environmental problems". This reflects the increasing importance placed on global cooperation on matters such as global warming and ozone layer depletion.

General principles of Community competence in the field of environmental protection are outlined in Article 130R (1) and (2). Article 130R(2), for example, repeats the principles of preventive action, rectification of damage at source and that the polluter should pay. In addition, it establishes a high level of environmental protection as one of the central aims of the Union. The Treaty also enshrines the precautionary principle and emphasizes the need to integrate environmental protection into the Community's other policies such as energy, transport and land-use.

Article 130R (4) introduces the subsidiarity principle, which, although implicitly already well-established, makes it explicit regarding the environment. To this end, the Treaty provides that "the Community shall take action relating to the environment to the extent which the objectives can be attained better at Community level than at the level of the individual member-states". One consequence of the introduction of this principle is that it enables member-states to challenge Community environmental measures by invoking Article 130R as a legal base (Redgwell 1994: 134).

Article 130S provides for environmental legislation to be passed by qualified majority voting. It also lists the specific areas on which unanimity of the Council is required. These include:

> Provisions of a primarily fiscal nature, measures concerning town and country planning, land-use with the exception of waste management and measures of a general nature, water management, and measures significantly affecting a member-state's choice between different sources of energy and the general structure of its energy supply.

These exclusions are vitally important areas and obviously dilute the importance of qualified majority voting for the other areas.

Article 130T allows member-states to introduce more stringent protective measures than those provided for in Community legislation, as long as they are compatible with the Treaty and the requirements of Article 36 on preventing damage to the Single Market from the proliferation of different environmental standards throughout the EU. However, the provision to introduce stricter legislation applies only to measures adopted under Article 130, and not to measures adopted in pursuance of internal market aims under Article 100A. Article 130R (5), finally, in recognition of the burden which the potentially high cost of pursuing the Community's environmental objectives under Article 130R may impose on less developed regions of the EU, allows for temporary derogations and/or financial support from the Community's Cohesion Fund (ibid. 136).

Overall, then, the amendments to the Treaty of Rome expand the powers of the EU in environmental, and hence climate-related, matters considerably. On the other hand, they provide for important exceptions which, as will be seen, are of particular relevance to the formulation and implementation of measures to reduce greenhouse-gas emissions.

5. The Instruments of Community Policy

EU environmental policy has traditionally been dominated by a regulatory approach based on legislative command-and-control instruments seeking to harmonize standards imposed on member-states. Legislative instruments have, in fact, proved to be an effective way of setting minimum levels of protection, of implementing international obligations and laying down EU-wide rules and standards necessary for the maintenance and integrity of the internal market. Examples of the use of regulatory measures for environmental policy include the introduction of Best Available Technology (BAT) or the Large Combustion Plant Directive (Directive 88/609/EEC), which aims to reduce the emissions of acid rain precursors from large combustion plants.

The most frequently used legislative instruments to achieve stated policy targets are *Directives, Regulations* and to a lesser extent *Decisions. Directives* and *Regulations* are designed to set fundamental levels of protection for public health and the environment as well as to implement wider international commitments and to provide EU-wide rules and standards necessary to preserve the integrity of the Single Market. *Decisions,* by contrast, are directed at a narrower group and can address companies as well as individuals.

A *Regulation* is of general application throughout the territory of the member-states. It is directly applicable and becomes automatically part of the law of the member-state without the need for further ratification. *Regulations* also prescribe the means to be employed to achieve given targets. *Directives,* by contrast, are binding insofar as the attainment of the goals is concerned but leave member-states some leeway in the implementation. *Decisions,* finally, are binding in their entirety. Other nonbinding legislative instruments include *Recommendations, Opinions* and *Communications.* Alongside these legislative instruments, the EU has at its disposal a range of *market-based instruments, horizontal support instruments* and *fiscal support mechanisms.* Particular emphasis has been placed on developing these policy tools in the context of the EU's evolving response to climate change.

Market-based instruments, such as the proposed carbon/energy tax, are designed to sensitize producers and consumers towards responsible use of natural resources, and the avoidance of pollution and waste by internalizing external environmental costs (through the application of economic and fiscal incentives and disincentives, civil liability, etc.). They are geared towards "getting the prices right" and ensuring that environmentally-friendly goods and services are not placed at a competitive disadvantage vis-à-vis polluting or wasteful competitors.

Horizontal support instruments, by contrast, are designed to improve background and statistical data as a basis for improved sectoral policies. They aim to support education, professional and vocational education and training. They also support the development and availability of "clean" technologies, primarily through scientific research and technological development.

Financial support mechanisms, finally, aim to contribute significantly to the financing of actions for the improvement of the environment, beside the budgetary lines. Projects falling under this category include programmes with direct environmental objectives, such as Life, the Structural Funds and notably Envireg. Also in this category is the new Cohesion Fund decided upon at the Maastricht Summit. Other relevant programmes include Thermie, Joule and ALTENER.

In evaluating the EU's response to climate change, the choice of instrument is of prime importance. It is possible to distinguish between instruments which are binding, such as Directives and Regulations, and those which are not. Yet the extent to which a particular instrument is or is not legally binding tells only part of the story. Equally important is the extent to which a measure is considered to be *politically* binding. A Directive which is legally binding but poorly implemented, for example, is likely to be less influential than a non-binding Resolution which carries political weight and which is actually complied with.

One of the principal weaknesses of EU environmental policy, perhaps not surprisingly, is the lack of implementation and/or poor enforcement of its environmental legislation. While there has been some progress in ensuring that EC legislation is implemented more effectively, it remains a fact that many directives are not properly applied by member-states, as witnessed in the number of cases brought by the European Commission before the ECJ for non-compliance with Community law. A special report concerning the environment prepared by the EC Court of Auditors in 1992, for example, lamented the slow progress made in the implementation of environmental Directives and recommended that the Commission take on a more pronounced role to ensure improved coordination of EU environmental policy (*Official Journal of the European Communities*, (1992) No. C245, 23 September). In the event, many Directives are the result of political compromise between member-states with fundamentally diverging interests. This can lead to a certain incoherence in the drafting of the texts – a fact which contributes to the problem of poor implementation and enforcement.

In the context of EU climate-change policy, the crucial test will be the extent to which stated policy objectives, such as the 1990 CO_2 emission-stabilization decision, are backed up by concrete measures to implement the policy and to monitor national compliance.

6. The Institutional Anatomy of the EU

Before turning to the evolution of EU climate-change policy, a few words on the institutional anatomy of the EU are in order. The principal EU institutions are the European Parliament, the European Commission, the Council of Ministers and the Court of Justice.

Despite numerous Treaty amendments which have progressively increased its powers, the European Parliament is still a largely consultative as opposed to a legislative body. Though it now has, under the Codecision Procedure, the power to amend or even veto a proposal for legislation relating to the functioning of the internal market that has been adopted by the Council, it has no direct influence over the final outcome. Despite its electoral legitimacy – the 626 representatives are popularly elected – it still suffers from a "democratic deficit" because it continues to play a secondary role to that of the Council (Visek 1995: 380–381).

The Commission, by contrast, occupies a central role. It drafts and initiates legislation, develops the action programmes and acts as the guardian of the treaties. Though it lacks any specific enforcement powers, it can and does initiate actions against member-states for breaches of the Treaty and failure to implement the obligations of environmental directives.[8] At the pinnacle of Community institutions stands the Council of Ministers. Made up of the relevant national ministers, for example of the Environment, the Council is the final decision-making authority. In this capacity it must approve all draft legislation before it can enter into force.

The ECJ, finally, as the highest judicial authority regarding Community law, is responsible for interpreting Community law and deciding on disputes between member-states and Community institutions. As such the ECJ can interpret the EU Treaty, review the legality of Community acts, decide whether a failure to act by one of the institutions constitutes a breach of Community law, answer questions regarding Community law from national courts of member-states, and rule on whether member-states are fulfilling their obligations under Community law (Visek 1995: 383).

7. The Evolution of EU Climate-change policy

To understand the context and evolution of EU policy towards climate change requires some brief reflections on the origins of EU environmental policy. Until the adoption of the Single European Act in 1987, the

---

8 The recently established European Environmental Agency is for the time being restricted to research and monitoring, though the intention is for it eventually to become involved in enforcement. In its present capacity it is of indirect assistance to the Commission in its enforcement role by providing relevant information.

basis for Community action on the environment was provided, somewhat tenuously, by a generous interpretation of the 1957 Treaty of Rome establishing the Community. Accordingly, the EC's environmental mandate was based on the commitment, contained in the preamble, to the constant improvement of the living and working conditions of EC citizens. An additional justification was found in the argument, as expounded during the 1972 Paris Summit of Heads of State of the EC, that differing national environmental standards could pose a threat to the creation and maintenance of the Common Market (Freestone 1991: 136).

Although the Treaty of Rome was silent on the environment, it enabled the Community to legislate on environmental protection issues. However, early measures were rather vague and generally had little impact on the behaviour of enterprises or individuals.

The EC's first piece of environmental legislation was the *Directive on Classification, Packaging and Labelling of Dangerous Substances* of 1967 (Directive 67/548/EEC) (*Official Journal of the European Communities*, (1967) No. 196, 16 August). Six years later, in 1973, in the wake of the 1972 Stockholm Conference on the Human Environment, Community environment policy was placed on a more formal footing with the adoption of the First Community Action Programme on the Environment. The Programme had three main objectives: the prevention and control of pollution and nuisance; improvement of the environment and quality of life; and Community action, or common actions by member-states, in international organizations dealing with the environment. The Programme also established the Polluter Pays Principle and reaffirmed the 1972 Paris statement that economic expansion was not an end in itself but that economic growth should be linked to improvement in the living and working conditions of EC citizens (Freestone 1991: 136).

The EC's First Community Environment Action Programme, which covered the period 1973–1977, has been succeeded by a further four programmes, each of which has reaffirmed and expanded the goals of the first. Each of these action programmes has been accompanied by a steady expansion in the scope of Community environmental legislation. Indeed, by mid-1995 over 300 pieces of legislation dealing with environmental issues had been adopted. Today, EC legislation covers a wide array of issue-areas ranging from the regulation of air, water and noise pollution to waste management, major accident hazards, dangerous substances, environmental impact assessment and nature conservation. Since the 1960s, moreover, the focus of EU environmental policy has shifted from local to global issues, with increasing emphasis on long-term pollution prevention rather than short term, end-of-pipe pollution control. Similarly, the single-media approach to pollution is increasingly being replaced by a multi-media, integrated permitting and pollution control approach. Thus, by the late 1980s the Community had become the main driving force of environmental policy in Western Europe, with Community legislation attaining a degree of influence which made it impossible to understand national policy without reference to EC policy.

7.1. The Beginnings of EU Climate-Change Policy

Despite the impressive expansion in the scope of EU action on the environment, climate change was for a long time dealt with primarily as a topic of research. The first Community institution to address the issue seriously was the European Parliament, which issued a Resolution on the subject in November 1986.[9] Yet, even as late as 1987, the Fourth Community Environment Action Programme for the period 1987–1992 made no mention of climate change except as a subject for further research. It was not until November 1988 that the Commission issued its first Communication on the subject (COM (88) 656 final). In this Communication, it reviewed scientific findings on the phenomenon and possible responses to the problem, but made no recommendations for immediate action.

Progressively, however, particularly following the establishment of the Intergovernmental Panel on Climate Change (IPCC) in 1988, Commission pronouncements on the subject of climate change began to identify the issue as one of the most pressing environmental problems of the energy sector (COM (89) 369). By March 1990, in fact, the Commission issued a statement on its policy targets on the greenhouse issue, in which it recommended the urgent need for a clear commitment by industrialized countries to stabilize CO_2 emissions by the year 2000 and to consider significant reductions by 2010 (Skjærseth 1993: 15).

Further steps in the direction of formulating a response strategy were taken in the spring of 1990 with suggestions made in a Draft Communication on the use of economic and fiscal instruments in EC environental policy. In it the Commission concluded that the only practicable solution in the short to medium term would be to limit the growing use of fossil fuels, improve energy efficiency and promote renewable energy sources. Accordingly, improvements in energy efficiency and the acceleration of energy-conservation efforts were to form the

9 To the knowledge of this author, this was the first time climate change was debated in an international political forum.

cornerstone of any energy policy capable of dealing with the greenhouse problem.

In June 1990, finally, the issue was taken up at the highest political level when the European Council meeting in Dublin pressed for early adoption of targets and strategies for limiting emissions of greenhouse gases, in particular CO_2. The Council also underlined the special responsibility of the Community and its member-states in encouraging and participating in international action to combat global environmental problems and providing leadership in this field. This Resolution paved the way for the target of stabilizing CO_2 emissions by the year 2000 at 1990 levels, adopted at a joint meeting of the Council of Energy and Environment Ministers in October 1990. The decision stated that:

> The EC and the member-states assume that other leading countries undertake commitments along (similar) lines and, acknowledging the targets identified by a number of member-states... are willing to take actions aiming at reaching stabilization of the total CO_2 emissions by 2000 at 1990 levels in the Community as a whole (COM(92) 226 final).

The declaration, which was adopted unanimously, "assumes" that other countries take measures, but does not make the goal explicitly conditional upon such action (Vellinga and Grubb 1993: 4). Though not a legally binding target, it was politically influential and provided the basis for the subsequent evolution of Community policy on climate change, particularly in the run-up to the 1992 UNCED in Rio de Janeiro. The declaration was also instrumental in pushing the Japanese government to take a stand on climate change (Fermann 1993).

In the two years since the publication of the 1988 Communication, then, EC policy had moved from the stage of problem recognition to making specific policy recommendations: a rapid and in retrospect relatively smooth process, particularly compared to the evolution of Community policy on ozone-layer depletion earlier in the 1980s. The political commitment to stabilize greenhouse-gas emissions, indeed, the very development of the EC's stance on climate change from 1988 onwards, was influenced by the desire to adopt a leadership role and the aspiration for a major presence at the Rio Earth Summit (Wynne 1993: 102). By this time several member-states had adopted national policies to curb, stabilize or reduce greenhouse-gas emissions and by making assumptions on what might happen in member-states which had adopted no national targets it appeared that stabilization was an attainable goal for the EC.

7.2. Preparing for UNCED: The Second Stage of EU Climate-Change Policy

The adoption of the 1990 stabilization target brought to an end the first phase of EU policy-making on climate change. The second stage of Community climate-change policy lasted from the adoption of the 1990 stabilization target to the signing of the FCCC in Rio in June 1992. During this period, the EC came to play a decisive role during the preliminary negotiations on the FCCC. Indeed, some observers feel that the latter might not have been so strongly worded had it not been for the position of the EC (Grubb et al. 1994).

Following the 1990 decision to stabilize greenhouse-gas emissions, the Commission started to formulate a specific response strategy. A confidential Draft Communication, prepared by the Directorates-General for Environment and Energy (DG XI and XVII) in late November 1990, stressed the importance of stabilizing CO_2 emissions by the year 2000. It suggested that to do so would in fact, in view of the projected business-as-usual increase in Community emissions to the year 2000, require a reduction of 10–20% in the growth of emissions. This Communication also indicated that existing measures aimed at increasing energy efficiency were insufficient to achieve the desired results.

Thinking on the issue was inspired by a belief that an adequate response strategy should comprise a variety of elements including regulatory measures, fiscal instruments, economic instruments, burden-sharing among member-states and complementary measures at national level. Measures to curb transport-related emissions such as speed limits and improved traffic and transport management were also contemplated (Skjærseth 1993: 16). On 28 November 1990, the Energy and Environment Commissioners jointly proposed to the Commission the adoption of A Community Action Programme to Limit Carbon Dioxide Emissions and to Improve the Security of Energy Supply (SEC(90) 2404). This programme comprised regulations to improve energy efficiency in houses, transport and commerce, industrial agreements on energy efficiency, least-cost planning and third-party financing of energy conservation, and a combined energy/carbon tax of around US$10/barrel of oil in the proportions 75/25.

For a while consideration was also given to the feasibility of adopting a Directive allocating different CO_2 emission targets, similar to the Large Combustion Plant Directive (Directive 88/609/EEC). In the end, the idea of a Directive allocating different targets was abandoned in favour of fiscal measures which subsequently became the focal point of Community strategy on climate change. One of the main reasons for abandoning the idea of setting national CO_2 emission targets at EC level was the reluctance to transfer the relevant powers over an issue which was likely to affect so many aspects of national life (Haigh 1994).

Community thinking on climate change was further refined when, on 14 October 1991, the Council adopted a Communication entitled A Community Strategy to Limit Carbon Dioxide Emissions and to Improve Energy Efficiency. The Communication reaffirmed the EC's desire to play a forward role in global negotiations leading up to the June 1992 UNCED Conference and stated that:

> The Community owes it to both the present and future generations to put its own house in order and to provide both leadership and example to developed and developing countries alike in relation to protection of the environment and the sustainable use of natural resources (SEC(91)1744 final).

As the date of UNCED in Rio drew closer, so the resolve of the EC to maintain the leadership role it had assumed at the Second World Climate Conference in November 1990 grew. In May 1992, the European Commission adopted a proposal to be made to the Council on an overall strategy to reduce CO_2 emissions in the EC. This strategy, adopted in response to a request by the joint Energy/Environment Council of December 1991, aimed to implement the decision to stabilize CO_2 emissions by 2000 at their 1990 level. This was to be achieved through reduced energy demand, increased energy efficiency and fuel-switching (IEA 1994: 184). To this end, the Community adopted a five-point strategy which was based on the following elements:

- Measures to conserve energy and improve energy efficiency through implementation of the existing SAVE proposals (Specific Actions for Vigorous Energy Efficiency) for a series of Directives on energy efficiency standards. These measures, it was felt, would achieve an estimated 3% reduction in CO_2 emissions by the year 2000.[10]

- A decision on renewable energies to support the development of renewable energy, which emerged as the ALTENER Directive. These measures, it was estimated, would reduce projected emissions by a further 1% by the year 2000.

---

10 The original conception of SAVE dates back to a paper published in 1987, entitled Towards a Continuing Policy for Energy Efficiency in the European Community, COM (87) 223. For a discussion, see Warren (1993).

- Strengthening of the Thermie Programme on the development and introduction of new energy conversion and use technologies, and the Joule Programme for energy R&D.

- The introduction of a combined energy/carbon tax on the carbon and energy content of fuels. This measure was projected to reduce emissions between 3% and 5.5%, depending on the policy stance on industry exemptions.

- A decision concerning a system for the monitoring of CO_2 emissions and other greenhouse gases within the EU.

Subsequently, in May 1992, the Commission introduced its proposal for a Council Directive introducing a tax on CO_2 emissions and energy, published in the *Official Journal* of 3 August 1992 (COM (92) 226 final). This proposal came to form the central plank in the EC's climate strategy, though it has been stalled ever since due to considerable resistance from a number of member-states, particularly Great Britain and some of the southern member-states such as Spain and Portugal.

The Draft Directive aimed to levy the tax in two main segments, 50% on the carbon content of fuel and 50% on the thermal value of fuels. The rates were to be introduced progressively from a base rate equivalent to US$3 per barrel of oil, increasing each year to a maximum rate equivalent to US$ 10 per barrel by the year 2000. The tax was to be fiscally neutral. Energy sources included within the scope of the proposal were fossil energies (gas, oil, coal and derivatives) and electricity produced by hydroelectric and nuclear power plants.

7.3. Towards Ratification of the Climate Change Convention

The signing of the FCCC by the Community in June 1992 ushered in the third stage in EU climate-change policy, a stage which ended with the ratification of the Convention by the Community on 15 December 1993. During this period, the Council had to make a decision on whether to accept the detailed package of proposals tabled by the Commission immediately prior to the UNCED, the most important of which in the eyes of the Commission was the proposed carbon/energy tax.

In a formal declaration the EC expressed its desire for a prompt start to the implementation of the Convention and reiterated the necessity for reinforcing the commitments entered into. On the same occasion, the EC stressed the need for an early preparation of protocols under the FCCC covering specific issues, in particular the limitation of CO_2 emissions. It also issued a statement in which it reaffirmed its target to stabilize emissions by 2000 at 1990 levels and in which it "urged all countries to translate the clear commitments established under the Convention into their national policies (and) to take steps and commitments similar to those adopted or foreseen by the Community and its member-states".[11]

Further steps towards implementing the EU strategy were taken during 1993 with the adoption of the following legislation: (i) Decision 93/389/EEC for a monitoring mechanism of Community CO_2 and other greenhouse-gas emissions (*Official Journal of the European Communities* (1993), No. L167, 9 July). (ii) Directive 93/76/EEC on Energy Efficiency (SAVE) (*Official Journal of the European Communities* (1993), No. L237, 22 September). (iii) Decision 93/500/EEC on Renewable Energy (ALTENER) (*Official Journal of the European Communities* (1993), No. L235, 18 September).

Of these, the monitoring mechanism represents the only substantive piece of EC legislation to have been adopted by the Council so far. It establishes a legal and institutional basis for working towards CO_2 stabilization in the EU and requires, *inter alia*, member-states to devise, publish and implement national programmes for limiting their CO_2 emissions in order to contribute to the Community objective. It also contains provisions for the evaluation of these national programmes by the Commission, establishing an inventory of greenhouse-gas emissions to the limitation of CO_2 and other greenhouse-gas emissions under the FCCC. The national programmes submitted so far under the Monitoring Mechanism suggest that EU stabilization will be difficult to attain if current trends persist (Grubb 1995a: 46; COM (94) 67 final).[12]

As for the Community's other measures in the field of energy efficiency, the SAVE Programme (Decision 93/500/EEC) and the ALTENER Directive (Directive 93/500/EEC) were reduced in both scope and content by the time they were adopted in 1993 (Skjærseth 1993: 19). The original SAVE Programme had envisaged a series of Directives on energy efficiency (including a promise for a new proposal on least-cost planning;

---

11 Statement by Mr Silva, Prime Minister of Portugal on the occasion of the signature by the EC of the FCCC, June 1992.

12 These conclusions have been reinforced by the Commission's Second Evaluation of Existing National Programmes under the Monitoring Mechanism of Community CO_2 and Other Greenhouse-Gas Emissions, published in March 1996. See also (IER 1996: 252).

industry; transport; and household/commercial measures), the combined effect of which was to contribute about a quarter to the reductions necessary to achieve the stabilization target. However, important measures were dropped from the original package and all detailed requirements were removed from the programme by the time of its adoption. As a result, far more responsibility than had been intended was left with the member-states (Grubb et al. 1994: 19). Moreover, it was not specified what member-states were to do in detail.

Similarly, the ALTENER Directive was strongly reduced in both scope and content, lacking in the end sufficient funding and substantive tools for implementation when it was adopted in January 1993. The result of these changes was, in the words of one commentator, to render EU level commitments towards stabilization of CO_2 emissions almost impotent (Grubb et al. 1994: 20).

The carbon/energy tax, for its part, which has represented the central plank in Community climate-change policy, has been moribund almost from the outset. In the face of sustained and deep-seated opposition to the tax since the proposal was first introduced in 1992, major concessions were made to sectors dependent on international trade and/or energy intensive industries such as the glass, steel, cement, non-ferrous metals, paper, pulp and coal sectors. In fact, the Commission agreed to grant these sectors partial or total exemption from the tax until the EU's main trading partners have followed suit. Similarly, in October 1993, the European Council recommended a deferral of the tax for the Cohesion Fund countries (Ireland, Spain, Greece, Portugal) for as long as their CO_2 emissions remain below 85% of the Community average, based on 1990 index figures. The reason for this form of "burden-sharing" was to allay fears in these countries of being penalized by the tax due to their lower energy consumption and CO_2 emissions (Sebenius 1994).

At the close of 1993, and throughout 1994, little progress was made on the carbon/energy tax. While efforts to find an acceptable way of implementing the tax in the EU continued, hostility in many member-states towards the perceived federal ambitions of the EU bureaucracy – sparked off by the Danish No vote in the 1992 referendum on the Maastricht Treaty – resulted in a nationalistic backlash to repatriate many elements of EU policy, including environmental policy under a redefinition of the legal principle of subsidiarity (Wynne 1993: 101).

The result of all this, some feel, has been to dampen, even fundamentally question the steady enlargement since the early 1970s of EC environmental policy influence over member-states. Even though formally no change took place in the EU's position on climate change following the 1992 UNCED, the EU has become more passive, both internally and at the international level (Haigh 1994). Despite these uncertainties, and the targeting of EU environmental policy-making by some member-states (e.g. Great Britain) as a particular candidate for repatriation, a substantial impetus towards concerted environmental and climate-change policy making has remained in the EU. On the other hand, much of the initial force and dynamism of EU action has invariably been undermined by these wider uncertainties.

7.4. Implementing the FCCC: The Further Development of EU Climate Change Strategy

The ratification of the FCCC on 15 December 1993 brought to an end the third phase of EU climate-change policy. A fourth stage started in the wake of the ratification by the EU of the Convention. Since then, indeed since the signature of the FCCC in June 1992, fundamental divisions over how to achieve the EU's targets for reducing emissions of greenhouse gases have come to the fore. The focal point of these divisions has been the proposed carbon/energy tax. Alongside these divisions, wider arguments over the very course of European integration in the post-Cold War era began to surface, particularly following the signing of the Maastricht Treaty. The controversy which ensued over the ratification of the Maastricht Treaty emphasized deep-seated scepticism, even fear, in many quarters of the EU over the extent and pace of European integration. While many of these fears were neither new nor concerned with environment or climate-related issues *per se*, EU environment and climate-change policy was invariably affected by what became almost a constitutional crisis over European integration.

Indeed, since 1992, there has been a marked slow-down in Brussels' legislative output in the field of environmental protection. The Commission work-programme for 1996, for example, which outlines the measures the Community aims to adopt during the course of the year, is much reduced, reflecting pressure on the European Commission to stem the flow of legislative proposals (*ENDS Report* No. 250, 1995: 35–36).[13] Notable among the measures not listed are the Commission's past

---

13 The reduction in the volume and scope of the Community's work programme continues a trend in evidence since 1992. In the event, Commission work programmes are not always reliable guides to its actual achievements. Many measures listed in the work programmes still await adoption, while other draft legislation may not have been announced in past work programmes.

commitments to produce EU-wide rules on CO_2 emissions from vehicles.[14]

Meanwhile, work on the carbon/energy tax has continued. During the Greek EU Presidency from January 1994 to June 1994, the Greek government presented new and substantially revised proposals which, instead of an automatic annual increase in the tax rate, provided for an initially low tax level of US3$/barrel of oil; this would be reviewed after three years, and not be implemented in all countries (Grubb et al. 1994: 25). Despite these and other concessions, such as exemptions for energy-intensive industries, which weakened the projected emission reductions, the Council of Ministers could not agree on the proposals.

Subsequent attempts by the German EU Presidency from July 1994 to December 1994 to breathe new life into the tax proposal did not succeed either in breaking up the log-jam. Although the focus shifted towards using the existing system of excise duties in order to circumvent the thorny prospect of a new type of EU-level taxation, opposition to the idea of the EU being given competence to develop tax proposals effectively blocked the proposal – primarily, though not exclusively, due to the attendant sovereignty implications (Grubb et al. 1994: 26).

By the end of 1994 it became clear that a fully fledged carbon/energy tax of the kind originally proposed was less and less likely, whatever its long-term justification (EC Committee 1995: 75). The unified tax on CO_2 emissions was finally buried when, after over three years of controversy, the Council of Ministers meeting in Essen on 15–16 December 1994 decided to replace it with a harmonized tax on an optional basis.

Despite this apparent setback, Germany received support from some countries (Denmark, Netherlands, the Association of Small Island States – AOSIS) for going even further and reducing emissions by 2005 and beyond. While national measures would be relied upon primarily in the future to achieve individual, national emission targets, it was agreed that the Commission should issue guidelines to assist member-states in applying a carbon/energy tax on the basis of common, though yet to be defined, parameters. Subsequently, on 10 May 1995, the Commission issued a revised proposal for a voluntary CO_2 tax which outlined the specifications of an EU-wide levy (*International Environment Reporter*, 11 January 1995: 3).[15]

---

14 At the time of writing, the Commission was on the verge of proposing an outline for reducing CO_2 emissions from cars. See *Environment Watch Western Europe*, Vol. 4, No. 23, 15 December 1995, p. 1.

15 The new tax is to be set at a level to be decided upon by each member-state. By 1998, the Commission is to assess the situation and try again to harmonize tax rates throughout the EU, with a target tax rate of US$19/bbl. Exemptions for energy-intensive industries are to be possible.

7.5 The First Conference of Parties: The Berlin Mandate

As the date of the First Conference of Parties (COP) of the FCCC in March 1995 drew closer, efforts to forge a common EU position for the Berlin Conference intensified. Apparently motivated by the fear that failure to reach agreement on the tax meant the risk of going to Berlin empty-handed, the European Council agreed in October 1994 to set up a group of experts to prepare the Community position for Berlin, in particular in relation to establishing Community emission-reduction targets for the post-2000 period. No definitive conclusions were reached, although the insufficient nature of existing measures to reduce emissions was acknowledged by the Council. It was furthermore admitted that this would necessitate the implementation of agreed measures and even new ones such as fiscal instruments (Grubb et al. 1994: 74).

Amidst the continued controversy over the EU's carbon/energy tax proposal, the first CoP took place in Berlin in March 1995. This UN-sponsored conference, the follow-up to the 1992 UNCED in Rio, focused attention and interest on climate change and measures taken by governments around the world to reduce greenhouse-gas emissions. The aim of the conference was to reaffirm and, where possible, extend the commitments made at Rio. In the event, while recognizing the need to cut CO_2 emissions, parties failed to agree to any substantive action, deciding instead to defer further action until 1997 (see Chapter 2). Accordingly, it was agreed that a two-year negotiation process with the mandate of approving a protocol to the FCCC would be started. The protocol would set specific, legally binding targets and timetables for reducing greenhouse-gas emissions beyond the year 2000, thereby following the model established under the Vienna Convention and its associated protocols on substances that deplete the ozone layer.

In what has subsequently been dubbed the "Berlin Mandate", it was agreed that measures taken by industrialized nations were inadequate, and that industrialized countries should set quantified emission-reduction targets with specified time-frames such as 2005, 2010 and 2020 for greenhouse gases not controlled by the Montreal Protocol (*Financial Times*, 8/9 April, 1995). The EU, for its part, was an active participant in preparing the conference. During the event, the Community worked with the so-called Green Groups that had formed uniting the G-77 and the industrialized North in negotiating the final mandate (*International Environment Reporter*, 131 May 1995: 439). The process led to an unusually open North–South debate and underlined the increased importance accorded climate change issues by many developing countries, including China and India. During the conference, a number of EU environment ministers stressed the need for a protocol, and resisted pressure from the United States, Japan, Canada, Australia and New Zealand to avoid substantive action. Despite the EU's efforts, however, the "Berlin Mandate" did not include specific commitments (*Financial Times*, 8/9 April, 1995).

8. Determinants of EU Climate-change Policy

In retrospect, despite considerable scientific uncertainty and the complexities of the climate change problem, the early phases of the Community's response to climate change progressed remarkably smoothly. Following initial delays and the overcoming of various hurdles, particularly in moving from treating climate change as a topic for research to a recognition of the greenhouse issue as a threat to the global environment requiring determined action, the Community made relatively speedy progress in formulating a response strategy. Initial successes included the announcement of the 1990 stabilization target and the preparation of the Community's position for the 1992 UNCED, the highlight of which undoubtedly was the proposal for a carbon/energy tax.

While the 1990 stabilization target amounted to little more than the aggregate of the separate national statements of intent publicized at the time, leaving unanswered the question as to whether there actually was a distinct EC policy in the first place, the fact that the EC managed to rally around a highly publicized response strategy and to adopt a leadership role at an early stage in the climate change debate did much to raise awareness around the world of the importance of the greenhouse issue (Wynne 1993: 108). Without the EC it is doubtful that the FCCC, despite its evident shortcomings, would have been as strongly worded as it is. Since the 1992 UNCED, however, EU climate-change policy has run into serious problems which question the very nature of its response strategy.

What factors underlie the initial success of EU climate-change policy, and what obstacles have since been encountered? Underlying the transformation from problem recognition to the formulation of a response strategy was undoubtedly the growing understanding of the potential negative effects which climate change and global warming could have for the Community. Initial studies carried out on behalf of the European Commission indicated that without appropriate action there would be significant negative consequences in a number of member-states. While underlining the high uncertainty regarding the consequences of climate change, these studies suggested that there were likely to be considerable variations, not only between the regions of the EU but also between economic sectors and member-states (Rotmans et al. 1994).[16]

Public opinion was a further important factor that shaped the early stages of EU climate-change policy formation. In the late 1980s and early 1990s, public concern throughout Europe was running high, following a series of hot, record-breaking summers. Adding to this, a number of environmental disasters of international scope (e.g. Chernobyl, Sandoz spill into the Rhine, Exxon Valdez) occurred during this period and contributed to public anxiety over environmental pollution.

The desire to play a leadership role at the 1992 UNCED was a further potent factor involved in determining the pace and scope of the Community's response to climate change. Similar to the 1972 Stockholm Conference on the Human Environment, which had a decisive impact on the development of EC environmental policy, the 1992 UNCED focused the minds of policy-makers in Brussels. Bolstered by the Single European Act and the drive to achieve the Single Market – a process which boosted the powers of the Community – the EC had clear ambitions to become a major player in the expanding field of international environmental politics.[17]

Moreover, the EC had by this time attained an unprecedented centrality with regard to the formation of national environmental policy. Widely considered as the motor of environmental policy in Europe, climate change appeared as a logical extension to the Community's environmental competence. Of course, as long as the environment ranked high on the political agenda, and the costs of cutting emissions were still considered manageable, it was relatively easy to make general policy commitments.

However, once it became clear what costs would be involved in cutting emissions and public opinion shifted to other issues, it was bound to be more difficult to stick to the commitments made. Since 1992, EU climate-change policy has been substantially watered down, particularly since it has moved from the declaration of the symbolic 1990 stabilization target to more substantive aspects of policy response. With hindsight,

---

16 See for example EC: DG XI, Development of a Framework for the Evaluation of Policy Options to deal with the Greenhouse Effect, Main Report: Assessment of Strategic Options, May 1992. Draft Report prepared by Climate Research Unit (CRU), University of East Anglia and Environmental Resources Limited (ERL).

17 As Wynne (1993: 102) points out, amid the heated conflicts and uncertainties over the scope of national sovereignty, the EC member-states share a common interest in creating a sufficiently united identity to be recognized as a global power in foreign policy, security and trade agreements in the new post-Cold War world order. Environmental issues increasingly pervade these larger agendas.

it is possible to say that the early phases of European climate-change policy were somewhat overambitious.

Among the numerous obstacles in the way of the EU achieving its climate-change policy objectives, uncertainty over the impacts of climate change ranks prominently. There are also serious economic and political stumbling blocks. Adding to this are the cultural, environmental and institutional differences between member-states. One principal difficulty derives from the fact that climate-change is a collective problem which does not lend itself to easy national fixes; every country is, in a sense, an emission source while, at the same time, everyone is impacted by climate change (Grubb 1995b: 476). Indeed, even though the polluter pays principle is accepted as a basis of Community policy, its application to the climate-change problem raises complex questions. How, for example, is one to measure contributions from different gases over time from different countries?

Unlike the more focused ozone issue in the 1980s, controlling greenhouse-gas emissions is far more difficult, not least because it affects practically all aspects of economic behaviour and is likely to have major impacts on established social habits. Proposals for equitable burden-sharing between member-states, based on the principle of proportionality (i.e. responsibility in proportion to emission contribution) or the possibility of introducing tradeable emission quotas, must ultimately overcome these obstacles.

Another problem derives from the fact that the EU represents, in a sense, a microcosm of the greenhouse problem at global level. Given the wide variations between member-states regarding economic development as well as cultural and institutional factors, in fact, there is a certain North–South dimension to EU climate-change policy. A given strategy will have different costs for different countries, depending on economic structure, tax provisions and resource availability: a fact which influences the disposition of the member-states to adopt more or less stringent actions to combat climate change. Political and institutional differences between member-states accentuate these problems (Grubb 1995a; Wynne 1993). Of course, opposition to a given policy for one set of reasons may often masquerade behind other, perhaps more politically palpable, rationales (Sebenius 1994: 296).

In considering the determinants of EU climate-change policy, it is useful to briefly dwell on some of the key obstacles facing the EU. These range from scientific uncertainty and problems over burden-sharing to subsidiarity and national sovereignty, waning public concern, industrial lobbying and institutional hurdles. Problems in integrating energy and environmental policy will also be briefly discussed here.

8.1. Scientific Uncertainty

First, there is the problem of modelling the impacts of climate change. Even though there is growing acceptance that different regions and sectors of the Community will suffer different impacts, and that the costs of climate change will be significant, attempts to simulate the likely consequences of climate change under various emission scenarios have proved to be unreliable. To date there simply is no way of reliably predicting specific impacts, be it at the regional or local level.

As many of the EU-sponsored studies on climate change have not been predictive, governments still tend to pay more attention to short-term costs rather than the long-term (Skjærseth 1993: 66). If past trends are anything to go by, scientific uncertainty is likely to continue to be used as an excuse for delaying substantive action.

8.2. Internal Differences and Burden-sharing

The EU faces distinct difficulties in allocating emission-abatement measures equitably between the member-states. These difficulties arise from the fact that the measures required to implement emission constraints involve action at a variety of levels, some of which will have more or less significant impacts on a country's economic policies. There is thus a lack of real incentive on the part of member-states to contribute to the collective goal of emission abatement (Grubb 1995a: 47).

Moreover, the EU is made up of countries at different stages of development. This becomes clear when comparing *per capita* CO_2 emissions which vary from 0.75 tons in Portugal to 6.46 in Luxembourg and 3.28 in Germany: disparities which, in turn, make it difficult to define Community interests. In essence this means that some member-states will have to reduce emissions more than others. Not surprisingly, this is difficult to achieve politically.

There is within the EU a distinct North–South divide with regard to ability to pay, GDP *per capita* emissions, primary energy requirements and anticipated greenhouse-gas emissions. Many northern member-states, for example, are constrained in terms of fuel choice due to a restricted resource base and the fact that they have already exhausted most low-fossil fuel options, thereby restricting the potential for further energy-efficiency measures (Skjærseth 1993: 67).

By contrast, nations such as Greece, Portugal and Spain, perceive climate-change measures as costly and detrimental to economic growth. In their view, the richer industrialized countries of northern Europe should shoulder much of the greenhouse-gas reduction load; these are countries, they argue, that are better equipped, financially and technologically, to take on the issue and that, in their view, face a responsibility for historic emissions.[18] In the end, the issue boils down to the question of who should pay for the necessary abatement measures.

Though latterly some northern member-states have become less worried about the economic consequences of emission abatement – Sweden, Denmark and the Netherlands have already taken steps towards introducing energy taxes and Germany is contemplating the issue (*International Environment Reporter*, 14 June 1995: 457–458) – dividing the burden on an equitable basis will be difficult to achieve. Fears of economic recession and the desire to avoid additional budgetary drains are likely to reinforce the reluctance to incur significant costs.

Despite an increasing understanding and awareness of the likely consequences and costs of future climate change, finding an equitable way to allocate emission-abatement measures will thus be difficult. For all the oft-quoted homogeneity of the Community, member-states vary considerably with regard to degree of affectedness, perception of environmental issues and the relative costs of abatement measures: a fact borne out by existing plans regarding national CO_2 emissions.

As Grubb (1995a: 49) points out, the roots of the problem are similar to those that underlie the global endeavour: how to implement collective commitments among diverse and jealously sovereign states in an area as fundamental as energy policy. Burden-sharing is therefore likely to remain a substantial challenge.[19]

8.3. Subsidiarity

Climate-change policy, like environmental policy more generally, does not take place in a political vacuum. Much of its success or failure will be determined by how it is linked to the political priorities of key actors.

---

18 The European Council recommended a deferral of the tax for the Cohesion Fund countries (Ireland, Spain, Greece, Portugal) for as long as their CO_2 emissions remain below 85% of the Community average, based on 1990 index figures. The reason for this form of "burden-sharing" was to allay fears in these countries of being penalized by the tax due to their lower energy consumption and lower CO_2 emissions.

19 A Council statement made on 18 December 1995 reflects this dispute. In it the Council states that the objective of equitable sharing within the EU should be discussed and agreed in parallel and that it should take into account cost-effectiveness and the elements defined in the "Berlin Mandate", such as differences in starting points and approaches, economic structures and resource bases, the need to maintain strong and sustainable economic growth, available technologies, and other individual circumstances, as well as the need that each member-state must contribute substantially to the fulfilment of the obligations of the Treaty.

Success and failure is also influenced by policy-makers' perceptions and the relationship between their priorities and the institutional structures in place at the time (Bergesen and Haigh 1994: 4).

The complex and evolving constitutional structure of the EU, for example, has deeply affected the establishment and implementation of a Community policy on climate change. As Wynne (1993: 125) points out, EU greenhouse-gas policies are threaded with deeply unresolved conflicts over sovereignty, institutions and social identity; a fact which complicates the EU's role in international negotiations and hampers internal implementation.

Indeed, since the signing of the Maastricht Treaty, efforts to craft a coherent climate-change policy have to a considerable extent been overshadowed by the politics of European integration. Fearing the loss of national sovereignty, some member-states (e.g. Great Britain) have attempted to repatriate many elements of EU decision-making, including environmental policy, by focusing discussion on the so-called subsidiarity principle, a far-reaching form of decentralization of the Community's integration process.[20]

One of the first victims of the subsidiarity debate were EU environment and climate-change policy (Bleischwitz 1993: 15). Given the pervasiveness and cross-sectoral implications of the EU's proposed climate-change policy measures (e.g. carbon tax), this is perhaps not surprising. Waning public concern over environmental issues, economic recession and the emergence of external political factors and priorities, such as the conflict in the former Yugoslavia, have reinforced this tendency.

Subsidiarity is not new to EU environmental policy. As early as 1973, the First Community Environment Action Programme underlined that the member-states play a primary role in implementing environmental policy (Axelrod 1994: 127). In the run-up to the ratification of the Maastricht Treaty, emphasis was increasingly placed on subsidiarity as a means of ensuring less centralized decision-making in Brussels. Faced with mounting Euro-scepticism, the European Commission adopted a distinct low profile and put many environmental policy initiatives on hold.

Highlighting the apparent power vacuum in Brussels, different member-states interpreted subsidiarity in different ways (Axelrod 1994: 122). Great Britain, for one, emphasized the importance of repealing many directives and devolving decision-making powers to the member-states.

---

20 In practice, subsidiarity aims to ensure that action is taken at the most appropriate level (i.e. national or Community level). With regard to environmental protection issues, subsidiarity means that the Community is to take action only where environmental objectives can be attained better at EU level than the level of the member-states.

Although this interpretation of subsidiarity did not command universal support, the European Commission, in a pre-emptive, conciliatory move, suggested the roll-back of some 24 pieces of environmental legislation, including the proposed Directive for labelling energy consumption of domestic appliances (ibid. 123). Similarly, the transfer to the national level of the SAVE Programme was considered.[21]

The change in mood also affected the prospects of the carbon/energy tax proposal and bolstered those member-states such as Great Britain who where opposed to the very notion of granting the EU the power to raise taxes. In fact, the proposed carbon/energy tax has been rejected and weakened, largely on the grounds of subsidiarity and competitiveness, to the point where it can make little real contribution to the declared objective or to longer-term reductions.

Whether subsidiarity will continue to influence EU climate-change policy depends in large measure on whether agreement can be reached on more fundamental matters concerning the relationship between the member-states and the EU. Under current circumstances at least, a greater degree of national influence is likely to diminish the prospects for success of EU climate-change policy (COM (94) 67 final).[22] The increased emphasis placed on subsidiarity will at a minimum lead to a temporary delay in the initiation, agreement, implementation and enforcement of EU environmental and hence climate-change measures (Axelrod 1994: 128).

On the other hand, it is doubtful that a subsidiarity-based challenge to environmental legislation is likely to prove successful, if only because the problems it addresses do not confine themselves to the boundaries of member-states. Indeed, the integration of the European economy has made it more difficult for member-states to act unilaterally. Even so, the mere existence of the subsidiarity principle may embolden some member-states to vote against, or to seek to water down, environmental or climate-change initiatives (Visek 1995: 389).

Of course, given the broader cultural, economic, environmental, insti-

---

21 Only two of the original 13 proposals which formed part of SAVE have been adopted to date. These concern boilers and the labelling of domestic appliances. The first daughter Directive to the SAVE Programme, on refrigerators and freezers, came into force in January 1995. Two new Directives introducing energy labelling requirements for household washing machines and electric tumble driers were adopted by the Commission in June 1995.

22 This point is unambiguously emphasized in the Commission's first evaluation of existing national programmes under the Monitoring Mechanism of Community CO_2 and other greenhouse-gas emissions. In it, the Commission states that "concrete measures based on complementary national programmes are unlikely to be sufficient to reach the stabilization target".

tutional and political differences between member-states, key energy-policy decisions such as those required to control greenhouse-gas emissions cannot and perhaps should not be taken centrally. Yet, curbing greenhouse-gas emissions is a collective endeavour with large potential political, economic and environmental benefits accruing from cooperation (Grubb et al. 1994: 33).

8.4. Public Opinion

While there is no systematic correlation between perception and seriousness of the climate-change issue in public opinion and the strength of climate-change policy, the absence or presence of public pressure to respond to environmental challenges is frequently an important factor in determining the formation and execution of public policy (Skjærseth 1993: 68). During the late 1980s and early 1990s, for example, public concern over environmental degradation and climate change strongly influenced initial response strategies to climate change.

Since then, economic uncertainty and the emergence of external political issues associated with the end of the Cold War, not least the deliberations over the future course of European integration, have contributed to a waning of public concern over climate change. With this, there has been less pressure to deal with the issue in a serious manner. EU climate-change policy, in other words, faces a twofold challenge: that of waning public pressure, and the uncertainties over the future course of European integration.

8.5. Industrial Lobbying

An important obstacle to EU climate-change policy, particularly the carbon/energy tax proposal, has been the opposition by organized business and industry interests. The Union of Industrial and Employers Confederation (UNICE), the European Chemical Industry Federation (CEFIC) and the European Petroleum Industry Association (EUROPIA), for example, have put up fierce resistance to the carbon/energy tax proposal, both at national and EU level, ostensibly due to fears that the tax will harm Europe's competitiveness (*Platt's Oil Gramm News*, 15 January 1996: 2). Similarly, many of the Commission's efforts to promote minimum energy-efficiency standards have foundered or been watered down and delayed due to industry opposition.[23] Pressure has

---

23 This opposition has been brought to bear both at national level vis-à-vis member-state governments and directly vis-à-vis the Commission.

come from outside Europe as well. Arguing that the carbon/energy tax could affect the volume of oil exports, the tax proposal has been opposed by OPEC and the Gulf Cooperation Council (GCC).[24]

8.6. Institutional-Structural Obstacles

Institutional-structural factors have played their part in the slow-down in EU climate-change policy. The EU's voting arrangements, particularly the requirement for unanimity over energy questions, represent an important obstacle in this respect, as evidenced in successive Council stalemates over the carbon/energy tax proposal. Differing interests and lack of unanimity over climate change strategy and the choice of policy instruments pervade even the Commission.[25]

While the integration of the European economy has made it more difficult for member-states to act unilaterally, EU climate-change policy is ultimately determined by the actions of member-states. Structural obstacles at national level invariably affect the prospects for the EU as a whole. Aside from member-states' relative ability to reduce greenhouse-gas emissions, basic political and economic considerations have influenced national policy formation and implementation. Of particular concern to government authorities have been issues such as macroeconomic costs, international trade effects, energy security, political ability, legal flexibility and political support.

According to Fish and South (1994: 32), the single greatest consideration in blocking carbon taxes appears to be fears about international trade impacts. The perceived effect that unilateral and even multilateral actions to mitigate greenhouse-gas emissions could have on the international competitiveness of a country's products, as well as specific sectors and regions, appears in fact to have become the overriding concern for OECD governments (ibid. 40). Even though the carbon/energy tax proposal incorporates a conditionality clause, making action by the EU contingent on other OECD governments taking similar steps, and the Commission has agreed to exempt export-dependent and energy-

---

24 The GCC comprises countries such as Saudi Arabia, Kuwait, Bahrain, Oman, Quatar and the United Arab Emirates (UAE). See *Oil and Gas Journal*, 7 August 1995 and Perrin (1995). Some of these countries, it is reported, even threatened to break off diplomatic relations following the announcement of the energy tax.

25 Some member-states, such as Germany, argue against unilateral fiscal measures like the carbon tax and advocate an EU-wide tax. Others, such as Great Britain, by contrast, reject the proposed EU carbon tax for precisely the opposite reason, namely that important measures such as the tax are a matter for national governments to decide. See *International Environmental Affairs*, 7 February 1996, p. 81.

intensive industries, these concerns have invariably affected EU climate-change initiatives.

8.7. Compliance and Implementation Deficit

An implicit assumption pervading many studies of international environmental affairs is the belief that international environmental law is in some way self-executing, that once treaties are negotiated and agreed, the political process comes to a halt. Implementation, from this perspective, is the responsibility of the administrative and bureaucratic machinery of the nation-state.

Yet, as any observer of EU environmental affairs knows, there is a world of difference between member-states' political statements, legislative commitments and obligations on the one hand, and how EC legislation is actually implemented and enforced (Visek 1995: 392). This is a fact highlighted by the European Commission's latest report on monitoring the application of Community law, which underlines the persistent problem of implementation of EC environmental legislation (COM (95) 500).[26]

The reasons for this poor performance are diverse and include structural problems with the Community and its institutions, problems with the way the measures are adopted, and a lack of commitment on the part of member-states as a whole, particularly where member-states have become concerned that vigorous implementation and enforcement will damage their economic interests (Visek 1995: 392, 401). Indeed, in many cases, domestic political commitments have collided with international obligations under the FCCC and the EU. The 1990 stabilization commitment, for example, is no closer to realization in 1997 than it was when it was first announced.

Adding to this, controversy, lack of cohesion and protracted conflict over the underlying boundaries of political legitimacy surrounding Community climate-change measures (e.g. the carbon/energy tax proposal) are unlikely to have lent EU climate-change policy a high level of authority and credibility. This, in turn, makes efficient implementation all the more difficult, particularly since the EU lacks strong powers to implement, monitor and enforce legislation and depends largely on the member-states to implement and enforce Community policy. Problems at

---

26 The report indicated that implementation had improved only slightly in 1994 compared with previous years.

national level, therefore, can seriously undermine Community policy on environment and climate change (Visek 1995: 393).[27]

8.8. Problems in Policy Integration: Climate Change and the Energy-Environment Interface

The relationship between energy and environmental policy is at the core of energy-policy decisions in the EU and is thus central to an understanding of EU climate-change policy and the attainment of the EU's CO_2 emission reduction targets. But just how energy and environmental policy relate to each other is a complex question. Without going into too much detail, some brief consideration of the Community's scope of action in the field of energy and the interface between energy and environmental issues is in order here.

To begin with, there is no general power to define a Community energy policy. Although Article 3t of the Maastricht Treaty lists measures in the domain of energy as one of the objectives of the Community, the EU's scope of action in the energy sector is limited primarily to ensuring that national obstacles to a single market in energy products and services are abolished (van der Esch 1995: 14). Particular attention is given to ensuring that energy-specific requirements such as security of supply and regularity of supplies within a given energy-resources mix are reconciled with single market rules (ibid. 15). In all other respects, Community powers are limited by national energy-political sovereignty.

Although there is the prospect that an energy title may be added when the Maastricht Treaty is revised during the 1996 Intergovernmental Conference, national energy-political sovereignty continues, as documented by the provisions of Article 130S(1), which requires unanimity voting for environmental measures of a fiscal nature or which concern the mix of energy sources in the member-states. Given the continued primacy of national sovereignty in the energy sector, it is hardly surprising that progress towards achieving a single market in energy services and on related issues has been slow (van der Esch 1995: 15).

In the words of one commentator, it is with a fairly "balkanized" and fragmented Community energy market that EU environmental and hence climate-change policy interfaces. Climate change, of course, has provided a powerful impetus for greater integration between Community

---

27 There are large variations among member-states in levels of compliance with Community legislation. Some, such as Denmark, have taken environmental obligations very seriously, whereas others, such as the southern member-states, have taken a more relaxed approach.

energy and environmental policy. Moreover, EU actions in the field of energy have often overlapped with its environmental policy. Power stations, refineries and other large energy-users, for example, are subject to emission limits specified in the Large Combustion Plant Directive. Moreover, the EU has introduced a number of energy efficiency measures under the SAVE Programme and, more controversially, has proposed a carbon/energy tax.

Formally, environmental policy has attained constitutional status and is thus, at a programmatic level at least, equal to other policy areas, such as energy and transport. This is reflected in current Community energy policy,[28] which aims to: (i) Foster the development of an internal market in energy; (ii) improve the development of external energy relations and ensure security of supply; and (iii) ensure environmentally sustainable production, distribution and use of energy products and services.

These goals are reinforced by the Maastricht Treaty, which mandates a "high degree" of environmental protection. They are also reflected in the Fifth Environmental Action Programme which stresses the need to integrate environmental concerns into other major Community policies such as industry, energy and transport. The existence of an EU White Paper on EU energy[29] and the prospect of inserting a new Energy Title in the Treaty on European Union both underline the importance attached to integrating environmental concerns into European energy policy.

The main challenge will be to reconcile the goal of boosting competitiveness and security of supply with protecting the environment. The task will not be an easy one, particularly as barriers to integrating energy and environmental policy remain high, not least due to continued emphasis on national energy-political sovereignty. This, in turn, affects the Community's prospects for meeting its goal of stabilizing emissions at the 1990 level by the year 2000. In fact, in a recent review of the Fifth Environment Action Programme, the Commission forecasts an overall increase in emissions of 1% per year after the year 2000 and estimates that the stabilization target will be missed by up to 5%. Failure of the EU's energy-efficiency programme to achieve a 20% improvement target between 1986 and 1996 is believed to be one factor underlying this (*ENDS Report* No. 253, 1996: 25).

---

28 See, for example, COM(94) 659 final, 11 January 1995 (for a European Union Energy Policy – Green Paper).

29 The White Paper was published in December 1995 and is based largely on the Green Paper on an EU energy policy (COM(94) 659 final) issued in early 1995.

The question is whether national approaches to energy policy will suffice to stabilize CO_2 emissions, let alone to reduce emissions beyond the year 2000. Current evidence suggests that this is unlikely to be the case (Grubb 1995a).[30] According to Commission estimates, CO_2 emissions within the EU will increase substantially over the next two decades in the absence of strong policy interventions – a trend which would be incompatible with the Community's international commitments (COM (95) 682). Little has actually been done to reduce dependence on fossil fuels. Energy consumption in the Community will thus grow at almost 1% per year, but the structure of demand could change in favour of oil and gas. Sustainability in the energy sector is thus an acute issue and this naturally affects prospects for EU climate-change policy. Problems in integrating Community energy and environment policies will add to these difficulties.

That said, climate change has served as a powerful moving force for integration. EU energy policy is increasingly concentrating on energy efficiency – partly due to security-of-supply reasons but also for environmental reasons (COM (94) 659 final). Despite progress in this field, however, the current drive to liberalize energy markets and reduce prices has resulted in a weakening momentum towards energy conservation and environmental protection (*ENDS Report* No. 252, 1996: 38). Environmental concerns, in other words, have been more visible in general policy statements than in the implementation of specific measures (van der Esch 1995: 15).

9. The Future of EU Climate-change Policy

The release of CO_2 and other gases into the atmosphere – and more specifically its consequential effect on the earth's climate system – is now generally acknowledged as a significant international environmental problem.[31] Despite uncertainties over the rate and spatial distribution of human-induced climate change, there is growing awareness of the need for concerted action to avert climate collapse. Known colloquially as the "greenhouse effect", the phenomenon is the subject of the 1992 FCCC.

---

30 One problem with relying solely on national targets is that there is no way to alert in the case of non-achievement. Indeed, here lies one of the chief benefits of the EU's Monitoring Decision, as it provides for a kind of early warning when national policies prove insufficient to meet declared emission goals (Grubb et al. 1994: 34).

31 For a summary of the latest IPCC assessment report on climate change, see *International Environment Reporter*, 20 September 1995: 746–748. See also (IPCC 1995).

The FCCC commits its signatories, including the EU, to specific actions directed at stabilizing emissions of greenhouse gases at levels which prevent dangerous anthropogenic interference with the climate system.

But how well-placed are key actors such as the EU in achieving the objectives of the FCCC? And will these actors be in a position to positively influence the political process of developing the FCCC into an effective international environmental regime? Answering these questions in relation to the EU is difficult, if only because of the time-lag between policy response and policy outcome. Moreover, international institutions such as the EU must of necessity respect the principle of state sovereignty which limits their scope of action. Finally, climate change in many ways typifies what is seen as a growing series of international environmental problems (ozone-layer depletion, dwindling biodiversity, transboundary air pollution, etc.) characterized, *inter alia*, by scientific uncertainty, separations over time and space between cause and effects, and linkages between different aspects of global environmental change, creating joint cost-accounting problems and making it difficult to establish legal and financial liability (Leonard and Mintzer 1994: 330).

Despite these difficulties, it is possible to evaluate the EU in relation to such criteria as agenda setting, international policy formulation and national policy development. In each of these aspects the EU has played an important, indeed vital, role. The EU was instrumental in setting the international agenda on climate change by taking an early lead in the negotiations leading up to the 1992 UNCED and the signing of the FCCC. The adoption of the 1990 CO_2 emission-stabilization target was equally important in this regard. Similarly, the EU was influential, both before and during the 1992 UNCED, in arguing for clear commitments and a strongly worded FCCC. Finally, the EU has had a vital impact on national environmental policy development.

Despite recent setbacks and a slow-down in legislative output, the EU still enjoys unparalleled centrality regarding national environmental policy formulation. While it is difficult to say whether European initiatives on climate change would be better or worse off without the stance of the EU, most observers would agree that without EU leadership in the decisive phases of the negotiation of the FCCC, the latter, despite its obvious shortcomings, would not have attained its present form.

However, notwithstanding the EU's undoubted importance in international environmental affairs, and the professed optimism by EU spokesmen that the EU is on target to meet its goal to hold CO_2 emissions in the year 2000 to the 1990 level, even without the carbon/energy tax, Community climate-change policy faces many challenges. If current trends persist, in fact, it is doubtful whether the Community will be able to meet its climate change targets, let alone achieve emission reductions beyond the year 2000. According to a review of the EC's Fifth Action Programme undertaken by the Commission, the recent recession helped to limit CO_2 emissions, but these are expected to rise as economic recovery takes hold. The Commission furthermore forecasts that the EU's goal of stabilizing emissions by the year 2000 at 1990 levels will be missed by as much as 5%, and that beyond the year 2000 emissions could rise by 1% annually (*ENDS Report*, No. 253 1996: 25).

These conclusions are further substantiated by the Commission's Second Review of National CO_2 Programmes under the Monitoring Mechanism published in March 1996. According to the report, CO_2 emissions have fallen only in three of the 15 member-states – Austria, Germany and the United Kingdom. In the event, the fall in emissions in Germany is attributed to the economic effects of reunification with the former GDR and the installation of energy-efficient factories and power plants. And in the United Kingdom, the report suggests the decline is due to a move from coal to natural gas. Not suprisingly, the results of the report, which point to the insufficient nature of member-state policies to reduce emissions, have sparked widespread criticism of the EU and its member-states' climate-change policies and have renewed calls within the Commission and in some member-states for a Community-wide carbon tax (*International Environment Reporter*, 13 April 1996: 252).

9.1. The Post-Berlin Agenda

Many of these problems were highlighted at the First Conference of Parties to the FCCC in Berlin in April 1995 (*International Environment Reporter*, 18 October 1995: 789–790). While the Conference left many questions unanswered, and only began to address the sizeable gap between climate science and policy, it nevertheless provided an important starting point and set into motion a process for climate protection in the 21st century. This is a process in which EU leadership will be required, particularly in the forthcoming negotiations on a protocol to the FCCC.

Community policy towards climate change over the next few years will be influenced by two broad factors. The first and most important is internal and relates to the outcome of the current debate on the scope and speed of future European integration. The second is external and is driven as much by the growing scientific evidence that climate change is a reality requiring urgent action as it is by the policy response to climate change adopted by other actors, notably the United States and Japan.

As for the first set of factors, considerable uncertainty currently surrounds the process of European integration. While the 1996 Inter-governmental Conference did resolve some of the outstanding issues, uncertainty has spread and now affects crucial areas of Community policy, notably EU environmental policy.

The second set of factors, by contrast, in particular the growing scientific evidence of human-induced climate change, may act as a catalyst to stimulate a renewed push towards an effective international regime on climate protection. On the other hand, the slow-down in EU environment and climate-change policy is, if anything, reinforced by the lack of action on the part of Japan and the United States. EU leadership in the forthcoming negotiations on a protocol to the FCCC will thus be essential.

Broadly speaking, then, the EU's climate-change agenda will be dominated by the forthcoming negotiations of a protocol to the FCCC. Alongside these negotiations, EU policy-makers will have to address questions relating to the rate, magnitude and regional distribution of future climate change; the potential impacts on eco-systems and human society of the probable range of future climate change in Europe; and the optimal mix of greenhouse-gas emission control measures. These may even include tradeable emission quotas based on national emission targets which could be "traded" between the member-states or their industries (Grubb et al. 1994: 36). Crucially, EU policy-makers will have to find ways of determining the impacts of existing control measures, of monitoring emissions and putting into place mechanisms to enforce effectively any agreement reached. Finally, the EU will face increasing pressure to lead by example internally while finding ways of assisting developing countries and following joint implementation measures externally.

Addressing these issues will not be easy for the EU, particularly in a climate dominated by greater budgetary restraint and sustained pressure to decentralize decision-making authority in environmental affairs to the member-states. Repeated calls for the simplification of EC legislation and administration, as voiced by the EU Heads of State at the Corfu Summit in June 1994 and the Essen Summit in December 1995, are indicative in this respect. A report commissioned by the Heads of State in 1994 and published in 1995 on the reform of EU environmental policy, the so-called Molitor Report,[32] in fact, sets out a minimalist agenda for the

---

32 Report of the Group of Independent Experts on Legislative and Administrative Simplification.

future of EU environmental policy and reinforces the recent slow-down in environmental policy making in Brussels (*International Environment Reporter*, 28 June, 1995: 492–493).[33]

This slow down in environmental policy-making has already affected Community measures on climate change. Both the SAVE and the ALTENER programmes, for example, have been reduced in scope, and the carbon/energy tax is now being contemplated on a voluntary basis. If current trends continue, and EU environmental policy is further weakened, the Community's ability to agree and implement effective measures to combat climate change will invariably be affected. This is a development which would fit into a broader pattern in Europe of questioning of collective authority and commitment, particularly, though by no means exclusively, in the environmental arena (Wynne 1993: 101).

On the other hand, the hardening of scientific evidence regarding climate change and the increased concern expressed by many developing countries may provide new impetus to European climate-change initiatives. The expansion of the EU in 1995 to include Austria, Finland and Sweden, the so-called Accession countries, may over time even strengthen EU climate-change policy. While the Accession countries are expected to account for an additional 180 million tonnes of CO_2, or about 4–5% of the total for the EU-12, this is not expected to make a large difference to the EU stabilization commitment. All three have declared very similar emission targets and have strong environmental sensitivities. What is more, all have proved to be strong proponents of international environmental agreements and consider fiscal instruments to be an important element of climate-change policy (Grubb et al. 1994: 29–31). That said, they too face difficulties in achieving their targets and are likely to add only fractionally to the EU's overall effort to achieve the 1990 stabilization target (ibid. 31).[34]

9.2. European Integration and Climate Change

Most European countries share a common interest in creating a sufficiently united identity to be recognized as a global power in foreign policy, security and trade agreements in the post-Cold War era (Wynne 1993: 102). Yet, to be a credible international actor, the EU is dependent

---

33 In the event, the report, which emphasized the value of deregulation, did not generate the necessary political momentum at the European Summit in Cannes in June 1995.

34 The impact of the Accession countries on EU emissions is expected to push EU emissions up by only about 1–2% to 15%.

on greater institutional and political cohesion. The future of EU environment and climate-change policy, and hence the role of the EU as an actor in international environmental affairs, is thus in large part contingent on the outcome of the wider debate on the pace and scope of future European integration.

Efforts to develop a coherent EU response to the greenhouse challenge have coincided with the intensifying conflict over monetary and political union and the extension of the Treaty of Rome. Yet, while the politics of European integration overshadow the climate change debate, the issue cannot simply be ignored. At a time when EU policy-makers and their counterparts at the national level ponder the EU's organizational form as it moves into the 21st century, they must respond to wider transnational challenges such as climate change. Indeed, if the EU manages to overcome the current disputes over national sovereignty, it will be much better placed to face such challenges as climate change. At present, however, the EU and Europe more generally are still far away from such a vision (Kennedy 1993: 256–257; Wynne 1993: 102).

10. Conclusions

Despite these uncertainties, and the noticeable deceleration in the pace and scope of environmental policy-making since 1992, there remains a substantial impetus towards concerted environmental action in the EU, particularly in the field of climate change. Among the measures which harbour the best potential for emission reductions, energy efficiency figures prominently.

Yet the recently proposed extension of the Community's SAVE Programme to cover the period 1996–2000 appears, if anything, to have become a victim to deregulation and subsidiarity. As currently proposed, the SAVE II programme would encompass labelling and standardization of energy appliances and, in contradistinction to the original SAVE Programme, which contained plans for legislation to support energy-efficiency efforts, relies primarily on voluntary agreements with equipment manufacturers (*ENDS Report* No. 253, 1996: 42; *International Environment Reporter*, 14 June 1995: 457–458).

Beyond these measures, the next few years will see the further development and refinement of national climate-change programmes and their evaluation under the monitoring mechanism. In addition to recent policy initiatives, which include the adoption of new energy-efficiency standards for refrigerators, the adoption of new Directives on energy-labelling requirements for household washing machines and electric tumble driers as part of the SAVE programme, and the preparation of a

Draft Directive on curbing demand for energy (least-cost planning), the Commission plans to introduce additional measures over the next three years. These include: (i) Making energy efficiency an essential requirement in new and existing harmonization directives on energy-consuming equipment; (ii) issuing a Communication by 1998 on the use of fiscal instruments to promote energy efficiency and renewable energy sources; (iii) adopting a strategy to promote third-party financing of investments in energy efficiency; (iv) adopting a Communication containing a strategy for easing the commercialization of renewable energy sources in 1997; (v) issuing a Communication in 1996 setting out a framework for voluntary environmental agreements with industry, notably the electricity industry (*ENDS Report* No. 252, 1996: 38–39).

Despite these plans, there is no room for complacency, particularly if current trends in energy consumption are taken into consideration.

To return to the question raised at the beginning of this chapter, whether the EU has been a catalyst or a hindrance regarding the implementation of the FCCC, it is of course too early to tell. Most observers would agree that without EU leadership in the decisive phases of the negotiation of the FCCC, the Convention, despite its obvious shortcomings, would not have attained its present form. Nevertheless, EU climate-change policy faces many challenges. Despite repeated reaffirmations of the 1990 stabilization target and the commitments made under UN auspices, it is doubtful whether the 1990 targets will be met. According to the Commission's Second Evaluation of National Programmes under the Monitoring Mechanism, CO_2 emissions could increase by as much as 5% by the year 2000 from 1990 levels. Of particular concern to the Commission are the growing emissions from the transport sector which increased by 7% from 1990 to 1993 (*International Environment Reporter*, 3 April 1996: 252).

Whether or not the EU achieves its targets and manages to provide the necessary leadership to ensure that the forthcoming negotiations on the protocol to the FCCC are successful, will depend in large measure on whether agreement can be reached on more fundamental matters concerning the relationship between the member-states and the EU on European integration. At a minimum, subsidiarity will delay the initiation, agreement, implementation and enforcement of EU climate-change measures. Equally, scientific uncertainty and economic recession are likely to continue to be used as excuses for delaying substantive action.

Public opinion, of course, may once again come to force decision-makers to take a more aggressive approach on climate change. In this context, non-governmental organizations can play an important part in mobilizing the public. Nevertheless, while Community environmental policy will be characterized by a greater emphasis on sustainability and enforcement in years to come, the recent trend towards consolidation of existing measures is likely to continue for some time yet. This, in turn, will influence EU climate-change measures. Opposition to more binding obligations on member-states and key industry sectors within the Community may manifest itself in the forthcoming negotiations on a protocol to the FCCC. Equally, resistance may rise up and block action to implement commitments made at international and Community level. Specifically, opposition within some member-states may grow if and when target emission limits become more specific.

Notwithstanding the many economic, institutional and political hurdles that have yet to be overcome, climate change has provided a powerful impetus for integration of energy and environment policy in the EU. However, since efforts to curb greenhouse-gas emissions have major implications for economic policy and social habits, it is likely that the scope and nature of Community action on climate change will remain hotly contested within the EU.

Chapter 13

Political Leadership and Climate Change: The Prospects of Germany, Japan and the United States

Gunnar Fermann[1]

1. Introduction

1.1. Brief Recapitulation

Observers have been inclined to judge the 1992 Climate Change Framework Convention (FCCC) as a promising start to the development of an efficient climate change problem-solving regime. However, as pointed out in the introductory chapter, the FCCC suffers from serious weaknesses and has so far failed to be instrumental in curbing the increasing emissions of major greenhouse gases, which, indeed, are at the core of any solution to the problem of global warming. As concerns the future capacity of the climate change problem-solving regime to induce reductions in the emissions of greenhouse gases, the recent estimate of the International Energy Agency (IEA 1995: 49–53), that world emissions of CO_2 are likely to increase by some 30–40% by the year 2010, is a reminder that the steady improvement of the climate change regulatory regime should not be taken for granted.

Indeed, there is good reason to reject the assumption that the process

1 This chapter is a result of my association with The Centre for Environment and Development at the Norwegian University of Science and Technology, and the Energy, Environment and Development Programme of the Fridtjof Nansen Institute. I gratefully acknowledges the valuable comments and interventions provided by Christiane Beuermann, Ben DeAngelo, Per Ove Eikeland, Jill Jäger, Raino Malnes, Olav Schram Stokke, Arild Underdal and Jay P. Wagner to previous versions of this chapter in full or in part. Remaining weaknesses, factual or interpretive, are, of course, my sole responsibility.

The chapter is an elaborated version of my report: Political Leadership and the Development of Problem-solving Capacity in the Global Greenhouse: Prospects of Germany, Japan and the United States Towards the 21st Century, Lysaker: The Fridtjof Nansen Institute – EED Report No. 3, 1994, which was financed by the research programme SAMMEN, run by the Norwegian Research Council.

of regime-formation is governed according to some teleological principle; that the process necessarily and by its own momentum will increase the problem-solving capacity of the present regime. There are few "necessities" in international politics. The future development of international institution-building processes should be considered an empirical question to be determined only in retrospect. *A priori*, however, any outcome – including the more pessimistic ones – should be kept open: Problem-solving regimes may improve, stagnate, reverse, or even be dismantled.

In the previous chapters, various factors impeding attempts at remedying the problem-solving deficit suffered by the present climate change regime have been analysed: Problems related to scientific uncertainty and the relationship between science and politics were dealt with in chapters 3, 4 and 5. In chapters 1 and 5 obstacles related to the nature of the state and the international state system were discussed, while in chapters 6 and 7 the ambiguities and difficulties involved in arriving at an equitable and efficient solution to the problem of climate change were accounted for. Finally, analyses in chapters 8 through 11 indicated that developing countries and transforming economies facing more urgent problems should not be relied upon to initiate a closing of the evident gap between the seriousness of the climate change problem and the inferior institutional capacity developed internationally to cope with it. Regions and countries like Africa, China, Brazil and Russia, relying upon increasing fossil-fuel energy consumption to facilitate economic development, are unlikely to favour the development of an international regime putting radical limits on their CO_2 emissions.

1.2. The Requirement of Political Leadership: Research Questions

Granted that a more efficient international regime is required to manage the problem of climate change and the unlikelihood that the development of such a regime will be achieved by means of some invisible and mysterious self-propelling mechanism, the question arises as to *how* and *who* might be able to facilitate an improvement of the 1992 climate change problem-solving regime?

Considering the *how*-question first, experiences from the negotiation of the 1992 FCCC and the 1987 Montreal Protocol indicate that extraordinary state behaviour had an impact on both the speed and the outcome of the two negotiating processes (Benedick 1991; Haas et al. 1993). Indeed, scholars have extrapolated that political leadership might be a necessary condition for the successful development of international environmental regimes in general (Underdal 1991a; Sprinz and

Vaahtoranta 1994; Young 1991). As for the case of climate change, there is so much indicating that the initiatives taken by certain European countries during the 1988–92 period were vital in transforming the issue of climate change from one of science to one of international politics (Fermann 1994: 12). Furthermore, it is probable that various forms of political leadership will become as essential a factor in creating incentives for the further operationalization and actual implementation of the principles and intentions embedded in the 1992 FCCC, as they were instrumental in developing and marketing these ideas prior to UNCED.

Assuming then, that political leadership constitutes a potentially important and continuing asset for the world community in coping with international environmental challenges, and, more specifically, that various forms of leadership will be essential for the elaboration of the 1992 FCCC into protocols and coordinated action, *who* are the most promising candidates for assuming leadership?

For leadership to materialize, promising candidates are required – that is, states with the resources and incentives to make a special effort to control, stabilize or reduce world greenhouse emissions. Germany, Japan and the United States all belong to the rather exclusive, but heterogeneous, group of *critical actors* capable of contributing substantially to either the solution or the worsening of the problem of global warming. This "club" also includes large, fast-growing and fossil-fuel intensive developing economies like China and India, and countries, like Brazil, possessing large, but dwindling, resources of tropical forest (sinks). Owing to their strong priority of "catch-up" industrial development and their limited scientific and financial resources, however, these developing countries are not in a position to act as catalysts for a more effective climate change problem-solving regime. As documented in previous chapters, several governments of developing countries view international regulation of greenhouse emissions with scepticism, since such regulations are widely perceived to contradict the industrial strategy they have chosen to promote economic growth. A regulatory regime is, it seems, in their short-term interest only to the extent that it provides a mechanism for the additional and, preferably, unconditional transfer of technology and funds to their economies.

Although governments of some major developing countries may have the weight to act as "veto powers" to the elaboration of the climate change regime,[2] and this is exactly the reason why they count as critical

---

2 While industrialized countries are responsible for about three-quarters of the world's accumulated emissions of greenhouse gases, developing countries account for two-thirds of the increase in emissions (Toichi 1993: 26).

actors, they seem to have neither the motivation nor the resources required to take on a positive *leadership role* in the decade ahead. This demanding role seems more fit for industrialized countries like Germany, Japan and the United States, all of which possess the resources, if not the political will, required to fuel the policy process towards strengthening the international problem-solving capacity – including the operational-ization and implementation of the abatement principles and goals embed-ded in the 1992 FCCC – with research, technology, funds, political influence and administrative capacity. In the last chapter of this book, I thus attempt to: (i) shed some light on the leadership potential of Germany, Japan and the United States in the field of climate change, and (ii) suggest how this potential may be utilized to improve upon the less than satisfactory climate change regime, making it more effective in regard to the malign nature of the climate change problem itself.

The specific predictions will be based upon an empirical investigation of structural factors as well as various policy dimensions there is good reason to believe will influence the ability and willingness of Germany, Japan and the United States in taking a leadership role in the future. These factors include emission responsibility, energy conservation and fuel-switching potential, and current climate change policies including abatement target, commitment, and measures.

The mapping of the leadership potentials of Germany, Japan and the United States, and the subsequent inferences made to particular future leadership roles, are provided in sections 3 and 4 respectively. Prior to the empirical investigation, however, some preparatory work needs to be done on the analytical framework and on the content of the leadership concept.

2. Analytical and Conceptual Elaboration

2.1. The Analytical Framework

The world community's potential for solving the problem of climate change may be conceived as dependent upon (i) the nature of the chal-lenge as well as (ii) the extent to which an international institutional capacity has been developed to cope with it. More generally, then, the *problem-solving potential* (PP) is a function of the *problem-severity* (PS) of the challenge in hand and the *problem-solving capacity* (PC) of the inter-national community in the specific issue area; $PP = f(PS, PC)$.

It follows from this that the PP increases to the extent that the PS is small and the PC is large. How does extraordinary state behaviour in the form of political leadership relate to these parameters?

Recall the evaluation of the nature of the climate change problem and the status of the climate change problem-solving regime made in Chapter 1. Here it was concluded that climate change is characterized by *high problem-severity* due to the broad scope, the long duration, and the extreme complexity of the issue, and *low problem-solving capacity* as reflected in the failure of the FCCC to commit the signatories to specific and binding climate change abatement targets, as well as the fact that world CO_2 emissions continue to increase. Thus, the climate change issue seems to fit quite well into category "B" in the matrix presented here:

Table 1: Problem-solving Potential – Typology

Problem-severity	Problem-solving Capacity	
	High	Low
High	A	**B**
Low	C	D

If this evaluation is correct, it implies, in the case of climate change, that the *problem-solving potential* at present is *low*. How, then, to rephrase my previous question, can political leadership contribute towards *increasing* the problem-solving potential in this important field of global environmental change? Or, stated somewhat differently: How may the problem-solving deficit currently suffered in the field of climate change be remedied by "injections" of political leadership?

The short answer is either by (a) changing our *perception* of climate change itself (PS), or by (b) increasing the international problem-solving capacity (PC) available to *manage* this particular transboundary problem. Some elaboration is required.

(a) Extraordinary efforts (leadership) in the realm of climate change science are likely to deepen our understanding of the problem. One cannot rule out the possibility that further research may radically alter our perception of how real and serious the climate change problem is. The high level of scientific uncertainty allows for this (see chapters 3 and 5). In particular, much work remains to be done regarding the distribution of impacts between various regions of the world. Leadership directed towards improving our knowledge of the causes, mechanisms, and the impacts of climate change may produce three outcomes influencing the problem-solving potential in three distinctly different ways: Further research inspired by some extraordinary initiative (leadership) may (i) *confirm* that our perception of climate change is largely correct; (ii) demonstrate that climate change is an even *more serious* problem than previously believed; or (iii) prove that global warming is *a lesser threat* than indicated in the recent IPCC assessment. While the first outcome does not alter the present problem-solving potential, the second and the third findings will respectively increase and decrease the current problem-solving deficit.

Scientific leadership may be executed by "epistemic communities"[3] both within the IPCC process and in individual countries, and may, through the production of new knowledge, influence the behaviour of governments. For instance, if additional evidence confirms the recent conclusion of the IPCC (1995) that the present trend of global warming is outside the range of natural variation, this may become an incentive for more radical abatement measures. On the other hand, if science confirms the impression that the industrialized North is less vulnerable to the impacts of climate change than the developing South, this could seriously weaken the former countries' incentives to invest money in developing the present problem-solving regime.

(b) While extraordinary efforts may be directed towards revealing the remaining "secrets" of climate change (scientific leadership) – secrets which impact on our perception of how serious the problem of climate change might be, are, by definition, unknown – political leadership may be applied as fruitfully to improve upon the international capacity to mitigate and adapt to climate change (PC). By influencing the direction and speed of the future development of the problem-solving regime, the present problem-solving deficit may be reduced.[4]

The idea that political leadership may contribute towards improving problem-solving capacity is no novelty: Underdal perceives leadership as an important asset for strengthening the institutional "tools" required for an effective solution of international problems (1991a: 139–140). Young emphasizes the role of political leadership even more strongly, arguing that "institutional bargaining is likely to succeed when effective leadership emerges ... (and) fail in the absence of such leadership" (1989:

3 The term "epistemic community" refers to a knowledge-based transnational network of experts whose members share common views about the causes of environmental problems and the policies to control them (Haas 1990).
4 It should be noted that scientific efforts, whether initiated on the basis of some extraordinary political impulse or not, may themselves contribute towards a strengthening of the international problem-solving capacity through the development of better and more effective technological means of mitigation and adaptation. What distinguishes basic from applied research in our context is the fact that the former may change our perception of how severe the climate change is, while the latter may produce technological tools that increase our capacity to deal with the problem. Hence, both kinds of research efforts are potentially capable of influencing the problem-solving potential, although by different avenues.

373). He reinforces this opinion elsewhere, concluding that "the presence of leadership is a necessary condition ... for success in reaching agreement on the terms of constitutional contracts in international society" (1991: 302).

The notion that leadership is a requirement for developing problem-solving regimes seems also to be rooted in empirical experience: The relative success of the international ozone negotiations as reflected in the 1985 Vienna Convention (general agreement on principles), the 1987 Montreal Protocol (agreement on specific CFC regulations), and the 1990 and 1992 amendments (tightening of CFC regulations), can, in part, be attributed to the leadership role executed by the United States and UNEP (Benedick 1989: 43–50; 1991; Parson 1993: 27–73; Sprinz and Vaahtoranta 1994: 77–105). Stating this, the ozone case also illustrates that one should not expect one single actor to execute leadership in every aspect and phase of the negotiations.

The key role of leadership also seems to be confirmed by a brief glance at the climate change issue as it has evolved from the late 1950s: During the 1960s and 1970s, the United States research community took the lead in developing and proliferating scientific knowledge on climate change. By scientifically acknowledging the problem, the United States research community contributed towards diffusing the issue from the science arena to the international political agenda. Even more crucial for making climate change an issue of major political concern, however, seem to have been the initiatives taken by certain West European countries in the late 1980s and the early 1990s: By unilaterally committing itself to the stabilization of CO_2 emissions, the Netherlands was instrumental in initiating a process within the European Union that led to the adoption of CO_2 stabilization as a common policy goal (see Chapter 12). Furthermore, realizing the low cost-efficiency of uncoordinated greenhouse gas abatements at the national level, Norway introduced the concept of "joint implementation"[5] into the political debate in 1991. These initiatives by two small West European states served to fuel the political process and made an imprint on the general principles adopted in the 1992 FCCC.[6]

The empirical investigation which follows later in this chapter is based on the assumption that political leadership may become equally, if not

---

5 The concept of "joint implementation" refers to the idea that one country, facing high marginal costs on abatement efforts at home, provides assistance – technological, financial, administrative – to less energy efficient and less developed countries securing a more efficient use of its resources and gaining some credit (emission permits) for its foreign abatement efforts. See also Chapter 7.

6 See the FCCC, Art. 4.2.

more, important in specifying and implementing the general principles envisaged in the FCCC, since it was instrumental in developing and marketing these principles in the first place. Prior to this, however, it is crucial to develop a clear and distinctive definition of the leadership concept.

2.2. Towards a Core Concept of Leadership: Conceptual Analysis

To ask how political leadership can be identified is to inquire into what the defining characteristics of the concept should be. Ideally, the combination of characteristics should be unique for the specific concept in question. In the repertoire of concepts found in political science literature, there are many that do not satisfy the demand that the criteria characterizing different concepts should be mutually exclusive. This is unfortunate since the overlap between the defining criteria of two or more "neighbouring" concepts invites analytical confusion and tends to reduce their discriminatory value.[7]

This seems to be the case for the neighbouring concepts of "political leadership" and "power politics": Although both address the phenomenon of extraordinary behaviour in international politics and negotiation, both also possess unique features not shared by the other. Thus, in order to distinguish clearly between leadership and other kinds of extraordinary state behaviour, a rather narrow definition of leadership is aimed for.[8]

2.2.1. Considering Underdal and Young's Conceptual Contributions

Let me start my conceptual analysis by considering a contribution of Arild Underdal, whose main aim is to identify various mechanisms for the execution of leadership in negotiations. These mechanisms are criteria for a typology that distinguishes between several types of leadership to be discussed later. To account for several types of leadership, the author invokes a rather broad definition of the concept:

---

7 Everything else being equal, the analytical sharpness and distinctness of a concept tends to decrease with a broadening of the empirical scope of the concept.
8 For important contributions to the conceptual leadership debate, see Calvert (1992: 7–24); Frohlich et al. (1971); Malnes (1995: 87–112); Underdal (1991a: 139–153); Young (1989: 349–375); and Young (1991: 281–308).

Leadership [is] an asymmetrical relationship of influence, where one actor guides or directs the behaviour of others towards a certain goal over a certain period of time (Underdal 1991a: 140).

While this definition accounts for the relational, asymmetrical, supervisory and goal-oriented aspects of leadership, it lacks criteria capable of distinguishing leadership from other kinds of extraordinary state behaviour aimed at influencing the actions of other states: Reading Underdal's definition and not knowing that it is leadership he defines, it can quite easily be confused with the neighbouring concept of *power*, which is, according to a standard definition, associated with "the ability of an actor to get others to do something they otherwise would not do" (Keohane and Nye 1977: 11).

Comparing the two definitions, it seems clear that Underdal's conception of leadership encompasses the definition of power referred to above. More specifically, the execution of power can be conceived of as one of several possible specifications of Underdal's broad leadership concept. This interpretation is confirmed when looking into Underdal's trichotomic typology of various modes, or kinds, of leadership: Besides "instrumental leadership" and "leadership through unilateral action", Underdal includes "coercive leadership" in his leadership typology. According to the author, the coercive mode of leadership works "through sticks and carrots affecting the incentives of others to accept one's own terms" (1991a: 143).

The blurring of differences between power politics and acts of leadership is also evident in the works of Oran Young. Emphasizing the ability of leaders to "solve or circumvent collective action problems ... [that complicate the realization of] joint gains" (1991: 285), like Underdal, Young distinguishes between three kinds of leadership: the structural, intellectual and entrepreneurial. While these categories only in part correspond with Underdal's typology differentiating between "coercive", "instrumental" and "unilateral" modes of leadership, the main task here[9] is to draw attention to the fact that Young's conception of "structural leadership" is very similar to Underdal's understanding of "coercive leadership". In particular, Young associates "structural leadership" with "the ability to deploy threats and promises in ways that are both carefully crafted and credible" (ibid.: 290).

---

9 The "intellectual", "entrepreneurial", "instrumental", and "unilateral" modes of leadership will be discussed later in section 2.2.4.

2.2.2. Empirical Costs of a (too) Broad Leadership concept

Both Underdal and Young accept bribery and blackmail as leadership strategies and, by doing so, invoke a broad leadership concept which includes attributes of the established concepts of power and power politics as well. Underdal's and Young's analytical choices involve certain costs which induce me to reconsider their conception of leadership:

(a) *Characteristics of issue area argument*: To begin with, it can be argued that the coercive elements of power politics (punishment and rewards to change preferences) are ineffective as means of solving collective action problems on international *environmental* issues (Keohane and Nye 1977; Benedick 1991: 143)[10] and should therefore be excluded from the leadership concept. However, even if it can be substantiated that coercion is a marginal element in developing and securing support for international environmental regimes, this argument is not in itself a strong enough incentive to reduce the empirical scope of the leadership concept. Whether an empirical phenomenon (coercion) makes a difference or not (for the development of problem-solving capacity), is, in principle, irrelevant for the *initial task* of constructing analytical tools capable of grasping and discriminating between these phenomena.

(b) *Discriminatory concepts argument – general version*: Much more serious is the argument that concepts referring to the same analytical level should, as a rule of thumb, be as narrow as possible in order to (i) avoid overlap and confusion with related (but not identical) concepts, and to (ii) increase the discriminatory capabilities of concepts so that phenomena otherwise being lumped together are distinguished as qualitatively different.[11] Although this general principle provides an *incentive* to delimit the concept of leadership from the power concept, the *means* of actually doing so must be based on concept-specific arguments such as the one presented below.

(c) *Discriminatory concepts argument – specific application*: Previously it was stated that to ask how acts of leadership can be identified is to inquire into what the defining characteristics of the concept should be. Now this question can be specified: *By what criteria can leadership be distinguished from power?* Or, put differently: What makes leadership unique in relation to acts of power politics? In responding to these questions, let

---

10 Young argues, for instance, that "it is virtually impossible to achieve high levels of implementation and compliance over time through coercion" (1994: 133).

11 If this principle were given more attention in conceptual work, the tendency towards duplicating the confusion surrounding so many concepts of political science could be minimized.

me shift the focus from leadership and power politics to the political actors that become subject to the guidance of leaders and power-executing states. The difference between leadership and power is most clearly accentuated when we look into the diverging mechanisms through which they influence the actions of other actors. As a starting point, let me briefly consider two sources of "authority".

"Authority" may not only be based on *power* conceived of as the capacity to make credible threats and promises to punish and reward subjects to secure their obedience. To the extent that "authority" is perceived as just and is voluntarily accepted, *legitimacy* provides an additional source of "authority". This is illustrated in the cases of "totalitarian" and "democratic" regimes: In "totalitarian" regimes, the execution of power is the sole source of "authority", and thus the two concepts are interchangeable. In "democracies", however, the "authority" of the government has a broader base – including both coercive elements and elements of voluntary acceptance (consent).

2.2.3. Distinguishing Two Kinds of Extraordinary State Behaviour: Leadership and Acts of Power Politics

The above exposition can now inspire the making of a reasoned distinction between leadership and power: States depending on power (coercion) to influence the behaviour of other actors base their strategy on threats and promises of punishment and reward. States giving in to such pressure are victimized and take the role of a "client" in relation to one or more "patron" states. A state influencing other actors through leadership does not rely on coercion, but on *persuasion* and *demonstration*. Its role as a leader is not based on the capacity to enforce policies on others, but rests rather on its ability to extract the *voluntary acceptance* (consent) from other actors for its goals and strategies in a specific issue area. States consent to the guidance of leaders not because they are forced to, but because they conceive the aims, ways and example set by the leader to be in their *mutual* interest. From this it follows that a leader does not create victims or engage in "patron–client" relations. States consenting to the leadership of another state on a specific issue should rather be conceived of as *followers*.

Table 2: Distinguishing "Leadership" from "Power Politics" – Defining Criteria

	"Leadership"	"Power Politics"
Goals/Motives	Common/mutual interest	Self-interest
Influencing Strategy	Demonstration, persuasion	Coercion, manipulation
Subjects	Consenting "followers"	Obedient "clients"

Turning the focus from the followers back to the leader, we are now in a position to grasp what distinguishes leadership from the concept of power and acts of power politics: In contrast to states exhibiting power politics, leaders influence other states through the mechanisms of persuasion and demonstration, and attract followers on the basis of common interest. The use of non-coercive means of persuasion and demonstration, and the mixed motives encompassing both special and common interests, are the characteristics (criteria) that distinguish leadership from power (see Table 2).[12]

Based on these considerations, I am inclined to propose a new definition of leadership. Revising Underdal's original formulation, the core definition presented below excludes the intentional and strategic elements of power from the concept of leadership, including the "coercive" mode of leadership that works "through sticks and carrots" and "structural" leadership which is associated with "the ability to deploy threats and promises in ways that are both carefully crafted and credible" (Young 1991: 290). Thus:

> Leadership is an asymmetrical relationship of influence, where one actor guides or directs the behaviour of others *by means of persuasion and demonstration* towards a certain *goal of mutual interest* over a certain period of time.

It should be emphasized that to qualify for leadership, good proposals are not enough. In order to attract followers, the goals and means forwarded by a would-be leader must be perceived of as acceptable or favourable by other actors. The core of the matter is that intentions and initiatives, by means of persuasion and demonstration, are actually transformed into collective action capable of bringing about the specification, implementation or institutionalization of these initiatives.

---

12 This conclusion is much in line with a recent work of Raino Malnes who, for the purposes of sharpening the term, excludes state action based on narrow "self-interest" (motivation) and executed by means of "threats and offers" (strategies) from the concept of leadership (1995: 105–106).

Moreover, by reducing the empirical scope of the leadership concept, I do not imply that leadership is the only, or even the most important, factor influencing the future development of the climate change problem-solving regime – or other environmental regimes for that matter. Power politics, including the subtle use of coercive strategies, will probably also play a role. However, in the present study I have no intention of studying the role of power politics in future climate policy-making explicitly or implicitly as part of a leadership concept that encompasses too much.

2.2.4. Modes of Leadership

Recalling that "coercive" and "structural" modes of leadership were excluded from the core concept of leadership elaborated above – that leadership is rather a way of influencing the actions of other states by means of persuasion and demonstration – what are the remaining mechanisms through which leadership is conveyed and expressed? Based on Underdal's and Young's contributions, I shall discuss and distinguish between four types of leadership: the "intellectual", "entrepreneurial", "instrumental", and "unilateral" modes of leadership. In the final predictive analysis, this typology will provide us with the analytical vocabulary required to conceptualize *how* the leadership potential of each one of the countries under consideration may be executed in the future (leadership roles).

2.2.4.1. Intellectual Leadership

The intellectual leader is "an individual who produces intellectual capital or generative systems of thought", and who "relies on the power of ideas to shape the way in which participants in institutional bargaining understand the issue at stake" (Young 1991: 288, 298). By defining the "options available to come to terms with the issue", the "intellectual leader" may, according to Young, play "an important role in determining the success or failure of efforts to reach agreements" (ibid.: 298).

Two "generative systems of thought" can be imagined: "Intellectual" leaders can provide *causal models* which, in Raino Malnes's formulation, "lie behind beliefs about the effectiveness and distributive implications of various solutions to international problems" (1995: 107). By altering cognitive perceptions, he may influence "national preferences and bargaining positions" (ibid.). But the "intellectual" leader may also influence the preferences of states by marketing *normative principles*, "which, together with self-interest, make up the evaluative component of preferences and positions" (ibid.). Young mentions "sustainable development" and "environmental security" as examples of moral principles (1991: 300).

The persuasive power of such intellectual frameworks of interpretation should not be underestimated. On issue-areas such as climate change, characterized by imperfect information and considerable uncertainty, the influence and persuasive power of intellectual frameworks may become significant, especially so, it would seem, if supported by influential "epistemic communities" (Haas 1990).

2.2.4.2. Entrepreneurial Leadership

According to Young, the "entrepreneur" is "an individual who relies on negotiating skill to frame issues in ways that foster integrative bargaining and to put together deals that would otherwise elude participants" (1991: 294). Adding to this, Malnes suggests that "entrepreneurs" also "can direct their effort ... at the institutional setting of negotiations" (1995: 106–107).

The framing of issues and institutional settings may be accomplished (i) by means of *agenda-setting* "shaping the form in which issues are presented"; (ii) through *popularization* and thus to "draw attention to the importance of the issue at stake" (Young 1994: 294); and (iii) by *inventing* "inclusive package deals" to overcome bargaining obstacles (Malnes 1995: 106). What distinguishes the "entrepreneur" from the "intellectual" leader is that while the latter provides negotiating premises by supplying "generative systems of thought", the former is a popularizer and communicator (agenda-setter) of such frameworks of interpretation. "Accordingly", Young notes, the "entrepreneur" often "becomes a consumer of ideas generated by intellectual leaders" (1991: 300).

2.2.4.3. Instrumental Leadership

The essence of instrumental leadership is finding means to achieve common ends. The demand for such services is greatest in situations and in issue-areas characterized by uncertainty: That is, in situations where actors have "incomplete and imperfect information and (only) vague preferences" (Underdal 1991a: 145). To the extent that negotiations involve searching, learning and innovation there is also scope for instrumental leadership. According to Underdal, in order to succeed in executing instrumental leadership, an actor needs to be well equipped with certain key capabilities:

(i) *Skill* is required whether the aim is to develop substantive solutions or engage in the political engineering of consensus. In the first case, the instrumental leader should be able to provide some answers to

problems of technological feasibility, economic efficiency and burden-sharing. As to the political engineering aspect, the leader is
capable of formulating politically feasible agreements.

(ii) Along with skill, *energy* is a necessary requirement for executing
instrumental leadership. "The amount of energy an actor will bring
to bear on a problem can be conceived of as a function of capacity
and interest" (Underdal 1991a: 146).

(iii) *Status* refers to an actor's formal and informal roles in a system, and,
according to Underdal, "generally serves as a key of access to decision-makers, arenas and issues" (ibid.: 147). Status is also a source
of legitimacy and respect.

It is evident that Underdal's notion of instrumental leadership to a considerable extent *overlaps* with Young's conceptions of intellectual and
entrepreneurial leadership. One way of integrating the leadership typologies of Underdal and Young would be to see entrepreneurial and intellectual leadership as sub-categories of instrumental leadership. Certainly,
both the intellectual and the entrepreneurial leader must bring much
energy to his role to succeed. What distinguishes the two leaders is the
kind of skill and status they possess.

2.2.4.4. Leadership through Unilateral Action

According to Underdal, "unilateral action may provide leadership
through at least two different mechanisms"; through its "substantive
impact" and as a means of persuasion. Concerning the mechanism of
"substantive impact", a distinction can be drawn between a direct and
an indirect version: In the former, the unilateral action of an actor is sufficient to provide the collective good in question (i.e. enhance security,
protect the ozone layer). In the latter, the unilateral action of one actor
"sets the pace to which other parties may find themselves more or less
compelled to adapt to" (Underdal 1991a: 141).[13]

---

13 At this point, two remarks should be made – one analytical and one substantial:
(i) As pointed out by Underdal, in the instance that other actors find themselves compelled to adapt to the example set by an actor, "unilateral action may very well be coercive in effect" (ibid.). In such a case, the extraordinary behaviour of an actor should
be classified as "power politics" rather than as "leadership" (see Table 2). (ii) Moreover,
in the instance where an actor provides a collective good all by itself, "(this) unilateral action by one actor may weaken rather than strengthen (the incentive of other
actors) to contribute", thus invite free-riding (ibid.). In this way, leadership may back-
fire and prove counter-productive for the purpose of directing collective action.

In contrast to the former mechanism (substantive impact), unilateral action as a means of persuasion may be executed also by small and weak countries: Underdal argues that "in situations characterized by high problem similarity (common problem), unilateral action may be used for the purpose of demonstrating that a certain cure does indeed work, or to set a good example for others to follow" (Underdal 1991a: 143). To the extent such a demonstration contributes to removing uncertainty about future actions, it can, as Underdal points out, *indirectly* contribute towards solving the problem in question "by helping to persuade others to follow" (ibid.).

It is evident from the above exposition that the intellectual, entrepreneurial, instrumental, and unilateral modes of leadership do not constitute a single typology satisfying the requirement of mutually exclusive categories.[14] Most notably, Underdal's conception of instrumental leadership overlaps with Young's categories of intellectual and entrepreneurial leadership, although the fit is not perfect. Moreover, the above list of leadership modes fails to be exhaustive. As Malnes points out, Young's conception of intellectual leadership "overlooks the influence that vivid descriptions of problems and dangers can have on people's readiness to do something about them, and makes no mention of how ... political positions [may] depend on potentially alterable attitudes to risks" (Malnes 1995: 107).

A somewhat similar criticism is directed towards Underdal's conception of leadership: Malnes finds that Underdal does not account for the possibility that leaders may be capable of "altering values and interests" of other actors (ibid.108). The "configuration of actor interests and preferences" is taken more or less for granted, making no mention of the possibility that moral appeals and sound judgement may make actors "think anew about their ultimate goals" (ibid.). However, this possibility is accounted for in Malnes's conception of *directional* leadership and could therefore be suggested as a fifth category of leadership.

To sum up: Although some scholarly work remains before the various typologies of leadership become harmonized, the contributions discussed above do provide a rich repertoire of ideas and insights on how leadership may be executed. In the final section, I draw upon this repertoire in conceptualizing how the future leadership role of Germany, Japan and the United States may be implemented. It should be emphasized, however, that in real politics one should not expect actors to manoeuvre strictly in accordance with analytical distinctions: Quite the contrary,

---

14 This is not surprising given the fact that Underdal and Young designed their typologies independently of each other.

successful leadership would seem to require more often than not a strategy including the mixing of several modes of leadership, and even the mixing of leadership (persuasion) and power politics (coercion).

3. The Leadership Potential of Germany, Japan and the United States: Structural and Political Dimensions

The essence of predictive analysis is to anticipate (forecast) the development (variation) of a specific phenomenon by means of identifying and describing the current status (value) of a set of factors there is good reason to believe will influence the future state of the phenomenon in question. While the main task of the explanatory study is to account for *past* events, the predictive study assumes causal relationships (in a probabilistic sense) between a set of independent variables and the dependent variable in order to *forecast* changes in the latter.[15]

This section is devoted to the selection and mapping of a set of independent variables believed to be crucial to the ability of Germany, Japan and the United States to take a role of leadership in the future. By mapping each country's potential for leadership, that is, their "performance" on certain key dimensions to be specified below, I should be in a position to suggest more specifically how Germany, Japan and the United States might contribute towards the improvement of the present climate change problem-solving regime (Section 4).

---

15 Since prediction depends so much on the explanatory variables chosen, this operation invites serious consideration. What constitutes, more specifically, a "good reason to believe" that a set of variables "Xn" will continue to influence the workings of a phenomenon "Y"? How are we to justify the assumption that some variables are capable of explaining the future variation in the phenomenon to be predicted? Two justifying strategies are available:

The choice of explanatory variables can either be based on the *experience* that "Xn" has influenced "Y" in the past, or derive its legitimacy from a more or less formalized body of *theory* which, by describing the mechanisms in operation, explains exactly why "Xn" should influence "Y". The two strategies are not mutually exclusive. Indeed, the more empirical evidence which is available to confirm the relationship, and the more clearly the mechanism communicating this influence can be described, the more convincing it is that the explanatory variables ("Xn") are capable of predicting future changes in the dependent variable ("Y"). Assuming that everything else remains constant (ceteris paribus), it is, for instance, reasonable to predict that consumption of oil will increase as long as the present price of oil is low or on the down-grade. The basis of such a forecast can be either axioms of economic theory (i.e. the mechanism of demand and supply), or simply the empirical experience that demand tends to increase when the price of a commodity decreases.

3.1. Identifying Aspects of Leadership Potential: Capacity and Willingness

Having argued that Germany, Japan and the United States are critical actors in international climate policy-making due to their scientific, technological, financial and administrative resources, what is, more specifically, meant by *leadership potential*? How is this term to be operationalized? In evaluating the respective potential for leadership of Germany, Japan and the United States, an elementary distinction must be made between the *capacity* and the *will* to take a leadership role in international climate change policy-making. This distinction corresponds to the distinction between military capacity and political intention made by strategists when estimating military threats, and accounts for the fact that strategic or political behaviour cannot be inferred from estimates of capacity alone.

Below, two aspects of the leadership potential corresponding to the distinction made between capacity and political will are introduced.[16]

a. Structural features (capacity):
  – Energy supply structure and substitution potential
  – Energy conservation potential

b. Political and perceptual features (political will):
  – Abatement targets
  – Abatement measures
  – Credibility of abatement commitment
  – Perception of climate change[17]

By studying each country along these dimensions, I hope to narrow the scope of political choice and identify the most promising avenue for taking a leadership role for each country under consideration.[18]

---

16 No positive correlation between capacity and political will is assumed. If it had been the case that favourable structural conditions systematically translate into strong climate policies, it would be of limited interest to study political and perceptual indicators empirically. They could simply have been inferred from the structural conditions revealed. Empirical studies indicate, however, that a strong positive link between objective conditions and policy-making cannot be taken for granted in environmental politics. For an illustration, see Dahl (1993).

17 This dimension is not treated separately, but integrated into the mapping of the other variables.

18 The relevance of each of these dimensions for the prospects of taking a leadership role are discussed prior to the documentation of each of the dimensions.

In addition to these six dimensions, a final variable is considered: responsibility – the moral dimension of leadership. It will become evident that the question of responsibility includes aspects of both structure and policy/perception. This is the reason the issue of responsibility is treated in a separate section prior to the other six comparative dimensions.

3.2. A Cautious Remark on Prediction[19]

Before setting out to describe and compare Germany, Japan and the United States along the mentioned dimensions, a fundamental epistemo-logic question shall be granted consideration: *How far into the future can we by means of predictive analysis expect to penetrate?* Or applied to the specific purpose of the present chapter. How detailed, specified or crisp a picture of the future leadership roles of Germany, Japan and the United States can we hope to arrive at without lapsing into pure speculation?

This question of predictive scope is closely related to the issue of *scientific uncertainty*: In the social sciences there seems to exist a merciless trade-off between the *precision* of predictions made and the *probability* that forecasts will be confirmed by future events. Although the relationship between scope and uncertainty is a complex one,[20] for present purposes it is reduced to the following question: How much uncertainty are we willing to accept before we consider ourselves crossing the line between scientifically argued prediction and imaginative/intuitive speculation?[21]

In the attempt to develop a reasoned attitude to this question, let me first point out two critical sources of scientific uncertainty restricting the reliability of predictive propositions.

(i) *Inadequate specification of explanatory/predictive model*: Predictions are

---

19 For a fuller account of the essence, scope and uncertainties inherent in predictive analysis, see Fermann (1994: 81–86).

20 Everything else being equal, the more precise and specific a forecast, the lesser the probability that the prediction will be confirmed by future events. This general point is illustrated in the following example: It is more likely that the rather cautious proposition that "the price of oil will increase within five years" will be fulfilled than the more committed forecast that "the oil-price will stabilize at US dollar 16–18 in the year 2000'.

21 In presuming that there are distinct limits to how far or specific we can forecast future events by means of predictive analysis, the present author is not particularly radical. In fact, some scholars both within the social and the humanistic sciences take the argument a step further by categorically rejecting the possibility of making reliable and non-trivial predictions about future social and political behaviour. They simply regard the ambitious aim of making reliable predictions on future social behaviour outside the reach of science.

made on the basis of an explanatory model. Two assumptions are made: first, that all the most influential explanatory variables are included; and, second, that the explanatory factors demonstrated to be influential in the past continue to be at work in the future. However, we have no guarantee that these ideal requirements are fulfilled. Not only may the explanatory model in question suffer from inadequate specification (either by including irrelevant explanatory dimensions or, worse, by excluding important variables), moreover, we have no guarantee that history will repeat itself. As pointed out by Paul Kennedy, "unforeseen happenings, sheer accidents, the halting of a trend, can ruin the most plausible forecasts; if they do not, then the forecaster is lucky" (1987: 438).

(ii) *The voluntary and creative nature of human choice*: In contrast to the natural sciences, the objects of the social sciences are not atoms that strictly obey the laws of physics or creatures whose actions are predetermined by instinct. Whether studying individuals, sub-national interest-groups, multinational corporations or states, the social scientist has to account for the fact that human beings, although influenced by genes and social norms, have the ability to overrule their own inhibitions and suppress their own inclinations. Thus, the social scientist cannot rely on laws to predict (or explain, for that matter) social phenomena (events, trends, actions); he has to content himself with probability estimates calculated on the assumption that the explanatory factors shown to be influential in the past (explanatory model) will continue to be so in the future. However, there is always the possibility that decision-makers may escape or overrule the structural setting and historical heritage surrounding and pressuring them, thus creating events and patterns not foreseen in the "most likely" scenarios of scientific futurists. While structural conditions and present policies may strongly *influence* future policymakers, they do not *determine* decisions made by man. The role of the creative (and therefore unpredictable) political action should not be underestimated.

Hence, the secrets of the future can be revealed only to a certain extent. The bolder our predictions – that is to say, the further into the future they extend and the more specific they are – the greater the probability they will prove to be wrong. But where do we draw the line between scientifically reasoned *predictions* assumed to be reasonably certain and *speculations* based more on the subjective intuition of the researcher? What is the scope of predictive research in a scientific sense?

Figure 1. Political Leadership – Scope of Predictive Analysis

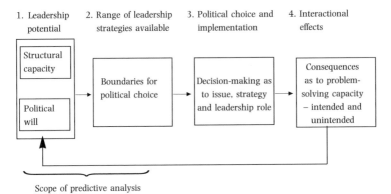

Scope of predictive analysis

There is no definitive answer to this question. However, on the basis of the two sources of scientific uncertainty explained above, I have taken a rather cautious position and decided to limit the scientifically argued predictions of my study to what I judge to be the most promising avenue for taking a leadership role for each of the countries under consideration. This "avenue" is assumed to lie within the boundaries for political action available for decision-makers (see Figure 1).[22]

Having argued that the *scope* of prediction (as a *scientific* project) should be limited to making inferences from the present (step 1 – presented in Section 3.3.–3.5.) to the future most promising avenue for taking a leadership role (step 2 – presented in Section 4), I still deem it interesting to *speculate* a step further as to actual leadership behaviour (step 3 – presented in the concluding Section 5), applying my subjective intuition. The latter is done with the understanding that, although both challenging and interesting, forecasting actual leadership *behaviour* lies outside the scope of *scientific* inquiry.[23]

---

22 The verification of predictions can only be achieved post hoc by means of explanatory studies.

23 The inability of energy economists to predict correctly the medium and long-term price of oil is but one example confirming that such a cautious position is justified.

3.3. Responsibility: The Moral Dimension of Leadership

Accounting for the responsibility of Germany, Japan and the United States is to describe each country's contribution to the extended greenhouse effect caused by anthropogenic emissions. As explained in Chapter 6, however, the task of assigning responsibility among countries for greenhouse gas emissions is an immensely complex issue, depending on the time frame applied, the sources accounted for, the greenhouse gases that are included, and the specific conception of national responsibility applied.

The problem of biased measurement is greatest when assessing the responsibility of countries belonging to different socio-economic categories (developing versus industrialized countries) and climate zones (temperate versus tropical and sub-tropical areas). When assessing the responsibility of Germany, Japan and the United States, however, the bias problem is minimized; especially so, it seems, when choosing the current emissions of energy-related CO_2 as indicator. When applying this indicator within the conventional nation-based approach for which data are available, I am reasonably confident that none of the industrial giants will become "victims" of the assessment strategy chosen.

According to this measure, Germany, Japan and the United States all feature among the five largest emitters in the world (number five, four and one respectively). Together, they are responsible for one-third of energy-related world CO_2 emissions. Having observed this, the differences between the three are nevertheless significant: While Germany (united) and Japan account for about one-twentieth of world CO_2 emissions each, the United States alone is responsible for nearly one-quarter of world emissions (see Table 3). The extraordinary position of the United States is duplicated on the *per capita* related indicator as well. The United States emits nearly 20 tons of CO_2 *per capita*, while the corresponding figures for Germany and Japan are 13.05 and 8.58 tons of CO_2 respectively. Finally, Japan compares favourably with Germany and the United States when observed according to the CO_2 emissions per unit of GDP indicator. The ratio of Japan is 0.63 compared to 1.09 and 1.34 of the United States and Germany.

Table 3: Assessing Responsibility – Germany, Japan and the United States[24]

1990	Germany	Japan	USA	OECD
% of energy-related world CO_2 emissions	4.82	4.91	**23.27**	48.21
Energy-related CO_2 emissions *per cap.*[25]	13.05	8.58	**19.97**	10.34
Energy-related CO_2 emissions/unit GDP[26]	**1.34**	0.63	1.09	1.15

Taking all three indicators into consideration, Japan and the United States clearly represent the two extremes of the spectrum, with Germany in a midway position. Japan scores favourably on all indicators, while the United States shows extremely high estimates on the share of energy-related world CO_2 emissions and on CO_2 emissions *per capita*. The unimpressive performance of Germany on CO_2 emissions per unit of GDP (and to some extent on the remaining indicators) is largly due to the unification of the coal intensive and inefficient East German economy with the much more efficient energy economy of the Federal Republic of Germany.

Germany, Japan and the United States are responsible for about one-third of the world's total emissions of greenhouse gases while being inhabited by less than one tenth of the world population. How may this assessment of responsibility induce Germany, Japan and the United States to take a leadership role in the future? In particular, what might their incentive be for taking this responsibility seriously?

The incentive for Germany, Japan and the United States to reduce their share of world CO_2 emissions can certainly not be inferred directly from the size of their shares. For instance, it should not be assumed that the United States' government, being responsible for nearly one-quarter of energy-related world CO_2 emissions, would feel more obliged to reduce this share than Germany and Japan being responsible for "only" 5% each. The relationship between responsibility and (future) political leadership is far more complex. One cannot expect altruism or a sense of guilt to become potent mechanisms that will induce Germany, Japan and the United States to transform great responsibility into constructive action. Much more convincing for decision-makers in Germany, Japan and the United States, each ultimately responsible towards its own

---

24 Based on OECD (1992a, 1992b). Least favourable scores emphasized.
25 Ton CO_2 per person.
26 Ton CO_2 per US$1000 at 1985 prices.

national electorates and interest groups, are arguments based on national self-interest – that their countries may suffer from the direct or indirect impacts of climate change, and/or are likely to benefit (directly or indirectly) from increasing their own abatement efforts.

Below, four *incentive mechanisms* are suggested which may induce Germany, Japan and the United States to translate their responsibility for huge emissions of greenhouse gases into progressive action in accordance with some kind of leadership role (see also Table 4).

A. *Vulnerability to the direct impacts of climate change*: The first category of incentive mechanisms covers instances of vulnerability to the direct impacts of climate change. While it is likely that the tropical and subtropical South is more vulnerable to climate change than the temperate North, this does not imply that countries like Germany, Japan and the United States are near to immune to the negative effects of global warming. They are likely to experience more turbulent weather, reduced water flows, depleted agriculture, and rising sea level. The United States alone would lose up to 18 000 square miles of wetlands and dry land – a larger area than Massachusetts – if the sea level rises one metre (Kennedy 1993: 112–114). As a mountainous island with a dense population, Japan is even more vulnerable to the rising sea level. With an increasing number of negative impacts of climate change in the northern hemisphere, decision-makers in Germany, Japan and the United States may increasingly perceive it in their interest to take their emission responsibility seriously. Such an incentive may be strengthened by the fact that on the issue of climate change, what the South does can hurt the North. If the North does not act on its historic and current responsibility, it has small moral leverage on the South which is steadily increasing its share of world greenhouse emissions, thus becoming more and more critical of the future management of climate change.

B. *Vulnerability to the indirect impacts of climate change*: The second kind of incentive mechanism is a reminder that decision-makers in Germany, Japan and the United States should not merely consider the immediate impacts of climate change, but also account for their potential vulnerability to possible *indirect* effects of global warming. Chief among those is the challenge of mass migration to rich countries such as Germany, Japan and the United States from developing countries suffering from climate change. Global warming will tend to increase the migratory flood caused by the "demographic imbalance between rich and poor societies" (Kennedy: 95) and threaten the social cohesion and political stability of industrialized countries. Another negative indirect effect of climate change may be disturbances in international trade upon which trading states like Japan and Germany very much depend. Finally, one cannot

Table 4: Incentive Mechanisms Transforming Responsibility Into Constructive Leadership

Impacts of climate change	Vulnerability	Opportunity
Direct impacts	A	C
Indirect impacts	B	D

rule out the possibility that the negative image resulting from a "policy of neglect" in the field of climate change may spill over into other issue-areas with damaging effects. Because of their huge economic resources and great responsibility for the greenhouse threat, Germany, Japan and the United States are particularly exposed to international criticism in this regard.

C. *Opportunities related to the direct impacts of climate change*: While the first two incentive mechanisms are related to the *costs* of *not* executing a progressive abatement policy in accordance with the huge emission responsibility of Germany, Japan and the United States, the third and fourth mechanisms address the potential *benefits* of taking this responsibility seriously. Perceived of as a problem, climate change may provide Germany, Japan and the United States with ample long-term incentives to act on their responsibility. The impulse to initiate abatement efforts is further strengthened to the extent that decision-makers and interest groups are visionary enough to see the (business) opportunities inherent in climate change: Germany, Japan and the United States all have the industrial base required to develop new markets for climate change abatement and adaptation technologies. The first one to exploit this opportunity will gain a competitive edge.

D. *Opportunities related to the indirect impacts of climate change*: This final mechanism is more obscure than the previous ones, but may become potent to the extent policy-makers acknowledge the numerous links between climate change and other issues. In particular, a progressive abatement policy is likely to strengthen the environmental image and enhance the general goodwill of Germany, Japan and the United States. This, in turn, may increase the moral standing and credibility of these countries, as well as their political leverage on related issue-areas.

While these incentives may have general relevance, their specific workings and role will have to be decided empirically for each country. In particular, the influence of responsibility on leadership very much depends on each government's perception of vulnerability (negative incentives) and its ability to take advantage of the challenge of climate change (positive incentives). This means that the influence of responsibility on the leadership prospects of Germany, Japan and the United States can be assessed only after these perceptions and capacity have been mapped.

3.4. Leadership Potential: Structural Dimensions

3.4.1. Energy Supply Structure and Potential for Fuel Substitution

How does energy structure relate to political leadership in the field of climate change? To the extent a country is able to restructure its energy supply (or assist other countries in doing so) so as to reduce the role of carbon-rich fossil fuels, this would provide an avenue for taking a leadership role. However, to the extent that fuel-switching is impracticable for technical, economic or political reasons, energy structure represents a *limitation* on the prospects for taking a leadership role. The crux of the matter is that the ability of a country to change the composition of energy supply influences the range of strategies available for developing climate change problem-solving capacity.

It is one of the ironies of energy economics that past efforts at energy substitution (and energy conservation) have tended to increase the marginal costs of current efforts.[27] Hence, a government possessing a good record in energy substitution and energy conservation may find continuation of its efforts into the future prohibitively costly.

This also means that a country with a small potential for energy substitution will hardly be likely to exhibit leadership by demanding that all countries stabilize energy-related CO_2 emissions by restructuring energy supply. The technical, economic or political obstacles to (further) energy substitution in such a country would more likely provide an incentive towards taking a leadership role by *other means* – by arguing, for instance, that international abatement efforts should be cost-effective and take into account the varying circumstances prevailing in each country. The establishment of wide consensus on such a principle would render it quite legitimate for a country suffering high marginal costs in the field of fuel-switching to follow less costly strategies. The lack of capacity or potential in one field may induce a country to take leadership by other means and on other issues.

Having explained the relevance of energy structure for leadership, I now proceed by considering what characterizes the energy structures of Germany, Japan and the United States, and what prospects there are for

---

27 This will continue to be true, provided that no "technological breakthroughs" are within reach.

increasing the share of non-fossil fuels in these countries? As a preliminary step towards answering these questions, let me recapitulate and
link up to the differences in CO_2 emissions (responsibility) between
Germany, Japan and the United States.

The differences in emissions of energy-related CO_2 between Germany
and Japan on the one hand, and the United States on the other, are only
partly explained by variations in the magnitude of the economy (GDP).
This fact is accentuated when the GDPs of Japan and the United States
are compared with energy-related CO_2 emissions. Japan has a GDP 55%
the size of that of the United States, but only 20% of the United States'
energy-related CO_2 emissions (WRI 1993: 236–237).

In order to account for this discrepancy, two additional factors must
be taken into consideration: One is the present energy structure and
prospects for fuel-switching (energy substitution), the other is energy
efficiency in various sectors of demand and the prospects for further
energy conservation.

Comparing the energy supply structure of Germany, Japan and the
United States with a focus on the share of fossil fuels in total energy supply, the differences are less than might be expected (see Table 5): Fossil
fuel accounts for 87.6%, 85.7% and 86.6% of the total energy supply of
Germany, Japan and the United States, respectively. The difference
between the fossil fuel intensive German economy and the less fossil fuel
intensive Japanese economy should not be exaggerated. It is, in fact, only
1.9%.

Thus, at this level of aggregation, the *similarity* between the three
countries is more striking: Germany, Japan and the United States all rely
more on fossil fuel in their energy supply than the OECD average of
83.8%. Switching from fossil to non-fossil fuels over time, Japan reduced
reliance on fossil fuel from 97.4% in 1973 to 85.7% in 1991. This
amounts to an 11.7% reduction. The comparable figures for Germany
and the United States are 10% and 8.4% respectively. The general
impression is that Germany, Japan and the United States still rely heavily on fossil fuel. The fuel-switching achievements of Japan are somewhat

Table 5: Energy Supply Structure – Fossil and Non-fossil Fuels in 1991 (1973)[28]

Fuel shares %	Germany	Japan	USA	OECD
Fossil fuels	87.6 (97.6)	85.7 (97.4)	86.6 (95.0)	83.8
Non-fossil fuels	12.4 (2.4)	14.3 (2.6)	13.4 (5.0)	16.2

---

28 Calculated from OECD (1992b).

more extensive than those of Germany and the United States – efforts which have made the Japanese economy slightly less dependent on fossil fuels than the other two.

At the aggregate level, variations in the share of fossil fuels in total energy supply go only part way in explaining the differing CO_2 emissions of Germany and Japan, on the one hand, and the United States, on the other. Even at a disaggregated level, distinguishing between coal, oil and gas, the differences observed cannot explain much of the variations in CO_2 emissions (see Table 6). Considering our two "extreme" cases, the share of coal in Japan's energy supply (18%) is considerably less than the role of coal in the American energy supply (24%).

Table 6: Energy Supply Structure in 1991(1973)[29]

Fuel shares %	Germany	Japan	USA	OECD
Coal	33.2 (31.6)	17.3 (18.0)	24.0 (18.1)	21.4
Oil	39.2 (56.2)	58.3 (77.8)	39.7 (47.0)	42.9
Gas	15.2 (9.8)	10.1 (1.6)	22.9 (29.9)	19.5
Nuclear	11.3 (1.2)	12.3 (0.8)	8.4 (1.3)	10.6
Hydro	0.4 (0.5)	2.1 (1.9)	1.7 (1.4)	2.8
Other	0.7 (0.7)	-	3.3 (2.3)	2.8

If the same pattern had been observed in the case of oil, the conclusion could have been drawn that the greater role of carbon-rich fossil fuels (coal and oil) in the American energy economy explains much of the discrepancy between Japan and the United States in relation to CO_2 emissions. This is not the case, however, since Japan's share of oil is much higher (58.3%) than that of the United States (39.7%).

Adding together the shares of coal and oil in total energy supply (Table 7), both of which have a considerably higher content of carbon than natural gas, the United States, in fact, scores unexpectedly lower (63.7% of total supply) than Japan (75.6%). From this, one may conclude that differences in the composition of fossil fuels cannot explain the relatively higher CO_2 emissions of the United States. To the extent that a correlation can be detected, it is, in fact, negative – the reverse of what could have been expected.

Having observed (i) that the total share of fossil fuels varies little between Germany, Japan and the United States, (ii) that the composition of fossil fuels explains little, if anything, of the divergent CO_2 emissions

---

29 (OECD 1992b).

of these countries, and (iii) that the share of fossil fuels in total energy supply has been reduced by 8.4–11.7% since 1973, what are the prospects for Germany, Japan and the United States continuing substitution towards low carbon and non-fossil fuels?

As a point of departure, let me first consider the projected changes in the share of fossil fuels in total energy supply towards the year 2000. Based on information provided by each country, OECD/IEA estimates that the role of fossil fuels in Germany (united) will remain stable from 1991 to 2000, accounting for about 87% of energy supply (Table 7). Comparing the projected estimate of united Germany in 2000 with that of West Germany in 1990, the share of fossil fuels will, in fact, increase by 2.3%. It should be noted, however, that the major part of this increase is accounted for by the growing role of natural gas, a fossil fuel with a relatively low content of carbon.

Table 7: Past, Current and Projected Energy Structure – Share of Fossil Fuels in Total Energy Supply[30]

Fossil fuels – share of TPES	Germany	Japan	USA
1973	97.6	97.4	95.0
1991	87.6 (84.9)[31]	85.7	86.6
2000	87.2	76.9	84.0

The pattern of stability is also apparent in the case of the United States, which expects no more than a small reduction in the role of fossil fuels in the energy supply. In the words of a close observer of the US energy sector:

> The history of US energy demand and the existing resources, infrastructure and institutions make the US economy as dependent on fossil fuels as a heroin addict is on the needle. Analysis of future projections, the institutional obstacles, and the US political process suggests that achieving major changes to break the addiction would be a momentous political task (Rayner 1993: 30).

The seemingly deviant case is Japan, which hopes to reduce the share of fossil fuels in energy supply by 8.8% – from 85.7% in 1991 to 76.9% in 2000. Although Japan has strong reasons for reducing the share of fossil fuels – for example, concerns about energy security – the estimated

---

30 Calculated from OECD (1992b).
31 West Germany only, 1990 figure.

reduction in the share of fossil fuels towards 2000 seems to be inflated by wishful thinking.

Firstly, the Japanese government would seem to have underestimated the incentive provided by the persistently low oil and coal prices. Although it is in Japan's long-term interest to reduce dependence on imported oil, cheap oil (relatively speaking) provides a strong incentive for economic growth the recession-stricken Japanese are unlikely to resist.

The primacy granted to economic growth over long-term concerns about energy security is aptly demonstrated by the historical record: Having experienced a downward trend from 1973 (77.8%) to 1985 (54.8%) during a period of high oil prices, the dramatic price fall in 1986 provoked an increase in the share of oil to 56.6% in 1988 and 58.2% in 1990 (OECD 1992b). In view of the enduring competitiveness of oil, one may doubt that oil will play any part in reducing the share of fossil fuels towards 2000. Recent price relations indicate that oil is liable to increase its share of energy supply.

Secondly, it is assumed that the share of nuclear power in the Japanese energy supply will increase from 12.3% in 1990 to 16.6% in 2000, thus constituting the single most important energy source filling the gap from the shrinking share of fossil fuels (oil). Meeting this target, however, will necessitate the building of many more nuclear plants than the eleven currently under construction. In fact, Japan's Nuclear Power Programme is already delayed due to the problem of finding appropriate sites in a country with volcanic geology, a dense population, and inhabitants keenly aware of the hazards of atomic power (Fermann 1993: 295).

Finally, Japan has, in contrast to both Germany and the United States, a reputation for being too optimistic about the prospects of reducing the share of fossil fuels (notably oil) in energy supply. This is demonstrated by comparing the real figures from 1991 with the projected figures for 1991 estimated in 1986: It turns out that the actual share of fossil fuels in 1991 was considerably higher (85.7%) than had been forecast five years earlier (81.9%).

The main conclusion to be drawn from the above comparison seems to be that substitution towards non-fossil fuels will slow down considerably, or even stop, in Germany, Japan and the United States towards the year 2000. While Japan's substitution potential *may* prove to be significant, fuel-switching is nevertheless unable to offset the effect of an ever increasing energy demand on CO_2 emissions. In absolute terms (Mtoe), the consumption of oil is expected to increase in all three countries towards 2000.

The implication of all this is that fuel-switching will contribute only marginally to the stabilization of CO_2 emissions in Germany, Japan and the United States, and, hence, cannot be relied upon as a strategy for taking a leadership role unless these countries contribute to energy substitution towards non-fossil fuels in other, less developed, countries (joint implementation).

3.4.2. Energy Efficiency and Energy Conservation

Energy conservation efforts contribute strongly to increasing the GDP output per unit of energy (energy efficiency), and may be initiated to reduce the burden of energy costs, enhance energy security (reduce dependence on imports), or as part of environmental policy. The fact that energy conservation can serve multiple aims ("no-regret") makes it an attractive strategy for decision-makers and an important factor to be considered in a study of climate change leadership. Which country, Germany, Japan or the United States, has traditionally taken the leading edge in the field of energy conservation and what are its prospects for leadership by means of conservation in the future? As a preamble to answering these questions, I first consider the contribution of energy conservation in explaining the divergent CO_2 emissions of Germany and Japan on the one hand, and the United States on the other.

The discrepancy between CO_2 emissions and GDP level became evident when comparing Japan with the United States: The former has only one fifth of the latter's CO_2 emissions, but more than half of the GDP of the United States. Furthermore, it was found that energy structure could not account for the weak correspondence between GDP and CO_2 emissions. The share of fossil fuels in energy supply was, in fact, very similar in Germany, Japan and the United States, and could be expected to remain so towards 2000. So, what about energy conservation – efforts aimed at increasing the output of GDP per unit of energy applied? Can differences in energy efficiency, as expressed by the energy intensity ratio, explain the relatively (and absolutely) higher CO_2 emissions of the United States?

It seems so, to a rather large extent: According to Table 8, Japan has a superior TPES/GDP ratio compared to that of Germany and the United States. In fact, Japan manages, along with Switzerland, the most energy efficient economy in the world, and can thus correctly be characterized as a "world leader in the effective use of energy" (OECD 1992b). It has been calculated that if other OECD countries were as energy efficient as Japan, energy consumption in the industrial, residential and transportation sectors could be cut by 54%, 9% and 13%, respectively (Tomitate 1992).

Table 8: Past, Current and Projected Energy Efficiency[32]

Energy intensity[33]	Germany	Japan	USA	OECD
1973	0.52	0.37	0.58	–
1991	0.47	0.25	0.42	0.39
2000	0.34	0.20	0.39	–

The rather low energy efficiency (0.47) of the German energy econo-my can in part be explained by the unification of Western and Eastern Germany. The energy economy of West Germany only is comparable to that of the United States in efficiency (0.42).

Although differing with regard to the present level of energy efficien-cy, Germany, Japan and the United States have all improved (decreased) their energy intensity ratio since 1973. This is largely due to transfor-mation of industry towards fewer energy-intensive industries and to extensive energy conservation efforts, especially in the industrial sector. So, what are the prospects for further energy conservation? To a vary-ing degree, all three countries expect to continue their energy intensity ratio towards the year 2000.

Energy conservation efforts in the United States are likely to be fairly modest, enough to secure an improvement in energy efficiency from 0.42 to 0.39. The obstacles to a more offensive energy conservation strategy in the United States are not so much technical as political: "Institutional and societal constraints are such as to limit the rate at which [energy] efficiency can be improved" (Rayner 1991: 277). Due mostly to the restructuring and modernization of the East German indus-try, the German energy economy is expected to improve considerably from the 0.47 of 1991 to 0.34 in the year 2000. However, this improvent is insufficient to bring Germany to a level comparable to that of Japan today (0.25), and, more importantly, unable to secure fulfilment of the German abatement target (25% reduction of CO_2 emissions). It is all the more impressive that Japan plans to intensify energy conservation efforts and expects to improve energy efficiency to 0.20 by the year 2000. Energy conservation efforts will continue in spite of increasing marginal costs, pressed forward by considerations about energy security, rising energy imports and various environmental issues, including cli-mate change. These incentives will most likely be strong enough to secure Japan's continued leadership in the field of energy conservation.

---

32 OECD (1992b).
33 TPES/GDP.

But can this lead be transformed into a leadership role in the field of *climate change?*

Japan will most likely retain leadership in energy conservation despite increasing cost, but this will not be enough to offset the emissions resulting from increased consumption of energy.[34] This implies that a Japanese leadership role cannot be based upon the contribution of energy conservation to the reduction of *national* CO_2 emissions. What options remain? It would seem that Japan can develop problem-solving capacity in the area of climate change, and thus take a leadership role by making its superior energy conservation technology available for other less technologically advanced countries. Japan seems to have both the capacity and the incentive to contribute towards the development of a *mechanism* for transfer of technology – whether conceived of as correcting market imperfections or facilitating the development of an international coordinating body: Among the incentives are the increasing marginal costs of conservation in Japan. While Germany and the United States also possess advanced energy conservation technology, they nevertheless do not seem to have the equally strong incentive of Japan to develop a mechanism for international transfer. This line of argument will be developed in Section 4.

3.5. Leadership Potential: Political and Perceptual Dimensions

Current policies are important parameters to consider when assessing a country's *willingness* to take a leadership role in developing the international problem-solving capacity to curb greenhouse gas emissions. In this section, the official climate policies of Germany, Japan and the United States are compared along dimensions (targets, time-tables, measures and legal status of commitment) believed to provide insight into the considerations (motives) and perceptions guiding these countries in their climate policy-making. Prior to this, however, I briefly re-examine the

---

34 While Germany (especially after the unification) and the United States have considerable potential for further energy conservation in the industrial sector, Japan's potential is almost exhausted because of the extensive energy conservation efforts already undertaken in the 1970s and 1980s. No technological "quick fixes" are likely to be introduced in the field of energy conservation during the next decade, and thus cannot be relied upon to reduce the high marginal costs of energy conservation measures in the industrial sector. Moreover, there is a tendency for conservation gains in the residential and transportation sectors being cancelled out by increased consumption (Fermann 1995: 118–119).

divergent profiles of Germany, Japan and the United States during the process leading up to the 1992 FCCC.

3.5.1. Leader, Follower, Laggard: Brief Review of Historical Profiles

During the 1980s, climate change underwent gradual transformation from being an issue of science to one of politics. The first concrete international proposals for the establishment of targets to reduce emissions came in June 1988 at the Toronto Conference on the Changing Atmosphere: Implications for Global Security. As a response to this call for action, the Dutch government organized a conference at ministerial level.

The November 1989 Noordwijk Ministerial Conference on Atmospheric Pollution and Climate Change proved to be something of a turning point: The Netherlands, along with Germany, Canada, France, Norway and Sweden, proposed the adoption of a binding international agreement to stabilize and ultimately reduce CO_2 emissions, thus propelling climate change to the top of the international political agenda (Hatch 1993: 13). Followed by key countries like China, Japan and the Soviet Union, the United States led the opposition against adopting quantitative emission targets and time-tables on the grounds that "further study was necessary before binding controls could be proposed" (Schreurs 1993: 17). Although these countries managed to water down the text of the conference declaration, they were unable to arrest the trend towards unilateral adoption of CO_2 emissions stabilization/reduction targets.[35]

Germany was well prepared to take up the challenge from the *Noordwijk Conference*: Spurred by a major debate in the German *Bundestag* in March 1987, the (first) *Enquete Commission* was constituted in October 1987.[36] Based on the recommendations of the Commission, in June 1990 the Federal Cabinet decided to adopt a 25% CO_2 emissions reduction target by the year 2005. This target was confirmed in November 1990 accounting for the prospects of German unification (Beuermann and Jäger 1993: 6–8).

---

35 In the compromise reached in the Noordwijk Declaration, the need to stabilize CO_2 emissions was recognized, and it was agreed that such stabilization should be achieved "as soon as possible", though, in the view of "many" industrialized nations, such stabilization should be achieved as a first step at the latest by the year 2000. See the Noordwijk Declaration on Climate Change, Ministerial Conference on Atmospheric Pollution and Climate Change, Noordwijk, the Netherlands, 6–7 November 1989.

36 Enquete Commission – "Vorsorge zum Schutz der Erdatmosphäre" (Preventive Measures to Protect the Earth's Atmosphere).

In view of Germany's strong support for quantitative CO_2 emissions targets at the November 1989 Noordwijk Conference, the subsequent German adoption of a CO_2 emissions reduction target did not come as a surprise. More surprising, however, was the October 1990 Japanese government decision to stabilize CO_2 emissions by the year 2000 on a per capita basis. This happened only two weeks prior to the November 1990 Geneva Climate Change Conference, and constituted a profound departure from the "no-action-wait-and-see-stand" advocated at the 1989 Noordwijk Conference.

The Japanese seemed determined to avoid a repetition of the barrage of criticism aimed at them by several EC countries at the Noordwijk Conference – a criticism felt to be unjustified since the Japanese perceived themselves to "supporting the EC stand in principle" while awaiting the results of "serious analysis into each sector of energy use to see if an emission target could be achieved" (Fermann 1993: 291).

By fall 1990, a majority of OECD countries had declared their support for the principle of setting emissions targets with time-tables and were in the process of developing climate change policies including such targets. A notable exception to this trend was the United States. Although George Bush in the course of his 1988 presidential campaign warned that the "White House Effect" (New York Time, 24 September 1988) – the notion that his new administration would take global warming seriously – should not be underestimated, and Secretary of State, James Baker, later emphasizing that "we ... cannot afford to wait until all of the uncertainties [of climate change science] have been resolved before we do act" (Hatch 1993: 11), subsequent developments proved that the Bush administration was "by no means ready for concrete and immediate action" (Andresen 1993: 6): The no-regret approach[37] advocated by the Bush administration at the Washington Conference and the Bergen Conference in the spring of 1990, and in various fora up to the 1992 UNCED Conference in Rio, made it clear that the Bush administration was continuing with the defensive policies of the Reagan administration; it strongly resisted emissions targets and time-tables, and was opposed to additional funds to the Third World (Andresen 1993: 4–10).

The historical record points to Germany as a *leader* in rendering

---

37 The notion of "no-regret" refers, in the context of scientific uncertainty, to measures which, if implemented, will have beneficial effects even if the threat of global warming turns out be false. Energy conservation measures are often perceived as "no-regret" since such measures are capable of simultaneously reducing CO_2 emissions, reducing energy demands and increasing energy security. The notion of "no-regret" can be conceived as competing with the "precautionary principle" which holds that you should act so as to minimize the probability of the worst possible outcome.

climate change a top international political issue: Not only was Germany among the first major industrialized country to adopt a climate change policy based on the principle of emissions targets and time-tables, it was also a driving force in the political process culminating in the October 1990 EU Council decision to stabilize overall CO_2 emissions within the Union at the 1990 level by the year 2000 (OECD 1992a: 120–123).[38]

Moreover, Germany was a strong advocate within the EU for the adoption of a Union CO_2/energy tax to secure the implementation of the CO_2 emissions stabilization target, but so far has not succeeded (Manne and Richels 1993). Finally, along with Norway, Germany has promoted the concept of "joint implementation" – the idea that a country could receive credits towards its own emission objectives by reducing emissions or enhancing sinks in another country.

Japan has generally taken a *reactive* posture towards international developments (Inoguchi 1991). Typically, Japan is as "careful to avoid acting unilaterally on sensitive international issues" as it is to "escape the political costs of being the last to follow suit" (Fermann 1992: 43). While there are signs that this attitude is in the process of being modified, Japanese reaction to the climate change issue confirms the old pattern: At the 1989 Noordwijk Conference it became clear to the Japanese delegation that they had underestimated the growing importance of environmental issues in post-Cold War international affairs. Acknowledging this, the Japanese relatively rapidly adapted to the new situation; by the 1990 Geneva Conference they had joined the expanding ranks of industrialized countries adopting greenhouse gas emissions targets.

However, Japan is reluctant to adopt a CO_2/energy tax as a means of achieving its CO_2 emissions stabilization target. In order for Japan to follow suit on the tax issue, it is not enough that five small western European countries have adopted a CO_2/energy tax (OECD 1993a).[39] In view of the prevailing understanding that climate change will have

---

38 While not a binding commitment, the EC Council decision to stabilize CO2 emissions was confirmed by the EC Commission on 27 May 1992. Its Community Strategy to reduce CO2 emissions includes:

* SAVE programme on energy efficiency.

* ALTENER programme on fuel-switching to renewable energy (including the THERMIE technology innovation programme).

* Community mechanism to monitor the national implementation of plans and programmes, greenhouse emissions.

* Community CO2/energy tax to stimulate energy conservation and substitution of fossil fuels.

With the notable exception of the tax, all the proposals have been adopted. For further elaboration, see chapter 12.

39 These countries are Denmark, Finland, the Netherlands, Norway and Sweden.

limited impact on Japan, and the fact that environmental taxes implemented unilaterally will reduce Japanese competitiveness and delay
economic growth, any change in the Japanese attitude to such taxes will
depend very much on the fate of the EC tax proposal, which has long
faced a political dead-lock. Japan is not likely to take a lead in the introduction of CO_2 taxes, but is likely to follow suit if the EC succeeds in
settling for a CO_2/energy tax.[40]

Despite the fact that the United States has for decades been a scientific leader in basic and policy-oriented climate change research, in terms
of policy-making it is perceived as a *laggard*. This impression has been
created mainly by the Reagan and Bush administrations' reluctance to
commit the United States to emission targets and time-tables. In contrast
to Germany, which has based its policy on the "precautionary principle",
the United States has consistently ruled out any measures that could not
be justified in terms of "no-regret".

The fact that the United States was the fourth country in the world to
ratify the FCCC does little to mitigate the image of the United States as
a laggard on the climate change issue: With its lack of commitment
regarding targets and time-tables, and its emphasis on no-regret measures, the 1992 FCCC is, in fact, very much "a blueprint of US (policy)
to deal with the global warming issue". While the American influence
on the FCCC was profound, the United States stood out as "the big loser
regarding the less tangible, but maybe equally important, environmental
image" (Andresen 1993: 9–10).

Having briefly sketched the policy profiles of Germany, Japan and the
United States up to the 1992 FCCC, I now turn to the *current* climate
change policies of these countries while emphasizing (i) the scope and
ambition of emissions targets, (ii) policy measures, and (iii) the strength
of policy commitment.

3.5.2. Germany's Climate Change Policy

3.5.2.1. Ambition and Scope

Not only was Germany among the first industrialized countries to adopt
a climate change policy, it has also adopted a very *ambitious* CO_2 emission target.[41] With 1987 as base year, the June 1990 Federal Cabinet

---

40 Recent reports indicate, however, that the EU carbon/energy tax may have received
a fatal blow (*GECR*, Vol. 5, No. 24: 4). See also chapter 12.
41 For two studies explaining the making of the German climate policy, see Cavender
and Jäger (1993) and Schreurs (1993).

decided to adopt a 25% CO_2 emissions reduction target to be attained by the year 2005.[42] As noted in the previous section, this target was roughly in accordance with the recommendations of the German Enquete Commission[43] and was confirmed by the German cabinet first in November 1990, and again in February 1992 (Beuermann and Jäger 1993: 6–8).[44]

As regards the *scope* of the emissions target, German climate policy focuses heavily on energy-related CO_2 emissions. "Despite statements that all greenhouse gases have to be reduced", political action (measures and announcements) "concentrates on the issue of reducing CO_2 emissions" (Beuermann and Jäger 1993: 1). The German cabinet has yet to specify (quantify) the reduction of emissions of CH_4, NO_x and N_2O. Thus, the scope of the German climate policy does not deviate from most OECD countries' practice of emphasizing energy-related CO_2 emissions, while paying less attention to other greenhouse gases.[45]

---

42 Especially in view of the forecasted rise in GNP of around 50% during the same period of time (an average annual rise of 2.5%), this target is without doubt ambitious. In fact, it implies that the Germans intend to produce one and a half times the goods and services provided today with just 70–75% of 1987 CO_2 emissions (Vellinga and Grubb 1992: 34).

43 In fact, the Commission found that a reduction of CO_2 emissions of about 30% by the year 2005 was feasible. However, this judgement was based on the rather ideal requirement that all recommended measures were implemented in full.

44 The Enquete Commission on Precautionary Measures to Protect the Earth's Atmosphere was established on 3 December 1987 by the German Bundestag with 22 members – eleven scientists/experts and eleven parliamentarians from the Bundestag. According to Konrad von Moltke an enquete commission is an "obscure body used to deal with complex and often politically sensitive issues that the regular committees do not want to address ... (or) where jurisdictional squabbles make the use of regular committees impossible" (von Moltke 1991). The Commission submitted three reports to the German Bundestag: The first offers not only a detailed account of current scientific knowledge about stratospheric ozone depletion and the anthropogenic greenhouse effect, but also recommendations on measures to protect the earth's atmosphere (emphasizing chlorofluorocarbons). In its second report the Commission studied the problems involved in tropical forests. The Commission's last report (October 1990) makes extensive investigations into the avoidance and reduction of releases of radiative trace gases due to energy use, and the possible content of an international convention (adopted at UNCED in Rio June 1992) for the protection of the atmosphere (OECD 1992a: 65–66).

45 A notable exception to this pattern is the climate policy of the Netherlands: The Dutch government has developed a "Gas-by-Gas strategy" designed to reduce emissions of CH_4, NO_x, N_2O by 20–25% by the year 2000 (OECD 1993a).

3.5.2.2. Policy Measures and Implementation

The German climate policy is expressively founded on the "precautionary principle" (Vorsorgeprinzip) – the notion that one should act so as to minimize the probability of the worst possible outcome (Beuermann and Jäger 1993: 3). Indeed, it would seem that this principle of insurance has made an impact on the German CO_2 emissions target since its policy opts for substantial reductions rather than mere stabilization. We should correspondingly expect the "precautionary principle" to inspire the adoption of extensive policy *measures* beyond those qualifying as "no-regret", and capable of matching the ambitious CO_2 emissions reduction target.

A conclusive analysis of this expectation can only be provided through a detailed examination of the multiple effects of each particular measure currently implemented as compared to abatement costs and emission targets. This, however, is beyond the scope of this chapter. Instead, I choose a cruder strategy, inquiring whether the *aggregate* effect of Germany's policy measures seems capable of fulfilling the 25% CO_2 emissions reduction target.

Initially, the German measures to reduce CO_2 emissions were based on three pillars:[46]

1. Increasing energy efficiency by means of *energy conservation*, particularly in domestic heating, cars and electricity.
2. Promoting *energy substitution* towards non-fossil fuels (i.e. renewable energies) and low carbon content fossil fuels (natural gas).
3. Giving priority to the use of *economic instruments* – most notably the application of CO_2/fossil fuels taxes.

What is the current status regarding the *implementation* of these measures? Considering points 1 and 2, trends in the allocation of federal research resources indicate that an effort has been made to increase energy efficiency and to switch energy demand towards non-fossil fuels. Although the total federal budget for energy research decreased by about 2% from 1990 (DM 856 million) to 1991 (DM 837.5 million), research programmes on energy conservation and renewable technologies *grew* by 33% and 9% respectively, and has continued (OECD 1992b: 402).

---

46 For four detailed accounts of German policy measures, see Vellinga and Grubb (1992: 34–38), Beuermann and Jäger (1993: 5–9, 13–22); Cavender and Jäger (1993); and BMU (1993: 107–109).

While these efforts give credibility to the German cabinet's *intention* of fulfilling its CO_2 emissions reduction commitment, the reallocation of federal research resources is insufficient to bring it about. The positive development reported in the German National Climate Report that the total CO_2 emissions of unified Germany decreased by 14% from 1987 (1086 Miot) to 1992 (933 Miot) (BMU 1993: 84) is due mainly to the restructuring and transformation of the economy and energy structure of the former GDR.

Indeed, a recent study concludes that "measures taken so far are not enough for achieving the required federal goal by 2005" (Beuermann and Jäger 1993: 48). This conclusion is confirmed by a leading expert within the German energy industry,[47] and by an extensive study researched by Prognos AG predicting that German CO_2 emissions will be reduced by no more than 10% by the year 2005 compared to 1987 levels. Broken down into East (GDR) and West Germany (FRG), the Prognos AG study finds that the former GDR CO_2 emissions will decrease by 38%, while in the FGR an increase of 3% is expected. This prediction roughly confirms the 1987–92 pattern when the CO_2 emissions in the GDR decreased by 47% while the FRG experienced a 2% increase (Vellinga and Grubb 1992).

The likelihood that unified Germany will fail to fulfil the 25% reduction target is due mainly to the fact that the recommendations of the Enquete Commission did not seriously consider resistance from the industry. The German policy just assumed that the measures pointed out by the Enquete Commission would be implemented. However, in the context of economic recession, the "implementers" themselves (industry) were resisting the enforcement of abatement measures. German industry associations prefer voluntary agreements – agreements which can be bent and adapted to the changing needs of the industry.

The prime illustration of this controversy is the issue of a CO_2 tax – arguably the single most efficient economic instrument available for the reduction of CO_2 emissions. Although Germany, along with Belgium, Denmark, Italy, Luxemburg and the Netherlands, has acted as an arbiter for a CO_2/energy tax within the EU,[48] because of the resistance of industry fearing loss of international competitiveness the German government is not prepared to apply economic market incentives unilaterally

---

47 Confidential source.
48 Germany favours the adoption of an EC CO_2/energy tax. The Commission's tax proposal has so far been blocked by the Ministerial Council, which has not been able to agree unanimously to even the principle of using fiscal instruments to reduce CO_2 emissions. See Chapter 12.

(Beuermann and Jäger 1993: 48). While this scepticism is shared by most OECD countries, it is still worth noting that the German government is presently unwilling to adopt the means necessary to reach the ambitious CO_2 emissions reduction target. Instead, it is counting on the European Commission's proposed CO_2/energy tax to help meet its emissions target (*Global Environment Reporter*, Vol. 16, No. 16, 1993: 4).

However, in view of the strong British opposition to a modified tax proposal developed by the European Commission, it seems clear that no such help will be provided for the time being. After a meeting in the EU Council in October 1993, the German Environment Minister Klaus Töpfer flatly stated that "the bottleneck [to EU carbon tax] remains the UK" (IER 1993: 707).

Germany claims that its climate change policy-making is based on the "precautionary principle". This is reflected in Germany's comparatively ambitious CO_2 emissions target. Much remains, however, for Germany to devise the measures capable of meeting the target.

3.5.2.3. Credibility of Abatement Commitment

How deep is the German commitment to its 1990 CO_2 emissions reduction target? The question of commitment to *target* cannot be decided without reference to the question of commitment to *measures* instrumental in fulfilling the target. Above, it was observed that Germany has yet to introduce a CO_2/energy tax and that enforcement of many policy measures is lacking. These are indications that Germany is not fully committed to its emissions reduction target.

That its commitment is limited is confirmed by the fact that, unlike the parliamentary decision to phase out CFC gases as envisaged by the 1987 Montreal Protocol, the CO_2 emissions reduction target is not confirmed by law.[49] Although a Federal Cabinet resolution counts as a *political* commitment, the lack of parliamentary endorsement means that the emissions target is not *judicially binding*.

A final indication of the German commitment to abatement targets is

---

49 The German CO_2 emissions reduction target was set independently of the decision to phase out CFC – the latter being seen primarily as an issue related to ozone depletion. For the first time, however, the German National Report of August 1993 report figures on the combined reduction of CO_2, CFCs and other greenhouse gases (measured in CO_2 equivalents), concluding that greenhouse emissions since 1987 have been reduced by about 50% – an impressive figure indeed! Stating this, the German government does not draw the conclusion that the German greenhouse abatement commitment is thereby fulfilled. Fulfilment of the Montreal Protocol (CFC phase-out) is still not mixed up with reductions of other greenhouse gases.

its willingness to ratify the FCCC. Germany long delayed its ratification pending the result of the time-consuming EU decision-making process. While this fact fails to demonstrate that Germany's commitment is weak, it indicates again that it is restrained and that the concern for concerted EU action was long deemed more important than the wish for immediate ratification.[50]

However, on 9 December 1993 German patience came to an end as Germany finally decided to ratifify the FCCC. With this action, Germany effectuated its warning that it would ratify independently of the EU if the member countries could not rapidly achieve agreement on a unified EU ratification (Beuermann and Jäger 1993: 9). The unilateral ratification of Germany had the intended effect: 21 December 1993, a unified EU ratified the FCCC as the fifty-second ratifying party, thus securing enforcement on 21 March 1994 (United Nations 1993).

3.5.3. Japan's Climate Change Policy

3.5.3.1. Ambition and Scope

Compared with the ambitious German CO_2 emissions *reduction* target, the Japanese abatement target is quite modest: Although the Japanese 1990 Action Programme (AP) states that greenhouse gases, like methane and nitrous gases, "should not be increased", the Japanese abatement commitment focuses heavily on CO_2 emissions, which are to be *stabilized* "on a *per capita* basis in the year 2000 and beyond at about the same level as in 1990".[51] [52] While it is correct that during the 1970s Japan became a "world leader in the efficient use of energy" and "is among the developed countries where the emissions of CO_2 *per capita* is lowest", the fact remains that the *per capita* linked CO_2 abatement target allows for at least

---

50 The German willingness to subordinate ratification of the FCCC to the EU decision-making process is also a reminder that membership in the EU is not necessarily an asset for a country eager to play a progressive role in international environmental politics; on the contrary, the slow decision-making process within the EU and the strong norm of concerted action may constrain progressive countries in their efforts.

51 The per capita-based stabilization target is supplemented by the much weaker statement that "efforts should also be made ... to stabilize the total amount of CO_2 emissions". This formulation is a reflection of the bureaucratic-politics compromise reached between the Environment Agency (EA) and the Ministry of International Trade and Industry (MITI). For further elaboration, see Fermann (1992: 40–45).

52 Like Germany, Japan is a party to the 1987 Montreal Protocol and is ahead of schedule in the phasing out of CFCs. Excluding CFCs, CO_2 accounts for about 80% of Japan's anthropogenic greenhouse emissions.

a 6% increase in emissions from the base year 1990 to the year 2000 (OECD 1992a; II/345; AP 1990: 1).

The Japanese abatement target is further weakened by the fact that 1990 is chosen as base year for CO_2 emissions: Looking at recent emissions trends, Japan experienced an incredible 30% increase in CO_2 emissions from 1987 (244 Mil.t-C) to 1990 (318 Mil.t-C). If the Japanese government wanted to take this steep rise in CO_2 emissions more seriously, it would have selected 1987 or 1988 as base year for stabilization. Instead, by choosing a *per capita* linked stabilization target and 1990 as base year, under the 1990 Action Programme Japan allowed itself a *40% increase* in CO_2 emissions between 1987 (244 Mil.t-C) and the year 2000 (337 Mil.t-C).[53]

Comparing the Japanese abatement target with the ambitious German CO_2 emissions target and the less impressive American overall stabilization target, the Japanese stabilization target is clearly the weakest. Comparing Japan's abatement target within the broader range of OECD countries, Japan receives a low rating along with countries like France, Greece, Ireland and Portugal (OECD 1992a; Fermann 1992: 51–52).

3.5.3.2. Policy Measures and Implementation

Reviewing the main elements of the Japanese strategy to fulfil its CO_2 emissions stabilization target, it will be found that to some extent it resembles the German strategy:[54] (i) Energy conservation in all sectors of demand to increase energy efficiency; (ii) energy substitution "by the development and use of nuclear energy, ...geothermal power, ...and natural gas" (AP 1990: 5–8); (iii) technological research to improve CO_2 emissions control technologies and develop techniques of disposal of CO_2 emissions (sinks); and (iv) international cooperation by means of "technology transfer" and "support to conservation and development of tropical forests and other sinks" (AP 1990: 11)

It is evident that both Germany and Japan emphasize the contribution of energy conservation and energy substitution in fulfilling the CO_2 abatement target. The Japanese strategy deviates, however, from the German strategy in its lack of any mention of economic instruments (CO_2/energy taxes), and the priority given to technological research and international cooperation.

Owing to the relatively low oil and coal prices, and Japan's reluctance

---

53 For such reasons, a recent study characterizes the Japanese CO_2 stabilization target as "ambiguous and misleading" (Izumi et al. 1994).

54 For a detailed account of Japanese policy measures, see AP (1990) and Fermann (1992: 44–50).

to compensate for this by adopting a CO_2/energy tax, it will become difficult for Japan to fulfil even its modest abatement target. The low price on fossil fuels will impede incentives to continue energy conservation and reverse the trend towards the substitution of (carbon-rich) fossil fuels. The latter will be reinforced by the delay of the Japanese Nuclear Power Programme.[55]

The fact that Japan is unwilling to introduce a CO_2/energy tax, which currently seems to be a necessary condition for fulfilment of the CO_2 abatement target, is a strong indication that economic growth and fear of skewed competition take priority over the CO_2 abatement target.

3.5.3.3. Credibility of Abatement Commitment

Recalling that the Japanese *per capita* stabilization target allows for a 6% increase in CO_2 emissions from 1990 to the year 2000, how credible is this commitment? To what extent do abatement efforts match the abatement target?[56]

If Japan were fully committed to its CO_2 emissions stabilization target, one would expect that Japan, recognizing the insufficiency of current abatement efforts, to adopt *additional* measures (i.e. a carbon tax) capable of fulfilling the target. Instead, when confronted with the trend towards increased CO_2 emissions, Japanese officials are eager to emphasize that "Japan is not committed to the stabilization of CO_2, but to policy measures *aiming at* stabilization" (my emphasis); that the 1990 abatement target should be considered "a political symbol, rather than an imperative"; and that "even if [the abatement target] was not accomplished, [the failure to meet] it is not unconstitutional".[57]

The legal validity of the latter argument is confirmed by the fact that the 1990 Action Programme, stating the CO_2 abatement target, was adopted through a government decision and not made into law.[58] In the

---

55 In 1991, Japan spent about 3 billion US dollars (375 billion Yen) on energy-related research. Nuclear research accounted for more than 80% of this sum – 2.5 billion US dollars (310 billion Yen) (OECD 1992b: 433-434).

56 Since both the 1990 Action Programme and Japan's report, Japan's Response to Global Warming, to the Intergovernmental Negotiating Committee (INC) fail to provide detailed data on the implementation of abatement measures, it is almost impossible to tell the effect of these efforts. This shortcoming is also evident in the German and American reporting practice.

57 Based on interviews with Japanese officials in November 1992.

58 The Japanese Law Concerning the Protection of the Ozone Layer Through the Control of Specified Substances was adopted by the Japanese Parliament in late 1989, following the belated signing of the Vienna Convention and the Montreal Protocol in September 1989.

Basic Environmental Law (12 November 1993), the Diet (Japanese parliament) excluded any references to abatement targets and taxes, thus failing to use this opportunity to confirm and upgrade the Japanese abatement commitment (IER 1993: 853).[59]

Somewhat in contrast to Japan's reluctance to make its CO_2 abatement commitment into law, Japan relatively rapidly (28 May 1993) ratified the FCCC as the 21st party to the convention. Although there are serious inconsistencies in the climate change policy of Japan, the rapid Japanese ratification of the FCCC can be interpreted as a reflection of the emphasis put on international cooperation in Japanese climate policymaking.[60]

Indeed, international abatement efforts will become necessary if Japan is to *compensate* for its probable incapacity to reach its *national* CO_2 abatement target: In spite of the Japanese "recession" persisting since 1991, recent emission trends indicate that current efforts are insufficient to stabilize the growth in CO_2 emissions since 1987. With the worsening abatement conditions, the insufficiency of measures applied (most notably, the "no to CO_2 tax policy"), and the failure to implement measures already adopted (the delay in the Nuclear Power Programme), it has become increasingly clear that the Japanese stabilization target will not be reached (Fermann 1992, 1993; IER 1993: 446). Since 1991, *per capita* CO_2 emissions in Japan have increased yearly by an average of 1.5%. "If this trend continues", a recent study concludes, "Japan will not be able to reach its *per capita* stabilization target" (Izumi et al. 1995).

---

59 Like several other countries, Japan has made fulfilment of the abatement target conditional on others acting likewise. In the English translation of the Japanese 1990 Action Programme this is somewhat vaguely expressed in the phrase: "The Government of Japan, based on the common efforts of the major industrialized countries to limit CO_2 emissions, establishes the following target ...". Japanese scholars have recently contested this official translation, characterizing it as "inaccurate". A proper translation of the Japanese version of the 1990 Action Programme should read: "... on condition that the major industrialized countries make common efforts to limit their CO_2 emissions" (Izumi et al. 1995). As can be observed, the latter version reads much more like a precondition, suggesting that Japan is reserving the option not to meet its target if other OECD countries fail to make efforts deemed sufficient by Japan.

60 Stating this, however, Japan's ratification of the FCCC is unlikely to radically change the judicial status of the Japanese abatement commitment, as the Convention at present is more a declaration of intent than a treaty specifying legally binding commitments (targets).

3.5.4. United States' Climate Change Policy

3.5.4.1. Ambition, Scope, Commitment

Statements by Bush and Baker in 1988–1989 indicated that the new American administration would reconsider the rather obstructivist policies of the Reagan administration on the issue of climate change. Subsequent developments showed, however, that the Bush administration did no more than offer a change in rhetoric rather than a change of actual policies. By spring 1992, the United States was the only OECD country apart from Turkey failing to commit itself to abatement targets and time-tables.

Moreover, the Bush administration made it clear that the United States would not accept an international treaty on climate change that *committed* the OECD countries to any abatement target. Acknowledging the impotency of an international convention the United States declined to support, West European states had to accept the omission of an explicit reference to judicially binding abatement commitments from the FCCC. It was therefore not surprising that the United States was one of the first countries to ratify the convention.[61]

During the autumn 1992 election campaign for the presidency, Bill Clinton and Al Gore stated that they would opt for a more "environment friendly" American policy. As to the climate change issue, expectations were that a Clinton/Gore administration would commit the United States to some sort of emissions stabilization target. Recalling the statements made by Bush and Baker four years earlier and acknowledging that promises are easily made during election campaigns, most observers adopted a waiting attitude.

However, on Earth Day, 21 April 1993, President Clinton confirmed the expectations nurtured during his presidential campaign by announcing American support for the target-oriented approach and declaring that the United States was committed to the stabilization of "emissions of greenhouse gases to their 1990 level by the year 2000". On the same occasion, Clinton promised the finalization of an American action plan by August 1993 specifying how this target was to implemented.

The American adoption of an abatement target with a time-table was favourably received by other OECD countries. Adaption to the national abatement target approach seemed to constitute a virtual turnabout in American climate change policy-making.

---

61 The United States ratified the FCCC on 15 October 1992 as the fourth party to the Convention.

It would seem irrelevant to note that the American abatement commitment is not legally binding, since this limitation characterizes the abatement commitment of most OECD countries. But in its various other aspects, how *demanding* is the American abatement target compared to those of Germany and Japan?

Everything else being equal, the earlier the *base year*, the harder it is to reach the target. Following this statement, it would seem that Germany (1987 base year) has made a more demanding abatement commitment than the United States (1990) and Japan (1990). Such a conclusion is premature, however, since Germany has provided for more *abatement time* (2005) than the United States and Japan (both 2000). Much more crucial than time-tables, however, is the question whether abatement policies opt for stabilization or reductions. The same can be said for the particular greenhouse gases included in the policy commitment.

The German emissions *reduction* target (25%) is much more demanding than the emission stabilization targets of the United States and Japan. While this conclusion is roughly correct, the American abatement target is not as modest as it seems.

(i) While both Japan and the United States have adopted a *stabilization* target, only Japan has done this on a *per capita basis*. A stabilization target linked to population growth is certainly less demanding than a stabilization target opting for *total* stabilization. This observation pushes Japan down the ladder (of ambition), but fails to move the United States up closer to the demanding target of Germany.

(ii) What moves the United States into a middle position and closer to Germany is the fact that the United States commits itself *explicitly* to "return US greenhouse gas emissions to 1990 levels by the year 2000 with cost effective *domestic* actions" (*my emphasis*) (CCAP 1993: i).[62]

These two features of the United States' abatement commitment make it more demanding than it appears at first glance. However, a third peculiarity should also be commented on.

(iii) While the German and Japanese abatement commitments focus heavily on CO_2 emissions, the United States' commitment is to the *overall* stabilization of greenhouse gases. *If* the American commitment had been interpreted to *include* the CFCs already regulated under the 1987 Montreal Protocol, the American abatement target would not only have been easier to reach, it would, moreover, have implied that the United States took credit twice for phasing out this particular gas. This, however, is not the case: In addition to the major greenhouse gases (CO_2, CH_4, NO_x), the US stabilization target includes only the gases introduced

---

62 The Climate Change Action Plan (USA), October 1993.

following the phase-out of the CFCs – the HFCs (hydrofluorocarbons). While the warming effect of the "old" CFCs is extremely high (in addition to being the prime cause of ozone depletion), the warming effect of the substituting HFCs is much less significant.

Stating this, the American overall stabilization target still allows for increasing the emissions of one particular greenhouse gas (for instance CO_2) if this is balanced by a comparative reduction of other greenhouse gases. While this kind of *flexibility* may contribute to the undermining of the American abatement commitment, it also invites more cost-effective abatement efforts. The comprehensive American approach provides decision-makers with the flexibility to first reduce emissions of greenhouse gases offering the lowest marginal costs on abatements.

To sum up: Joining in with the predominant national abatement target approach, the new American climate policy nevertheless contains some unique features. Apart from the fact that the United States opts for a stabilization target compared to the Japanese *per capita* stabilization commitment and the German *reduction* target, the American commitment is strengthened by the fact that the stabilization target will be reached through domestic efforts alone. These elements contribute towards moving the United States up the ladder of ambition to a position somewhere between the demanding abatement target of Germany and the more modest Japanese target.

3.5.4.2. Policy Measures and Implementation

What is impressive about the American climate policy is neither the ambition of the abatement target nor the magnitude of abatement efforts, but rather the likelihood that the abatement target actually will be reached. In an effort to substantiate the latter evaluation, I first consider the main elements of the American effort to fulfil the abatement commitment.

The conception that the Clinton administration constituted a *break* with the climate policies of the Bush administration is not fully justified. As concerns policy *measures*, the Bush administration took several important initiatives: The 1990 Clean Air Act includes provisions strengthening emissions standards, thus contributing to the abatement of greenhouse gases. Moreover, the 1992 amendment to the Energy Policy Act facilitates the use of renewable energies and the improvement of energy efficiency in the transport sector.[63] Furthermore, the Bush administration was responsible

---

63 For studies accounting for the 1990 Clean Air Act and the 1992 Energy Policy Act, see Eikeland (1993a, b), Aarhus and Eikeland (1993: 108–117).

for a significant increase in federal funding of research on renewable energies which had been neglected during the Reagan administration.[64]

Whether one chooses to see the Clinton administration's October 1993 Climate Change Action Plan (CCAP)[65] as a *break* or a *continuation* of Bush policies, it nevertheless represents a considerable *expansion* of the American abatement effort. The CCAP contains more than fifty new or expanded projects. Energy conservation is calculated to account for about 70% of the abatement effort while energy substitution stands for the remaining 30% (IER 1993: 797–799).

While the CCAP provides a stronger medicine than previous policies, it still abides by the "no-regret" profile of the Bush administration's policy measures. This becomes evident in the assumption that the 1.9 billion US dollars of federal money invested in the plan is designed to increase federal revenues by 2.7 billion US dollars in the same period. A similar assumption is made concerning the private investment tag of 61 billion US dollars between 1994 and 2000, which is said to be recovered through energy savings over the same period and to garner another 207 billion US dollars between 2001 and 2010 (IER 1993: 797).

Furthermore, the CCAP relies mainly on *voluntary action* by industry. To the extent that the energy savings reported above are deemed credible by industry and business, government enforcement should not be required. However, commenting on this, Greenpeace spokesman Steve Kretzmann states:

> The industries the president expects to volunteer to combat climate change are some of the same ones pooh-poohing the notion that global warming is a threat. Either this administration is amazingly naive or Bill Clinton has once again flip-flopped on an environmental commitment to please big business. (IER 1993: 788).

The weakness of the CCAP, however, is not so much its content as what is lacking. It fails to introduce taxation on the consumption of fossil fuels or emissions of carbon. The symbolic energy tax adopted in the summer of 1993 (4 cents per gallon) is not capable of rectifying this state of affairs.

Despite these shortcomings, the CCAP and measures introduced under the Bush administration seem to be sufficient to secure the fulfilment of the

---

64 From 1989 to 1991, the federal funding of research on renewable energies was increased by 46% (OECD 1992b: 492–493).
65 In his Earth Day speech on 21 April 1993, President Clinton declared the action plan to be ready for August. As it went, submission of the report was delayed until 19 October.

American abatement target (Bleischwitz et al. 1993: 31; IER 1993: 798).
This is due mainly to the modest ambition of the American abatement target, modest whether related to the abatement targets of other countries or
to the fact that the marginal costs of American abatement efforts are relatively low due to the comparatively low energy efficiency of the American
economy. The latter increases the mitigation effect of policy measures.

4. Japan, Germany and the United States: Prospects for Taking a Leadership Role

In the previous section, Germany, Japan and the United States have been
compared along dimensions believed to be essential for these countries'
potential for taking a leadership role in the development of the problem-
solving capacity of the climate change regime. The task of the present
section is to make predictive inferences regarding *whether* the three
countries under consideration may execute leadership in the future, and,
if so, *how*. The judgements and predictions made must fulfil three
requirements. Predictions shall: (i) comply with the defining characteristics (attributes) of a core concept of leadership,[66] (ii) focus on how each
country may contribute to the enhancement of its problem-solving
capacity, taking the provisions of the FCCC into consideration, and (iii)
not contradict empirical patterns identified during mapping of the leadership potential of the respective countries.

The analysis proceeds in two steps: Firstly, the potential for leadership
of each country is restated, the crucial relationships between abatement
target, abatement *measures* and abatement *conditions* being emphasized.
Then, predictions are generated on the basis of the leadership potential
mapped. As functions of their respective *vulnerabilities* and *opportunities*,
it will become evident that Japan, Germany and the United States vary
considerably as to what kind of leadership role they might be capable of
executing in the years to come, even though they share the status of
being large economies and highly industrialized countries.

4.1. Japan

4.1.1. Current Leadership Potential: Strengths and Limitations

Although the Japanese abatement target is modest compared to that of
the United States, and even more so in relation to that of Germany,

---

66 See Table 2, Section 2.2.3.

Japan is unlikely to fulfil its *per capita* based CO_2 emissions stabilization by the year 2000. This is due partly to the worsening abatement conditions of the 1990s,[67] and partly to the fact that the Japanese government remains unwilling to adopt the *measures* required to fulfil the target.

The primary expression of the latter is the Japanese reluctance to apply a energy/CO_2 tax: While the adoption of such a tax has been discussed continuously in Japan since the EU countries raised the issue in 1991, and moreover is recognized as one of the most effective ways of mitigating CO_2, such considerations have been overruled by the fear that a tax would reduce the international competitiveness of Japanese exports and restrict economic activity (growth), thus prolonging the recession. Reinforcing such sentiments is the perception that the direct socioeconomic impacts of global warming on Japan will be small (EA 1992).

Japan is already one of the most energy efficient economies in the world. Unfortunately, this implies that an energy/CO_2 tax would have to be set quite high to have any substantial impact – that is, to further contribute to fuel-switching and energy conservation.[68] The fact that Japan, due to *prior* achievements in energy substitution and conservation suffers *high marginal costs* on *further* abatement efforts explains to a considerable extent why the Japanese national abatement target is not within reach.

However, this line of argument (high marginal costs) seems to be much too defensive to function as a basis for a future leadership role on the climate change issue. Nor is it enough to make Japan a "world leader in the efficient use of energy" (OECD 1992a: 345) if this cannot contribute to further abatements and secure the implementation of the Japanese abatement commitment for which Japan has accepted political credit ever since its adoption in October 1990.

4.1.2. Avenues for Taking a Leadership Role

Japan was not among the first to adopt a national abatement target; neither will it be the only major OECD country failing to implement its target. In fact, most OECD countries are currently struggling against high

---

67 Most notable among the worsening abatement conditions is the negative incentive provided by the relatively low prices for fossil fuels (oil and coal). Also important is the delay of the Japanese Nuclear Energy Programme, which further inhibits substitution towards non-fossil fuels. These developments are explained in Fermann (1992, 1993, 1995) and the conclusion that Japan is unlikely to fulfil its abatement commitment has recently been confirmed by the Japanese themselves.

68 The impact of an energy/CO_2 tax would also depend on the combination and magnitude of other abatement measures implemented.

odds to fulfil their abatement commitments. Nevertheless, Japan seems to command a unique combination of incentives making it both more interested in avoiding political embarrassment on environmental issues and more capable of compensating for the probable failure to fulfil the national abatement target by other means.

In view of the requirement of the FCCC that industrialized OECD countries should stabilize and ultimately reduce greenhouse emissions so as to provide margins for the expected increase in emissions from the developing economies, the Japanese failure to achieve its stabilization target is likely to prompt foreign criticism. Especially so, it seems, since the Japanese "stabilization" target – owing to the *per capita* qualification – already allows for a 6% increase in CO_2 emissions from 1990 to the year 2000.

Not only has Japan traditionally been eager to avoid political embarrassment, the impact of foreign criticism will furthermore be reinforced by the great expectations created by the repeated statements that Japan intends to play a leadership role on international environmental issues. Finally, recalling the Japanese (and German) criticism of past American lack of commitment to abatement targets, the United States is unlikely to avoid notifying Japan of any failure to fulfil its abatement commitment.

Thus, although the Japanese at present do not conceive themselves as vulnerable to the probable socio-economic impacts of global warming, Japan nevertheless is vulnerable to the political consequences of not fulfilling its 1990 abatement target. This vulnerability provides an incentive for the Japanese to seek abatement approaches capable of compensating for their inability to stabilize greenhouse gas emissions at home. Below, it is argued that Japan exhibits both the incentives and the resources required to develop the abatement approach of *joint implementation* – the idea that one country facing high marginal costs on abatement efforts at home provides assistance (technological, financial, administrative) to less energy-efficient and developed countries. Through joint implementation, Japan may secure both a more efficient use of its resources and gain some credit for its foreign abatement efforts.

The idea of joint implementation of abatement efforts was brought into the political discussion by Norway and Germany in 1991. However, this abatement mechanism has been practised *within* the United States for some years in the form of tradeable emission permits. Joint implementation is encouraged in the FCCC,[69] and presently discussed in follow-up working groups. What makes Japan well prepared to take a leadership role in developing the abatement mechanism of joint implementation?

---

69 Paragraph 4.2.a.

Aside from anticipating the criticism resulting from the probability that Japan will not be able to stabilize emissions at home, the strongest incentive to initiate cooperation with developing countries in mitigating greenhouse gas emissions is the extremely high marginal costs of abatement efforts within the Japanese economy. To the extent Japan contributes toward mitigation in other countries, it may receive political credits for these efforts. In the short run this credit could be provided tacitly, but in the longer run credit may be provided formally according to some credit-allocating scheme which Japan may assist in developing. Such pioneering work may well contribute to deflecting the criticism that arises from Japan's failure to stabilize emissions at home.

In addition to having an incentive to take a lead in elaborating the idea of joint implementation into a practical mechanism of joint abatement efforts, Japan seems to be well equipped to facilitate such a development. Not only is Japan one of the world's leaders in energy efficiency and environmental technologies, more importantly it commands the financial resources required to help transform the abatement *needs* of developing countries into actual *purchasing power*.

This important asset of investment-available money is often ignored when Germany, Japan and the United States are compared. While both Germany and the United States are technologically advanced, they have, for various reasons and to varying degrees, also become net importers of capital and have decreased their ODA in recent years. Japan, on the other hand, not only shares a co-world leadership in technological innovation, it also commands surplus money (trade) in need of recycling.

By allocating some of this surplus to abatement efforts in developing countries, Japan may simultaneously (i) solve some of its own financial recycling problem, (ii) create new markets for its energy conservation technologies, (iii) secure the transfer of technology to developing countries, (iv) enhance its own environmental image, (v) deflect criticism arising from its failure to fulfil its *national* abatement target, and finally (vi) increase the capacity of the present climate change regime to challenge the problem of global warming. Stated thus, facilitating abatement efforts abroad may be conceived of as a "no-regret" measure not available at home.

Against this backdrop, I consider international cooperation generally and the abatement approach of joint implementation specifically to be the most promising avenue available in which Japan can take a leadership role on the issue of climate change. The fact that Japan is the largest single contributor to GEF[70] as well as in terms of ODA, coupled with

---

70 A financial mechanism within the World Bank established to allocate additional funds to abatement efforts in developing countries.

recent Japanese statements expressing a positive attitude towards international cooperation and the idea of joint implementation, are indications that Japan has seen this opportunity. The next move could be for Japan to make joint implementation an integral and important part of its official climate change policy.

4.2. Germany

4.2.1. Current Leadership Potential: Strengths and Limitations

Although the German CO_2 abatement target was well researched (Enquete Commission) prior to its adoption in June 1990 and generally perceived as attainable, Germany presently seems unlikely to fulfil its ambitious abatement target of reducing CO_2 emissions by 25% in the year 2005.

This state of affairs is *not* due to the German unification which, on the contrary, should have made the abatement target *easier* to attain: A study predicts that the CO_2 emissions from the new Bundesländer (former GDR) will have decreased by 38% by 2005 due to a decrease in production and a restructuring of industry and energy production. However, this will not be enough to compensate for the estimated 3% increase in CO_2 emissions from the old Bundesländer (former West Germany) (Prognos AG 1992: 435f.). Taken together, unified Germany is expected to decrease its CO_2 emissions by 10% – which is 15% short of its ambitious abatement target.

Fulfilment of the German abatement target was based on the assumption that an energy/CO_2 tax would be adopted. Having failed to persuade reluctant EU members of the benefits of a EU-wide tax, Germany has not been willing to adopt a tax unilaterally. Furthermore, negotiations with German industry on voluntary agreements securing a more effective implementation of abatement measures have so far been largely unsuccessful.

4.2.2. Avenues for Taking a Leadership Role

Having taken a leadership role in marketing the national abatement target approach and instituting it as a key principle of the FCCC, Germany now faces the prospect of failing to fully comply with its own abatement commitment. The failure of Germany to fulfil its ambitious abatement target is likely to cause some political embarrassment. This is the more likely since Germany, as a prominent member of the EU, has often taken the opportunity to criticize the United States for not adopting a national abatement target. Having done this in 1993 and enjoying reasonable chances of fulfiling its abatement commitment, the United States is not likely to refrain from noting German failure.

One way of deflecting such criticism would be for Germany to direct attention to the fact that although Germany is unlikely to fulfil its abatement target, it will nevertheless manage to *reduce* the emissions of CO_2. Indeed, this should be considered a greater achievement than the probable overall emissions stabilization achievement of the United States. In the opinion of this author, Germany's best prospect for taking a leadership role is by intensifying the work presently being done in reducing the German CO_2 emissions. Few OECD countries are likely to manage to reduce, or even stabilize, their greenhouse emissions. Thus, Germany stands out as *an example* actually capable of fulfilling the FCCC's requirement that the industrialized countries stabilize their greenhouse emissions. Germany is likely to fulfil this demand with a good margin.

Germany has demonstrated considerable interest in the abatement approach of joint implementation since 1991. While Germany to some extent shares Japan's need for deflecting foreign criticism and has developed advanced energy conservation technology attractive to developing countries, it is currently directing its financial resources to the restructuring of the former GDR and is the leading provider of hard currency to Russia and Eastern Europe. Much of the money has been provided by foreign investors attracted by the high interest rate in Germany.

This implies that Germany's financial resources are already stretched to the limit. This, in turn, makes Germany less capable than Japan of providing developing countries with the funds required to create a market for energy conservation technology and simultaneously facilitate more efficient mitigation in developing countries. While this limitation certainly does not exclude a German contribution to the elaboration of the joint implementation approach, it nevertheless reduces the probability of Germany taking a *leadership* role in this field.

Another alternative would be for Germany to increase the pressure on reluctant members of the EU to adopt a community-wide energy/CO_2 tax. If such a tax was adopted, it would both increase the chances of fulfilling the German abatement target and put heavy pressure on Japan and the United States to do likewise. If, through the EU, Germany could trigger the OECD-wide adoption of an energy/CO_2 tax this would be a major achievement and a substantial contribution to the development of the present climate change regime. It would provide Germany with a *second pillar of leadership* in addition to the abatement achievement already secured at home.

Against this, it could be argued that the EU might be just as likely to function as an obstacle to states wanting to take a leadership role. To realize the full leverage of the EU, unanimous agreement is most often required. To reach such agreement, strong proposals are watered down, thus reducing the impact of decisions. The original tax proposal has for several years now been modified and stalled by reluctant member-states, and Germany seems much too preoccupied with the former GDR, Eastern Europe and Russia to bring sufficient pressure to bear on reluctant EU members (see Chapter 12).

To conclude, Germany's most promising avenue for taking a leadership role is likely to be found within the national abatement target approach. By reducing CO_2 emissions by 10%, Germany sets an unprecedented example, fulfils the requirement of the FCCC that OECD countries shall stabilize (and ultimately reduce) greenhouse emissions to provide margins for increases in emissions from the developing countries, and thus strengthens the legitimacy of the current abatement regime. This is likely to bring pressure to bear on other countries to intensify their abatement efforts. It should be emphasized, however, that future success is dependent on the implementation of additional measures, since continued reductions of emissions based on the restructuring of the energy sector and industry in the former GDR cannot be relied upon in the future.

Where alternative avenues of achievement are concerned, Germany currently seems to lack the financial surplus required to compete with Japan on the abatement approach of joint implementation. I also have some doubt whether Germany is willing or able to convince reluctant members of the EU to adopt an EU-wide CO_2 tax. The growth rate must increase and unemployment fall considerably before such a tax is likely to be given serious political consideration. The only action that would make some impression under present circumstances seems to be the unilateral German adoption of a CO_2 tax. So far, the German government has not been willing to risk losing the continuing cooperation of German industry by enforcing such a measure.

4.3. The United States

4.3.1. Current Leadership Potential: Strengths and Limitations

It is somewhat ironic that the United States, the last OECD country to commit itself to the abatement target approach, seems to be one of the few countries capable of actually fulfilling its target. Why is this so? Is it due to favourable abatement conditions, effective abatement measures or the fairly low ambition of the American abatement commitment?

In answering this question let us recall Section 3, where the abatement commitments of Germany, Japan and the United States were compared. Here it was concluded that the abatement targets of the United States and Japan were considerably less ambitious than that of Germany. Why is it then that the United States seems likely to fulfil its abatement target while Japan is likely to fail?

A proper explanation requires a shift of focus from univariate cross-country comparison to the comparison of relationships between abatement target, abatement measures and abatement conditions. What matters, then, is not so much the ambition of the target (which is similar for Japan and the United States), but the ambition *relative* to the abatement conditions and measures of each country.

Previously it was argued that Japan is unable to fulfil its CO_2 emissions stabilization target because of severe abatement conditions raising the marginal costs of abatement efforts, thus reducing the effect of abatement measures.[71] In the United States, the structural abatement conditions are much more favourable owing to the fact that the energy efficiency of the American economy is considerably lower than that of Japan. This makes abatement measures like energy conservation less costly, or, the converse, increases the abatement effect of conservation measures.

While Japan seems to be "punished" (in the form of high marginal costs on abatement efforts) for its past energy conservation achievements, the United States is currently "rewarded" for its relatively minor prior efforts in this field. Having removed the political and institutional obstacles to committing itself to the abatement target approach, the United States now enjoys the benefit of low marginal costs rendering no-regret measures sufficient to reach its abatement target. What can be inferred from this state of affairs concerning the prospects for an American leadership role?

4.3.2. Avenues for Taking a Leadership Role

One could jump to the conclusion that the United States will take a leadership role within the national abatement target approach since the Americans are likely to fulfil their overall stabilization target. This would

71 It was also argued that in addition to unfavourable abatement conditions, Japan's failure to fulfil its abatement target is due to Japan's reluctance to adopt an energy/CO_2 tax.

be premature, however. Recall that the American abatement target is fairly modest, whether compared to countries like Germany or to the OECD average. Although Germany is unlikely to fulfil its ambitious abatement target (25% CO_2 reduction), a German effort at reducing CO_2 emissions by 10% would still be considered a greater achievement than an American stabilization of greenhouse gas emissions. In the end, a country aspiring to environmental leadership should be judged according to how large the reductions of greenhouse gases it is capable of securing, rather than to whether it is capable of fulfiling a modest abatement target.

The United States is likely to fulfil its stabilization target, but this is not enough to secure leadership within the national abatement target approach. This avenue is much better suited to countries like Germany, the Netherlands and Britain. Moreover, the United States has little incentive to adopt a significant CO_2 tax (or equivalent) since it is likely to fulfil its modest abatement target without the utilization of economic abatement instruments.

This is not the case for most other countries, however. In Germany and Japan, a tax seems to be a necessary condition for fulfilling the national abatement targets. If the United States continues to reject the adoption of a substantial energy/CO_2 tax, Germany and Japan would decline to do it unilaterally, pleading skewed competition. Hence, to the extent the United States uses its modest achievement at home (overall stabilization) as an argument for not adopting a tax on fossil fuels, the American achievement would become *counter-productive* in a wider context.

Furthermore, it can be argued that the probable fulfilment of the overall stabilization target provides Americans with a disincentive to elaborate on the abatement approach of joint implementation. This may be reinforced by the fact that the United States has for years been a net importer of capital and is in the process of decreasing its ODA budget, thus limiting the subsidies available for developing countries in need of American energy conservation technology. Against this, it can be argued that the United States has for several years *practised* the idea of joint implementation domestically by establishing markets for tradeable emissions permits.

Nevertheless, I would say that the most promising avenue for exercising leadership is for the United States to continue its huge effort in the field of climate change basic science, and applied economic and policy research (i.e. tradeable emissions permits). The United States' scientific achievements have been somewhat neglected by observers eager to criticize the Bush administration for its failure to commit the United States to an abatement target. The impact of past and future American research efforts should not be underestimated, however: the American research community will still be crucial in reducing the uncertainties prevailing in our knowledge of climate change. Proper assessment of the challenge facing us remains a necessary condition for the effective design of response strategies capable of mitigating the problem of global warming.

This, however, seems to be the only kind of leadership the United States is capable of, or willing to, provide within the field of climate change in the years to come. Domestically, institutional lobbyists, political pressure groups and the strong emphasis put on the continuation of the "American way of life" are likely to function as obstacles to the realization of the United States' otherwise great potential for substantially reducing the emissions of greenhouse gases.

5. Conclusion: Recapitulation and Some Final Speculation

At the outset of this chapter, it was argued that renewed leadership would be required if the climate change regime was to mature, thus reducing the evident mismatch between the severity of the challenge facing us and the limited problem-solving capacity currently available. On this assumption, the aim of this chapter was to (i) identify the leadership potential of Germany, Japan and the United States, and from this (ii) infer what might become the most promising avenue for taking a leadership role for each country.

Prior to the empirical analysis, an attempt was made to distinguish leadership from other kinds of extraordinary state behaviour (power politics). Emphasizing the non-coercive and consensual nature of leadership, as well as the mutual interest prevailing in a leader/follower relationship, Underdal's original definition of the term was accordingly revised:

> Leadership is an asymmetrical relationship of influence, where one actor guides or directs the behaviour of others *by means of persuasion and demonstration* towards a certain *goal of mutual interest* over a certain period of time.

By inserting two additional qualifications into the definition (emphasized), the empirical scope of leadership concept was significantly narrowed. This core conception of leadership was found to exclude two of Underdal's and Young's specifications of leadership; one was the "coercive" mode of leadership which was to affect "the incentives of others to accept one's own terms" ... "through sticks and carrots" (Underdal 1991a: 143); the other was "structural" leadership associated with "the ability to deploy threats and promises in ways that are both carefully

crafted and credible" (Young 1991: 290). It was found that these strategies for exerting influence belonged to the realm of power politics, not to
that of leadership. The remaining modes of leadership described in
Section 2.2.4. – the "intellectual", "entrepreneurial", "instrumental",
and "unilateral" – will now be applied in a final attempt to conceptualize
the future leadership behaviour of Germany, Japan and the United States.

Historically, Germany, Japan and the United States stand out respectively as the leader, the reluctant follower, and the laggard on the issue
of climate change. This picture changed somewhat in 1993 when the
United States finally decided to follow suit with the large majority of
OECD countries, upgrading its climate change policy to the national
abatement target approach, and moreover in July 1996, when the US
delegates at the CoP-2 in Geneva announced that the United States government was in favour of judicially binding abatement targets. Great differences can still be observed between the three countries considered in
this chapter (see Table 9).

Germany has clearly adopted the most ambitious abatement target;
Japan the least ambitious; the United States remains somewhere in
between. In other respects the climate change policies of Germany, Japan
and the United States resemble each other. None of them has adopted a
significant CO_2 tax (or equivalent) to compensate for the low current
prices of fossil fuels. The effect of this common failure to include economic instruments as an integral part of their climate change policies is
diverse, however. Owing to relatively low marginal costs of abatement
efforts, Germany still manages to reduce CO_2 emissions by 10%. Japan
for its part, however, is unlikely even to fulfil the modest *per capita*-based
stabilization target which allows for a 6% increase in CO_2 emissions. As
for the United States, the Americans should be able to fulfil their overall
stabilization target. A minor increase in CO_2 emissions is nevertheless to
be expected.

Finally, how are we to *conceptualize* the "avenues for taking a leadership role" identified in Section 4 by means of the typology described in
Section 2.2.4.?

Germany's most promising avenue for taking a leadership role is likely to be found within the national abatement target approach. Even
without the adoption of a CO_2 tax, Germany is likely to reduce its emissions of CO_2 by 10% by the year 2005 compared to the 1987 level.
While this achievement is considerably less than the very ambitious
German abatement target of 25% reductions, the German abatement
effort might nevertheless prove to be a "world record" (at least within
the OECD area), and satisfies the requirement of the FCCC (stabilization)
with a good margin.

Table 9: Policies and Prospects – Germany, Japan and the United States

1–2: Policy 3–5: Reality/prospects	Germany	Japan	United States
1. Abatement target	Reducing CO_2 emissions by 25% from 1987 to 2005.	Stabilizing *per capita* CO_2 emissions from 1990 to the year 2000 (implies 6% increase in total CO_2 emissions).	Overall stabilization of greenhouse gas emissions from 1990 to the year 2000.
2. Measures suggested	Substantial CO_2 tax suggested by the Enquete Commission.	No CO_2 tax suggested in 1990 Action Programme.	Substantial energy tax proposed by the Presidency as part of policy to reduce the budget deficit.
3. Abatement conditions	1. Relatively low marginal costs of further abatement efforts due to the unification of Germany (low energy efficiency of the former GDR) 2. Strong resistance from the German coal-lobby against the unilateral adoption of a CO_2 tax.	1. High marginal costs of abatement efforts owing to past achievements within energy conservation (energy efficient economy). 2. Export oriented and energy-intensive industries hostile towards adoption of a CO_2 tax.	1. Relatively low marginal costs of abatement efforts – economic instruments can be applied as "no-regret" measures. 2. Strong opposition from "oil states", the car industry and other lobbyists against adoption of energy/CO_2 taxes.
4. Measures applied	1. No unilateral CO_2 tax adopted. 2. German attempts to introduce a EU-wide CO_2 tax stalled by reluctant member-states.	No CO_2 tax likely to be adopted in the medium term.	Presidency's energy tax proposal reduced to insignificant gasoline tax (4 cents per gallon) by Congress.
5. Estimated change in emissions	CO_2 emissions reduced by 10% in 2005 compared to 1987 levels.	CO_2 emissions likely to increase within the range of 8–12% from 1990 to the year 2000 – substantially more than the 6% increase allowed for within the *per capita* stabilization target.	Estimated plus/minus 0% change in overall greenhouse emissions from 1990 to the year 2000. CO_2 emissions likely to increase by 1–3% in the same period.

In what way might this achievement contribute towards making Germany a leader of the future? And how might the specific quality of German leadership reduce the current mismatch between problem-severity and problem-solving capacity in the field of climate change?

Firstly, it should be acknowledged that a 10% German reduction of CO_2 emissions would translate into 0.5% reduction in total world CO_2 emissions compared to the business-as-usual scenario. While such an achievement certainly would not be enough to *solve* the collective problem of climate change, the "substantive impact" of the German CO_2 reduction could be multiplied to the extent that this achievement would trigger more extensive abatement efforts from other countries. Thus, Germany may execute "leadership through unilateral action" (Underdal).

But Germany need not content itself with the example set by substantial reductions in national CO_2 emissions: The impact and effectiveness of Germany's climate change policy would increase considerably to the extent that a CO_2 tax would be adopted. This would not only increase the "substantive impact" and persuasive power of the German abatement effort, it would, moreover, point out Germany for "entrepreneurial" leadership (Young) along with the five small European states already having adopted such a tax. This kind of leadership would put pressure on both the United States and Japan to act likewise. Equally as important, however, the adoption of a German CO_2 tax would bring tremendous pressure to bear on reluctant EU member-states to accept an EU-wide CO_2 tax. The unilateral adoption of a CO_2 tax should, of course, be followed up by a German diplomatic offensive within the EU, demonstrating that Germany was in earnest.

In my view, this would be the best way for Germany to broaden its current leadership role. However, it would be naive to expect leadership to be the sole ingredient in such a venture. "Power politics" in the form of side-payments and some arm-twisting is likely to be required to break the present dead-lock within the EU. Germany's potentially greatest achievement is not what it can manage by itself. German leadership rests even more on the extent to which it is able to *promote* its own achievements so as to trigger collective action.

While Germany's continuing leadership may be secured within the national abatement target approach, Japan's most promising avenue for taking a leadership role is to be found in international cooperation in the form of joint implementation. The main *incentive* for exchanging money and technology for emission permits is the probable failure of Japan to fulfil its modest abatement target and the political embarrassment arising therefrom.

The main merit of the abatement approach of joint implementation is that it provides an incentive for countries with surplus capital and technology (but facing high marginal abatement costs at home) to utilize their resources to reduce emissions in a country where the marginal costs of abatement efforts are low. In return, the donating country would receive emission permits – allowances to emit at home. The aggregate outcome would, at least in theory, be more abatements for less money.

In my opinion, Japan has both the incentive (political embarrassment) and the resources required (surplus money and technology) to study and explore the abatement approach of joint implementation seriously. Although the Americans have done pioneering work by establishing markets of tradeable emission permits *within* the United States, the market mechanism has yet to be practised internationally.

This means that the abatement approach of joint implementation is open for both "instrumental" leadership and "leadership through unilateral action": "Instrumental" leadership can be executed if Japan succeeds in its attempt to persuade potential partners in the developing world (South-East Asia or other regions where the marginal costs of abatement efforts are considerably lower than in Japan) that the establishment of a bi-, tri- or multinational market for tradeable emission permits is mutually beneficial – both in terms of the global environment and for national economic and developmental reasons.

Having established such a market through trial and error, the next promising step for Japan would be to execute "leadership through unilateral action". To the extent that Japan is able to demonstrate that joint implementation is a practical and cost-effective abatement approach capable of accounting for both the comparative advantages and the needs of industrialized and developing countries, Japan will stand out as an example to follow for other pairs of countries struggling with high marginal abatement costs and a lack of capital and technology. A successful Japanese exploration of joint implementation would not only deflect foreign criticism of Japanese abatement failure at home, more importantly, it would trigger collective action securing substantial *additional* abatements.

While Germany's and Japan's most promising avenues for taking a leadership role (respectively, the national abatement target approach and the abatement approach of joint implementation) may contribute toward increasing the problem-solving capacity of the present climate change regime, the American leadership potential was untill recently to be found mainly in its superior capacity to enhance knowledge of the problem itself.

The United States scientific community currently accounts for a substantial share of the total research effort in the field of climate change.

Much of this research is basic science directed towards reducing the uncertainties prevailing as to how severe and real the problem of climate change is. Proper assessment of the challenge facing us remains a necessary condition for the effective design and calibration of response strategies capable of mitigating the problem of global warming.

The United States is likely to continue executing "intellectual" leadership (Young) by producing "generative systems of thought". Moreover, American top scientists are likely to play a vital role within the IPCC process assessing and popularizing basic research on climate change (Working Group I).

American "intellectual" and "entrepreneurial" leadership is, moreover, likely to be continued in the research on impacts of global warming (advanced modelling) and effective means of mitigation (Working Group II). However, because of the lack of incentives (the US is likely to fulfil its modest abatement target) and strong institutional and political obstacles, the United States is unlikely to extend its "intellectual" and "entrepreneurial" leadership to costly political action. Although the American delegates at the CoP-2 in July 1996 announced a considerable shift in the US climate change policy in arguing in favour of an international codification of binding and equal greenhouse gas emission targets among OECD countries, US officials have made it clear that a strengthened commitment to regulate the emissions of industrialized countries is dependent on the willingness of developing countries to accept a mechanism for tradeable emission permits (joint implementation) to be part of the new climate change regime to be agreed upon at the CoP-3, December 1997 (see chapters 1 and 2). This condition significantly reduces the potential costs of the American policy shift for the United States, since a joint implementation approach to emission reduction would allow the United States to invest in emission reductions in other countries where marginal costs are comparatively low, while avoiding more costly abatements at home which could challenge the domestic political consensus.

If this interpretation is correct, we can discern the pattern of a potential American leadership based on *two* components: One is the continuation of American *scientific* leadership, the other is the incentive of the United States' government to invest considerably more energy *in developing a more practical mechanism of joint implementation*. In the latter effort, American and Japanese interests to a considerable degree converge: Because of the high marginal costs of abatement in Japan (high energy efficiency) and the political barriers to implementing CO_2 and/or energy taxes in American society (e.g. conservative reactions to any threat to the "American way of life"), there are – at the present stage of technological development – clear limits to how far the Japanese and the United States are capable or willing to go in reducing emissions at home. The remaining alternatives are to intensify research on sinks technology, and to work for the adoption of tradeable emission permits as part of a more effective international climate change regime. In order to succeed in the latter strategy, Japan and the United States must not only convince the developing countries that joint implementation might actually work as a cost effective way of reducing global greenhouse gas emissions (Chapter 7), they must also demonstrate that joint implementation does not conflict with the developing countries' conception of fair burden-sharing (Chapter 6). Finally, the proponents of joint implementation must convince Third World leaders that the notion of tradeable emission permits does not conflict with the priority given to economic growth in developing countries. A monumental task indeed!

Appendix 1:

United Nations Framework Convention on Climate Change

The Parties to this Convention,

Acknowledging that change in the Earth's climate and its adverse effects is a common concern of humankind,

Concerned that human activities have been substantially increasing the atmospheric concentrations of greenhouse gases, that these increases enhance the natural greenhouse effect, and that this will result on average in an additional warming of the Earth's surface and atmosphere and may adversely affect natural ecosystems and humankind,

Noting that the largest share of historical and current global emissions of greenhouse gases has originated in developed countries, that per capita emissions in developing countries are still relatively low and that the share of global emissions originating in developing countries will grow to meet their social and development needs,

Aware of the role and importance in terrestrial and marine ecosystems of sinks and reservoirs of greenhouse gases,

Noting that there are many uncertainties in predictions of climate change, particularly with regard to the timing, magnitude and regional patterns thereof,

Acknowledging that the global nature of climate change calls for the widest possible cooperation by all countries and their participation in an effective and appropriate international response, in accordance with their common but differentiated responsibilities and respective capabilities and their social and economic conditions,

Recalling the pertinent provisions of the Declaration of the United Nations Conference on the Human Environment, adopted at Stockholm on 16 June 1972,

Recalling also that States have, in accordance with the Charter of the United Nations and the principles of international law, the sovereign right to exploit their own resources pursuant to their own environmental and developmental policies, and the responsibility to ensure that activities within their jurisdiction or control do not cause damage to the environment of other States or of areas beyond the limits of national jurisdiction,

Reaffirming the principle of sovereignty of States in international cooperation to address climate change,

Recognizing that States should enact effective environmental legislation, that environmental standards, management objectives and priorities should reflect the environmental and developmental context to which they apply, and that standards applied by some countries may be inappropriate and of unwarranted economic and social cost to other countries, in particular developing countries,

Recalling the provisions of General Assembly resolution 44/228 of 22 December 1989 on the United Nations Conference on Environment and Development, and resolutions 43/53 of 6 December 1988, 44/207 of 22 December 1989, 45/212 of 21 December 1990 and 46/169 of 19 December 1991 on protection of global climate for present and future generations of mankind,

Recalling also the provisions of General Assembly resolution 44/206 of 22 December 1989 on the possible adverse effects of sea-level rise on islands and coastal areas, particularly low-lying coastal areas and the pertinent provisions of General Assembly resolution 44/172 of 19 December 1989 on the implementation of the Plan of Action to Combat Desertification,

Recalling further the Vienna Convention for the Protection of the Ozone Layer, 1985, and the Montreal Protocol on Substances that Deplete the Ozone Layer, 1987, as adjusted and amended on 29 June 1990,

Noting the Ministerial Declaration of the Second World Climate Conference adopted on 7 November 1990,

Conscious of the valuable analytical work being conducted by many States on climate change and of the important contributions of the World Meteorological Organization, the United Nations Environment Programme and other organs, organizations and bodies of the United Nations system, as well as other international and intergovernmental bodies, to the exchange of results of scientific research and the coordination of research,

Recognizing that steps required to understand and address climate change will be environmentally, socially and economically most effective if they are based on relevant scientific, technical and economic considerations and continually re-evaluated in the light of new findings in these areas,

Recognizing that various actions to address climate change can be justified economically in their own right and can also help in solving other environmental problems,

Recognizing also the need for developed countries to take immediate action in a flexible manner on the basis of clear priorities, as a first step towards comprehensive response strategies at the global, national and, where agreed, regional levels that take into account all greenhouse gases, with due consideration of their relative contributions to the enhancement of the greenhouse effect,

Recognizing further that low-lying and other small island countries, countries with low-lying coastal, arid and semi-arid areas or areas liable to floods, drought and desertification, and developing countries with fragile mountainous ecosystems are particularly vulnerable to the adverse effects of climate change,

Recognizing the special difficulties of those countries, especially developing countries, whose economies are particularly dependent on fossil fuel production, use and exportation, as a consequence of action taken on limiting greenhouse gas emissions,

Affirming that responses to climate change should be coordinated with social and economic development in an integrated manner with a view to avoiding adverse impacts on the latter, taking into full account the legitimate priority needs of developing countries for the achievement of sustained economic growth and the eradication of poverty,

Recognizing that all countries, especially developing countries, need access to resources required to achieve sustainable social and economic development and that, in order for developing countries to progress towards that goal, their energy consumption will need to grow taking into account the possibilities for achieving greater energy efficiency and for controlling greenhouse gas emissions in general, including through the application of new technologies on terms which make such an application economically and socially beneficial,

Determined to protect the climate system for present and future generations,

Have agreed as follows:

ARTICLE 1

DEFINITIONS

For the purposes of this Convention:

1. "Adverse effects of climate change" means changes in the physical environment or biota resulting from climate change which have significant deleterious effects on the composition, resilience or productivity of natural and managed ecosystems or on the operation of socio-economic systems or on human health and welfare.

2. "Climate change" means a change of climate which is attributed directly or indirectly to human activity that alters the composition of the global atmosphere and which in addition to natural climate variability is observed over comparable time periods.

3. "Climate system" means the totality of the atmosphere, hydrosphere, biosphere and geosphere and their interactions.

4. "Emissions" means the release of greenhouse gases and/or their precursors into the atmosphere over a specified area and period of time.

5. "Greenhouse gases" means those gaseous constituents of the atmosphere, both natural and anthropogenic, that absorb and re-emit infrared radiation.

6. "Regional economic integration organization" means an organization constituted by sovereign States of a given region which has compe-

tence in respect of matters governed by this Convention or its protocols and has been duly authorized, in accordance with its internal procedures, to sign, ratify, accept, approve or accede to the instruments concerned.

7. "Reservoir" means a component or components of the climate system where a greenhouse gas or a precursor of a greenhouse gas is stored.

8. "Sink" means any process, activity or mechanism which removes a greenhouse gas, an aerosol or a precursor of a greenhouse gas from the atmosphere.

9. "Source" means any process or activity which releases a greenhouse gas, an aerosol or a precursor of a greenhouse gas into the atmosphere.

ARTICLE 2

OBJECTIVE

The ultimate objective of this Convention and any related legal instruments that the Conference of the Parties may adopt is to achieve, in accordance with the relevant provisions of the Convention, stabilization of greenhouse gas concentrations in the atmosphere at a level that would prevent dangerous anthropogenic interference with the climate system. Such a level should be achieved within a time-frame sufficient to allow ecosystems to adapt naturally to climate change, to ensure that food production is not threatened and to enable economic development to proceed in a sustainable manner.

ARTICLE 3

PRINCIPLES

In their actions to achieve the objective of the Convention and to implement its provisions, the Parties shall be guided, inter alia, by the following:

1. The Parties should protect the climate system for the benefit of present and future generations of humankind on the basis of equity and in accordance with their common but differentiated responsibilities and respective capabilities. Accordingly, the developed country Parties

should take the lead in combating climate change and the adverse effects thereof.

2. The specific needs and special circumstances of developing country Parties, especially those that are particularly vulnerable to the adverse effects of climate change, and of those Parties, especially developing country Parties, that would have to bear a disproportionate or abnormal burden under the Convention, should be given full consideration.

3. The Parties should take precautionary measures to anticipate, prevent or minimize the causes of climate change and mitigate its adverse effects. Where there are threats of serious or irreversible damage, lack of full scientific certainty should not be used as a reason for postponing such measures, taking into account that policies and measures to deal with climate change should be cost-effective so as to ensure global benefits at the lowest possible cost. To achieve this, such policies and measures should take into account different socio-economic contexts, be comprehensive, cover all relevant sources, sinks and reservoirs of greenhouse gases and adaptation, and comprise all economic sectors. Efforts to address climate change may be carried out cooperatively by interested Parties.

4. The Parties have a right to, and should, promote sustainable development. Policies and measures to protect the climate system against human-induced change should be appropriate for the specific conditions of each Party and should be integrated with national development programmes, taking into account that economic development is essential for adopting measures to address climate change.

5. The Parties should cooperate to promote a supportive and open international economic system that would lead to sustainable economic growth and development in all Parties, particularly developing country Parties, thus enabling them better to address the problems of climate change. Measures taken to combat climate change, including unilateral ones, should not constitute a means of arbitrary or unjustifiable discrimination or a disguised restriction on international trade.

ARTICLE 4

COMMITMENTS

1. All Parties, taking into account their common but differentiated responsibilities and their specific national and regional development priorities, objectives and circumstances, shall:

    (a) Develop, periodically update, publish and make available to the Conference of the Parties, in accordance with Article 12, national inventories of anthropogenic emissions by sources and removals by sinks of all greenhouse gases not controlled by the Montreal Protocol, using comparable methodologies to be agreed upon by the Conference of the Parties;

    (b) Formulate, implement, publish and regularly update national and, where appropriate, regional programmes containing measures to mitigate climate change by addressing anthropogenic emissions by sources and removals by sinks of all greenhouse gases not controlled by the Montreal Protocol, and measures to facilitate adequate adaptation to climate change;

    (c) Promote and cooperate in the development, application and diffusion, including transfer, of technologies, practices and processes that control, reduce or prevent anthropogenic emissions of greenhouse gases not controlled by the Montreal Protocol in all relevant sectors, including the energy, transport, industry, agriculture, forestry and waste management sectors;

    (d) Promote sustainable management, and promote and cooperate in the conservation and enhancement, as appropriate, of sinks and reservoirs of all greenhouse gases not controlled by the Montreal Protocol, including biomass, forests and oceans as well as other terrestrial, coastal and marine ecosystems;

    (e) Cooperate in preparing for adaptation to the impacts of climate change; develop and elaborate appropriate and integrated plans for coastal zone management, water resources and agriculture, and for the protection and rehabilitation of areas, particularly in Africa, affected by drought and desertification, as well as floods;

    (f) Take climate change considerations into account, to the extent feasible, in their relevant social, economic and environmental policies and actions, and employ appropriate methods, for example impact assessments, formulated and determined nationally, with a view to minimizing adverse effects on the economy, on public health and on the quality of the environment, of projects or measures undertaken by them to mitigate or adapt to climate change;

(g)   Promote and cooperate in scientific, technological, technical, socio-economic and other research, systematic observation and development of data archives related to the climate system and intended to further the understanding and to reduce or eliminate the remaining uncertainties regarding the causes, effects, magnitude and timing of climate change and the economic and social consequences of various response strategies;

(h)   Promote and cooperate in the full, open and prompt exchange of relevant scientific, technological, technical, socio-economic and legal information related to the climate system and climate change, and to the economic and social consequences of various response strategies;

(i)   Promote and cooperate in education, training and public awareness related to climate change and encourage the widest participation in this process, including that of non-governmental organizations; and

(j)   Communicate to the Conference of the Parties information related to implementation, in accordance with Article 12.

2. The developed country Parties and other Parties included in Annex I commit themselves specifically as provided for in the following:

(a)   Each of these Parties shall adopt national policies and take corresponding measures on the mitigation of climate change by limiting its anthropogenic emissions of greenhouse gases and protecting and enhancing its greenhouse gas sinks and reservoirs. These policies and measures will demonstrate that developed countries are taking the lead in modifying longer-term trends in anthropogenic emissions consistent with the objective of the Convention, recognizing that the return by the end of the present decade to earlier levels of anthropogenic emissions of carbon dioxide and other greenhouse gases not controlled by the Montreal Protocol would contribute to such modification, and taking into account the differences in these Parties' starting points and approaches, economic structures and resource bases, the need to maintain strong and sustainable economic growth, available technologies and other individual circumstances, as well as the need for equitable and appropriate contributions by each of these Parties to the global effort regarding that objective. These Parties may implement such policies and measures jointly with other Parties and may assist other Parties in contributing to the achievement of the objective of the Convention and, in particular, that of this subparagraph;

(b)   In order to promote progress to this end, each of these Parties shall communicate, within six months of the entry into force of the Convention for it and periodically thereafter, and in accordance with Article 12, detailed information on its policies and measures referred to in subparagraph (a) above, as well as on its resulting projected anthropogenic emissions by sources and removals by sinks of greenhouse gases not controlled by the Montreal Protocol for the period referred to in subparagraph (a), with the aim of returning individually or jointly to their 1990 levels these anthropogenic emissions of carbon dioxide and other greenhouse gases not controlled by the Montreal Protocol. This information will be reviewed by the Conference of the Parties, at its first session and periodically thereafter, in accordance with Article 7;

(c)   Calculations of emissions by sources and removals by sinks of greenhouse gases for the purposes of subparagraph (b) above should take into account the best available scientific knowledge, including of the effective capacity of sinks and the respective contributions of such gases to climate change. The Conference of the Parties shall consider and agree on methodologies for these calculations at its first session and review them regularly thereafter;

(d)   The Conference of the Parties shall, at its first session, review the adequacy of subparagraphs (a) and (b) above. Such review shall be carried out in the light of the best available scientific information and assessment on climate change and its impacts, as well as relevant technical, social and economic information. Based on this review, the Conference of the Parties shall take appropriate action, which may include the adoption of amendments to the commitments in subparagraphs (a) and (b) above. The Conference of the Parties, at its first session, shall also take decisions regarding criteria for joint implementation as indicated in subparagraph (a) above. A second review of subparagraphs (a) and (b) shall take place not later than 31 December 1998, and thereafter at regular intervals determined by the Conference of the Parties, until the objective of the Convention is met;

(e)   Each of these Parties shall :

(i)   Coordinate as appropriate with other such Parties,

(ii)   relevant and administrative instruments developed to achieve the objective of the Convention; and

(ii)   Identify and periodically review its own policies and practices which encourage activities that lead to greater levels of anthropogenic emissions of greenhouse gases not controlled by the Montreal Protocol than would otherwise occur;

(f)   The Conference of the Parties shall review, not later than 31 December 1998, available information with a view to taking decisions regarding such amendments to the lists in Annexes I and II as may be appropriate, with the approval of the Party concerned;

(g)   Any Party not included in Annex I may, in its instrument of ratification, acceptance, approval or accession, or at any time thereafter, notify the Depositary that it intends to be bound by subparagraphs (a) and (b) above. The Depositary shall inform the other signatories and Parties of any such notification.

3. The developed country Parties and other developed Parties included in Annex II shall provide new and additional financial resources to meet the agreed full costs incurred by developing country Parties in complying with their obligations under Article 12, paragraph 1. They shall also provide such financial resources, including for the transfer of technology, needed by the developing country Parties to meet the agreed full incremental costs of implementing measures that are covered by paragraph 1 of this Article and that are agreed between a developing country Party and the international entity or entities referred to in Article 11, in accordance with that Article. The implementation of these commitments shall take into account the need for adequacy and predictability in the flow of funds and the importance of appropriate burden-sharing among the developed country Parties.

4. The developed country Parties and other developed Parties included in Annex II shall also assist the developing country Parties that are particularly vulnerable to the adverse effects of climate change in meeting costs of adaptation to those adverse effects.

5. The developed country Parties and other developed Parties included in Annex II shall take all practicable steps to promote, facilitate and finance, as appropriate, the transfer of, or access to, environmentally sound technologies and know-how to other Parties, particularly developing country Parties, to enable them to implement the provi-

sions of the Convention. In this process, the developed country Parties shall support the development and enhancement of endogenous capacities and technologies of developing country Parties. Other Parties and organizations in a position to do so may also assist in facilitating the transfer of such technologies.

6. In the implementation of their commitments under paragraph 2 above, a certain degree of flexibility shall be allowed by the Conference of the Parties to the Parties included in Annex I undergoing the process of transition to a market economy, in order to enhance the ability of these Parties to address climate change, including with regard to the historical level of anthropogenic emissions of greenhouse gases not controlled by the Montreal Protocol chosen as a reference.

7. The extent to which developing country Parties will effectively implement their commitments under the Convention will depend on the effective implementation by developed country Parties of their commitments under the Convention related to financial resources and transfer of technology and will take fully into account that economic and social development and poverty eradication are the first and overriding priorities of the developing country Parties.

8. In the implementation of the commitments in this Article, the Parties shall give full consideration to what actions are necessary under the Convention, including actions related to funding, insurance and the transfer of technology, to meet the specific needs and concerns of developing country Parties arising from the adverse effects of climate change and/or the impact of the implementation of response measures, especially on:
    (a)   Small island countries;
    (b)   Countries with low-lying coastal areas;
    (c)   Countries with arid and semi-arid areas, forested areas and areas liable to forest decay;
    (d)   Countries with areas prone to natural disasters;
    (e)   Countries with areas liable to drought and desertification;
    (f)   Countries with areas of high urban atmospheric pollution;
    (g)   Countries with areas with fragile ecosystems, including mountainous ecosystems;
    (h)   Countries whose economies are highly dependent on income generated from the production, processing and export, and/or on consumption of fossil fuels and associated energy-intensive products; and

ARTICLE 6

EDUCATION, TRAINING AND PUBLIC AWARENESS

In carrying out their commitments under Article 4, paragraph 1(i), the Parties shall:

(a) Promote and facilitate at the national and, as appropriate, subregional and regional levels, and in accordance with national laws and regulations, and within their respective capacities:

   (i)   The development and implementation of educational and public awareness programmes on climate change and its effects;
   (ii) Public access to information on climate change and its effects;
   (iii) Public participation in addressing climate change and its effects and developing adequate responses; and
   (iv) Training of scientific, technical and managerial personnel.

(b) Cooperate in and promote, at the international level, and, where appropriate, using existing bodies:

   (i) The development and exchange of educational and public awareness material on climate change and its effects; and
   (ii) The development and implementation of education and training programmes, including the strengthening of national institutions and the exchange or secondment of personnel to train experts in this field, in particular for developing countries.

ARTICLE 7

CONFERENCE OF THE PARTIES

1. A Conference of the Parties is hereby established.

2. The Conference of the Parties, as the supreme body of this Convention, shall keep under regular review the implementation of the Convention and any related legal instruments that the Conference of the Parties may adopt, and shall make, within its mandate, the decisions necessary to promote the effective implementation of the Convention. To this end, it shall:

(a) Periodically examine the obligations of the Parties and the institutional arrangements under the Convention, in the light of the objective of the Convention, the experience gained in its implementation and the evolution of scientific and technological knowledge;

(b) Promote and facilitate the exchange of information on measures adopted by the Parties to address climate change and its effects, taking into account the differing circumstances, responsibilities and capabilities of the Parties and their respective commitments under the Convention;

(c) Facilitate, at the request of two or more Parties, the coordination of measures adopted by them to address climate change and its effects, taking into account the differing circumstances, responsibilities and capabilities of the Parties and their respective commitments under the Convention;

(d) Promote and guide, in accordance with the objective and provisions of the Convention, the development and periodic refinement of comparable methodologies, to be agreed on by the Conference of the Parties, inter alia, for preparing inventories of greenhouse gas emissions by sources and removals by sinks, and for evaluating the effectiveness of measures to limit the emissions and enhance the removals of these gases;

(e) Assess, on the basis of all information made available to it in accordance with the provisions of the Convention, the implementation of the Convention by the Parties, the overall effects of the measures taken pursuant to the Convention, in particular environmental, economic and social effects as well as their cumulative impacts and the extent to which progress towards the objective of the Convention is being achieved;

(f) Consider and adopt regular reports on the implementation of the Convention and ensure their publication;

(g) Make recommendations on any matters necessary for the implementation of the Convention;

(h) Seek to mobilize financial resources in accordance with Article 4, paragraphs 3, 4 and 5, and Article 11;

(i) Establish such subsidiary bodies as are deemed necessary for the implementation of the Convention;

(j) Review reports submitted by its subsidiary bodies and provide guidance to them;

(k) Agree upon and adopt, by consensus, rules of procedure and financial rules for itself and for any subsidiary bodies;

(l) Seek and utilize, where appropriate, the services and cooperation of, and information provided by, competent international organizations and intergovernmental and non-governmental bodies; and

(m) Exercise such other functions as are required for the achievement of the objective of the Convention as well as all other functions assigned to it under the Convention.

3. The Conference of the Parties shall, at its first session, adopt its own rules of procedure as well as those of the subsidiary bodies established by the Convention, which shall include decision-making procedures for matters not already covered by decision-making procedures stipulated in the Convention. Such procedures may include specified majorities required for the adoption of particular decisions.

4. The first session of the Conference of the Parties shall be convened by the interim secretariat referred to in Article 21 and shall take place not later than one year after the date of entry into force of the Convention. Thereafter, ordinary sessions of the Conference of the Parties shall be held every year unless otherwise decided by the Conference of the Parties.

5. Extraordinary sessions of the Conference of the Parties shall be held at such other times as may be deemed necessary by the Conference, or at the written request of any Party, provided that, within six months of the request being communicated to the Parties by the secretariat, it is supported by at least one third of the Parties.

6. The United Nations, its specialized agencies and the International Atomic Energy Agency, as well as any State member thereof or observers thereto not Party to the Convention, may be represented at sessions of the Conference of the Parties as observers.

Any body or agency, whether national or international, governmental or non-governmental, which is qualified in matters covered by the Convention, and which has informed the secretariat of its wish to be represented at a session of the Conference of the Parties as an observer, may be so admitted unless at least one third of the Parties present object. The admission and participation of observers shall be subject to the rules of procedure adopted by the Conference of the Parties.

ARTICLE 8

SECRETARIAT

1. A secretariat is hereby established.

2. The functions of the secretariat shall be:

   (a) To make arrangements for sessions of the Conference of the Parties and its subsidiary bodies established under the Convention and to provide them with services as required;

   (b) To compile and transmit reports submitted to it;

   (c) To facilitate assistance to the Parties, particularly developing country Parties, on request, in the compilation and communication of information required in accordance with the provisions of the Convention;

   (d) To prepare reports on its activities and present them to the Conference of the Parties;

   (e) To ensure the necessary coordination with the secretariats of other relevant international bodies;

   (f) To enter, under the overall guidance of the Conference of the Parties, into such administrative and contractual arrangements as may be required for the effective discharge of its functions; and

   (g) To perform the other secretariat functions specified in the Convention and in any of its protocols and such other functions as may be determined by the Conference of the Parties.

3. The Conference of the Parties, at its first session, shall designate a permanent secretariat and make arrangements for its functioning.

ARTICLE 9

SUBSIDIARY BODY FOR SCIENTIFIC AND TECHNOLOGICAL ADVICE

1. A subsidiary body for scientific and technological advice is hereby established to provide the Conference of the Parties and, as appropriate, its other subsidiary bodies with timely information and advice on scientific and technological matters relating to the Convention. This body shall be open to participation by all Parties and shall be multidisciplinary. It shall comprise government representatives com-

petent in the relevant field of expertise. It shall report regularly to the Conference of the Parties on all aspects of its work.

2. Under the guidance of the Conference of the Parties, and drawing upon existing competent international bodies, this body shall:

   (a) Provide assessments of the state of scientific knowledge relating to climate change and its effects;
   (b) Prepare scientific assessments on the effects of measures taken in the implementation of the Convention;
   (c) Identify innovative, efficient and state-of-the-art technologies and know-how and advise on the ways and means of promoting development and/or transferring such technologies;
   (d) Provide advice on scientific programmes, international coopera- tion in research and development related to climate change, as well as on ways and means of supporting endogenous capacity- building in developing countries; and
   (e) Respond to scientific, technological and methodological ques- tions that the Conference of the Parties and its subsidiary bodies may put to the body.

3. The functions and terms of reference of this body may be further elaborated by the Conference of the Parties.

ARTICLE 10
SUBSIDIARY BODY FOR IMPLEMENTATION

1. A subsidiary body for implementation is hereby established to assist the Conference of the Parties in the assessment and review of the effective implementation of the Convention. This body shall be open to participation by all Parties and comprise government representa- tives who are experts on matters related to climate change. It shall report regularly to the Conference of the Parties on all aspects of its work.

2. Under the guidance of the Conference of the Parties, this body shall:

   (a) Consider the information communicated in accordance with Article 12, paragraph 1, to assess the overall aggregated effect of the steps taken by the Parties in the light of the latest scien- tific assessments concerning climate change;

(b) Consider the information communicated in accordance with Article 12, paragraph 2, in order to assist the Conference of the Parties in carrying out the reviews required by Article 4, paragraph 2(d); and

(c) Assist the Conference of the Parties, as appropriate, in the preparation and implementation of its decisions.

ARTICLE 11

FINANCIAL MECHANISM

1. A mechanism for the provision of financial resources on a grant or concessional basis, including for the transfer of technology, is hereby defined. It shall function under the guidance of and be accountable to the Conference of the Parties, which shall decide on its policies, programme priorities and eligibility criteria related to this Convention. Its operation shall be entrusted to one or more existing international entities.

2. The financial mechanism shall have an equitable and balanced representation of all Parties within a transparent system of governance.

3. The Conference of the Parties and the entity or entities entrusted with the operation of the financial mechanism shall agree upon arrangements to give effect to the above paragraphs, which shall include the following:

(a) Modalities to ensure that the funded projects to address climate change are in conformity with the policies, programme priorities and eligibility criteria established by the Conference of the Parties;

(b) Modalities by which a particular funding decision may be reconsidered in light of these policies, programme priorities and eligibility criteria;

(c) Provision by the entity or entities of regular reports to the Conference of the Parties on its funding operations, which is consistent with the requirement for accountability set out in paragraph 1 above; and

(d) Determination in a predictable and identifiable manner of the amount of funding necessary and available for the implementation of this Convention and the conditions under which that amount shall be periodically reviewed.

4. The Conference of the Parties shall make arrangements to implement the above-mentioned provisions at its first session, reviewing and taking into account the interim arrangements referred to in Article 21, paragraph 3, and shall decide whether these interim arrangements shall be maintained. Within four years thereafter, the Conference of the Parties shall review the financial mechanism and take appropriate measures.

5. The developed country Parties may also provide and developing country Parties avail themselves of, financial resources related to the implementation of the Convention through bilateral, regional and other multilateral channels.

ARTICLE 12

COMMUNICATION OF INFORMATION RELATED TO IMPLEMENTATION

1. In accordance with Article 4, paragraph 1, each Party shall communicate to the Conference of the Parties, through the secretariat, the following elements of information:

   (a) A national inventory of anthropogenic emissions by sources and removals by sinks of all greenhouse gases not controlled by the Montreal Protocol, to the extent its capacities permit, using comparable methodologies to be promoted and agreed upon by the Conference of the Parties;

   (b) A general description of steps taken or envisaged by the Party to implement the Convention; and

   (c) Any other information that the Party considers relevant to the achievement of the objective of the Convention and suitable for inclusion in its communication, including, if feasible, material relevant for calculations of global emission trends.

2. Each developed country Party and each other Party included in Annex I shall incorporate in its communication the following elements of information:

   (a) A detailed description of the policies and measures that it has adopted to implement its commitment under Article 4, paragraphs 2(a) and 2(b); and

   (b) A specific estimate of the effects that the policies and measures referred to in subparagraph (a) immediately above will have on anthropogenic emissions by its sources and removals by its sinks of greenhouse gases during the period referred to in Article 4, paragraph 2(a).

3. In addition, each developed country Party and each other developed Party included in Annex II shall incorporate details of measures taken in accordance with Article 4, paragraphs 3, 4 and 5.

4. Developing country Parties may, on a voluntary basis, propose projects for financing, including specific technologies, materials, equipment, techniques or practices that would be needed to implement such projects, along with, if possible, an estimate of all incremental costs, of the reductions of emissions and increments of removals of greenhouse gases, as well as an estimate of the consequent benefits.

5. Each developed country Party and each other Party included in Annex I shall make its initial communication within six months of the entry into force of the Convention for that Party. Each Party not so listed shall make its initial communication within three years of the entry into force of the Convention for that Party, or of the availability of financial resources in accordance with Article 4, paragraph 3. Parties that are least developed countries may make their initial communication at their discretion. The frequency of subsequent communications by all Parties shall be determined by the Conference of the Parties, taking into account the differentiated timetable set by this paragraph.

6. Information communicated by Parties under this Article shall be transmitted by the secretariat as soon as possible to the Conference of the Parties and to any subsidiary bodies concerned. If necessary, the procedures for the communication of information may be further considered by the Conference of the Parties.

7. From its first session, the Conference of the Parties shall arrange for the provision to developing country Parties of technical and financial support, on request, in compiling and communicating information under this Article, as well as in identifying the technical and financial needs associated with proposed projects and response measures under Article 4. Such support may be provided by other Parties, by competent international organizations and by the secretariat, as appropriate.

8. Any group of Parties may, subject to guidelines adopted by the Conference of the Parties, and to prior notification to the Conference of the Parties, make a joint communication in fulfilment of their obligations under this Article, provided that such a communication includes information on the fulfilment by each of these Parties of its individual obligations under the Convention.

9. Information received by the secretariat that is designated by a Party as confidential, in accordance with criteria to be established by the Conference of the Parties, shall be aggregated by the secretariat to protect its confidentiality before being made available to any of the bodies involved in the communication and review of information.

10. Subject to paragraph 9 above, and without prejudice to the ability of any Party to make public its communication at any time, the secretariat shall make communications by Parties under this Article publicly available at the time they are submitted to the Conference of the Parties.

ARTICLE 13

RESOLUTION OF QUESTIONS REGARDING IMPLEMENTATION

The Conference of the Parties shall, at its first session, consider the establishment of a multilateral consultative process, available to Parties on their request, for the resolution of questions regarding the implementation of the Convention.

ARTICLE 14

SETTLEMENT OF DISPUTES

1. In the event of a dispute between any two or more Parties concerning the interpretation or application of the Convention, the Parties concerned shall seek a settlement of the dispute through negotiation or any other peaceful means of their own choice.

2. When ratifying, accepting, approving or acceding to the Convention, or at any time thereafter, a Party which is not a regional economic integration organization may declare in a written instrument submitted to the Depositary that, in respect of any dispute concerning the interpretation or application of the Convention, it recognizes as

compulsory ipso facto and without special agreement, in relation to any Party accepting the same obligation:

(a) Submission of the dispute to the International Court of Justice, and/or

(b) Arbitration in accordance with procedures to be adopted by the Conference of the Parties, as soon as practicable, in an annex on arbitration.

A Party which is a regional economic integration organization may make a declaration with like effect in relation to arbitration in accordance with the procedures referred to in subparagraph (b) above.

3. A declaration made under paragraph 2 above shall remain in force until it expires in accordance with its terms or until three months after written notice of its revocation has been deposited with the Depositary.

4. A new declaration, a notice of revocation or the expiry of a declaration shall not in any way affect proceedings pending before the International Court of Justice or the arbitral tribunal, unless the parties to the dispute otherwise agree.

5. Subject to the operation of paragraph 2 above, if after twelve months following notification by one Party to another that a dispute exists between them, the Parties concerned have not been able to settle their dispute through the means mentioned in paragraph 1 above, the dispute shall be submitted, at the request of any of the parties to the dispute, to conciliation.

6. A conciliation commission shall be created upon the request of one of the parties to the dispute. The commission shall be composed of an equal number of members appointed by each party concerned and a chairman chosen jointly by the members appointed by each party. The commission shall render a recommendatory award, which the parties shall consider in good faith.

7. Additional procedures relating to conciliation shall be adopted by the Conference of the Parties, as soon as practicable, in an annex on conciliation.

8. The provisions of this Article shall apply to any related legal instru-

ment which the Conference of the Parties may adopt, unless the instrument provides otherwise.

ARTICLE 15

AMENDMENTS TO THE CONVENTION

1. Any Party may propose amendments to the Convention.

2. Amendments to the Convention shall be adopted at an ordinary session of the Conference of the Parties. The text of any proposed amendment to the Convention shall be communicated to the Parties by the secretariat at least six months before the meeting at which it is proposed for adoption. The secretariat shall also communicate proposed amendments to the signatories to the Convention and, for information, to the Depositary.

3. The Parties shall make every effort to reach agreement on any proposed amendment to the Convention by consensus. If all efforts at consensus have been exhausted, and no agreement reached, the amendment shall as a last resort be adopted by a three-fourths majority vote of the Parties present and voting at the meeting. The adopted amendment shall be communicated by the secretariat to the Depositary, who shall circulate it to all Parties for their acceptance.

4. Instruments of acceptance in respect of an amendment shall be deposited with the Depositary. An amendment adopted in accordance with paragraph 3 above shall enter into force for those Parties having accepted it on the ninetieth day after the date of receipt by the Depositary of an instrument of acceptance by at least three fourths of the Parties to the Convention.

5. The amendment shall enter into force for any other Party on the ninetieth day after the date on which that Party deposits with the Depositary its instrument of acceptance of the said amendment.

6. For the purposes of this Article, "Parties present and voting" means Parties present and casting an affirmative or negative vote.

ARTICLE 16

ADOPTION AND AMENDMENT OF ANNEXES TO THE CONVENTION

1. Annexes to the Convention shall form an integral part thereof and, unless otherwise expressly provided, a reference to the Convention constitutes at the same time a reference to any annexes thereto. Without prejudice to the provisions of Article 14, paragraphs 2(b) and 7, such annexes shall be restricted to lists, forms and any other material of a descriptive nature that is of a scientific, technical, procedural or administrative character.

2. Annexes to the Convention shall be proposed and adopted in accordance with the procedure set forth in Article 15, paragraphs 2, 3 and 4.

3. An annex that has been adopted in accordance with paragraph 2 above shall enter into force for all Parties to the Convention six months after the date of the communication by the Depositary to such Parties of the adoption of the annex, except for those Parties that have notified the Depositary, in writing, within that period of their non-acceptance of the annex. The annex shall enter into force for Parties which withdraw their notification of non-acceptance on the ninetieth day after the date on which withdrawal of such notification has been received by the Depositary.

4. The proposal, adoption and entry into force of amendments to annexes to the Convention shall be subject to the same procedure as that for the proposal, adoption and entry into force of annexes to the Convention in accordance with paragraphs 2 and 3 above.

5. If the adoption of an annex or an amendment to an annex involves an amendment to the Convention, that annex or amendment to an annex shall not enter into force until such time as the amendment to the Convention enters into force.

ARTICLE 17

PROTOCOLS

1. The Conference of the Parties may, at any ordinary session, adopt protocols to the Convention.

2. The text of any proposed protocol shall be communicated to the Parties by the secretariat at least six months before such a session.

3. The requirements for the entry into force of any protocol shall be established by that instrument.

4. Only Parties to the Convention may be Parties to a protocol.

5. Decisions under any protocol shall be taken only by the Parties to the protocol concerned.

ARTICLE 18

RIGHT TO VOTE

1. Each Party to the Convention shall have one vote, except as provided for in paragraph 2 below.

2. Regional economic integration organizations, in matters within their competence, shall exercise their right to vote with a number of votes equal to the number of their member States that are Parties to the Convention. Such an organization shall not exercise its right to vote if any of its member States exercises its right, and vice versa.

ARTICLE 19

DEPOSITARY

The Secretary-General of the United Nations shall be the Depositary of the Convention and of protocols adopted in accordance with Article 17.

ARTICLE 20

SIGNATURE

This Convention shall be open for signature by States Members of the United Nations or of any of its specialized agencies or that are Parties to the Statute of the International Court of Justice and by regional economic integration organizations at Rio de Janeiro, during the United Nations Conference on Environment and Development, and thereafter at United Nations Headquarters in New York from 20 June 1992 to 19 June 1993.

ARTICLE 21

INTERIM ARRANGEMENTS

1. The secretariat functions referred to in Article 8 will be carried out on an interim basis by the secretariat established by the General Assembly of the United Nations in its resolution 45/212 of 21 December 1990, until the completion of the first session of the Conference of the Parties.

2. The head of the interim secretariat referred to in paragraph 1 above will cooperate closely with the Intergovernmental Panel on Climate Change to ensure that the Panel can respond to the need for objective scientific and technical advice. Other relevant scientific bodies could also be consulted.

3. The Global Environment Facility of the United Nations Development Programme, the United Nations Environment Programme and the International Bank for Reconstruction and Development shall be the international entity entrusted with the operation of the financial mechanism referred to in Article 11 on an interim basis. In this connection, the Global Environment Facility should be appropriately restructured and its membership made universal to enable it to fulfil the requirements of Article 11.

ARTICLE 22

RATIFICATION, ACCEPTANCE, APPROVAL OR ACCESSION

1. The Convention shall be subject to ratification, acceptance, approval or accession by States and by regional economic integration organizations. It shall be open for accession from the day after the date on which the Convention is closed for signature. Instruments of ratification, acceptance, approval or accession shall be deposited with the Depositary.

2. Any regional economic integration organization which becomes a Party to the Convention without any of its member States being a Party shall be bound by all the obligations under the Convention. In the case of such organizations, one or more of whose member States is a Party to the Convention, the organization and its member States shall decide on their respective responsibilities for the performance of their obligations under the Convention. In such cases, the organization and the member States shall not be entitled to exercise rights under the Convention concurrently.

3. In their instruments of ratification, acceptance, approval or accession, regional economic integration organizations shall declare the extent of their competence with respect to the matters governed by the Convention. These organizations shall also inform the Depositary, who shall in turn inform the Parties, of any substantial modification in the extent of their competence.

ARTICLE 23

ENTRY INTO FORCE

1. The Convention shall enter into force on the ninetieth day after the date of deposit of the fiftieth instrument of ratification, acceptance, approval or accession.

2. For each State or regional economic integration organization that ratifies, accepts or approves the Convention or accedes thereto after the deposit of the fiftieth instrument of ratification, acceptance, approval or accession, the Convention shall enter into force on the ninetieth day after the date of deposit by such State or regional economic integration organization of its instrument of ratification, acceptance, approval or accession.

3. For the purposes of paragraphs 1 and 2 above, any instrument deposited by a regional economic integration organization shall not be counted as additional to those deposited by States members of the organization.

ARTICLE 24

RESERVATIONS

No reservations may be made to the Convention.

ARTICLE 25

WITHDRAWAL

1. At any time after three years from the date on which the Convention has entered into force for a Party, that Party may withdraw from the Convention by giving written notification to the Depositary.

2. Any such withdrawal shall take effect upon expiry of one year from the date of receipt by the Depositary of the notification of withdrawal, or on such later date as may be specified in the notification of withdrawal.

3. Any Party that withdraws from the Convention shall be considered as also having withdrawn from any protocol to which it is a Party.

ARTICLE 26

AUTHENTIC TEXTS

The original of this Convention, of which the Arabic, Chinese, English, French, Russian and Spanish texts are equally authentic, shall be deposited with the Secretary-General of the United Nations.

IN WITNESS WHEREOF the undersigned, being duly authorized to that effect, have signed this Convention.

DONE at New York this ninth day of May one thousand nine hundred and ninety-two.

ANNEX I

Australia
Austria
Belarus[a]
Belgium
Bulgariaa
Canada
Czechoslovakia[a]
Denmark
European Economic Community
Estonia[a]
Finland
France
Germany
Greece
Hungary[a]
Iceland
Ireland
Italy
Japan
Latvia[a]
Lithuania[a]
Luxembourg
Netherlands
New Zealand
Norway
Poland[a]
Portugal
Romania[a]
Russian Federation[a]
Spain
Sweden
Switzerland
Turkey
Ukraine[a]
United Kingdom of Great Britain and Northern Ireland
United States of America

[a] Countries that are undergoing the process of transition to a market economy.

Source: Text and Annex have been downloaded from UNEP/WMO Information Unit on Climate Change as an electronic document.

Bibliography

Aarhus, Knut and Eikeland, PO (1993) *Virkemidler i forurensningspolitikken: Sverige, Tyskland, Nederland, Storbritannia og USA.* Lysaker: The Fridtjof Nansen Institute – Report No. 4.

Adams, J.A.S., Mantovani, M.S.M. and Lundell, L.L. (1977) "Wood Versus Fossil Fuel as a Source of Excess Carbon Dioxide in the Atmosphere: A Preliminary Report", *Science*, No. 196, pp. 54–56.

Agarwal, A. and Narain, S. (1991) *Global Warming in an Unequal World – A Case of Environmental Colonialism.* New Delhi: Centre for Science and Development.

Albert, Bruce *(1992)* "Indian Lands, Environmental Policy and Military Geo-politics in the Development if the Brazilian Amazon: The Case of the Yanomani", Development and Change, Vol. 23, No. 1, pp, 35–70.

Albin, Cecilia (1993) "The Role of Fairness in Negotiation", *Negotiation Journal*, July, pp. 223–244.

Allegretti, M. H. (1988) "Natureza e politica externa brasileira", Tempo e Presença, Vol. 10, No. 330, pp. 14–15.

Allen, Elisabeth (1992) "Calha Norte: Military Development in Brazilian Amazonia", Development and Change, Vol. 23, No. 1, pp. 71–99.

Altshuller, Aubrey P. (1958) "Natural Sources of Gaseous Pollutants in the Atmosphere", Vol. 10, No. 4, pp. 479–492.

Altvater, E. (1987) *Sachzwang Weltmarkt.* Hamburg: VSA-Verlag.

Ambio (1994) *Integrating Earth Systems* (special issue), Vol. 23, No. 1.

Andersen, Elisabeth (1993) *Value in Ethics and Economics.* Cambridge, MA: Harvard University Press.

Andrea, Meinrat O. (1996) "Raising Dust in the Greenhouse", *Nature*, Vol. 380, 4 April, pp. 389–390.

Andresen, Steinar (1990) *USAs drivhuspolitikk: Bakstreversk eller realistisk?* Lysaker, Norway: The Fridtjof Nansen Institute – EED Report No. 18.

Andresen, Steinar (1993) *US Climate Policy: Ideology versus Pragmatism.* Lysaker, Norway: The Fridtjof Nansen Institute – EED Report No. 3.

Andresen, Steinar and Wettestad, Jørgen (1990) "Climate Failure at the Bergen Conference?" *International Challenges*, Vol. 10, No. 2, p. 23.

AP (1990) "Action Programme for Arresting Climate Change". Tokyo: Japanese Environment Agency.

Arnold, James R. and Anderson, Ernest C. (1957) "The Distribution of Carbon-14 in Nature", *Tellus*, Vol. 9, No. 1, pp. 28–32.

Arnt, Ricardo Azambuja and Schwartzman, Stephen *(1992) Um Artificio Organico*

Transição na Amazônia e Ambientalismo. Rio de Janeiro: Rocco.

Arrhenius, Erik (1992) "Population, Development and Environmental Disruption: An Issue on Efficient Natural-Resource Management", *Ambio*, Vol. 21, No. 1, pp. 9–13.

Arrhenius, Svante (1896) "On the Influence of Carbonic Acid upon Temperature at the Ground", *Philosophical Magazine*, Vol. 41, No. 237.

Aunan, K., Seip, H.M. and Aaheim, H. A. (1993) *A Model Framework for Ranking of Measures to Reduce Air Pollution with a Focus on Damage Assessment*. Oslo: Center for International Climate and Environmental Research, Oslo – CICERO Working Paper No. 12.

Ausabel, J. H. and Victor, D. G. (1992) "Verification of International Environmental Agreements", *Annual Review of Energy and the Environment*, Vol. 17, No. 1, pp. 1–43.

Axelrod, Regina S. (1994) "Subsidiarity and Environmental Policy in the European Community", *International Environmental Affairs*, Vol. 6, No. 2, pp. 115–132.

Barrow, J. D. (1991) *Theories of Everything*. London: Oxford University Press.

Beijing Declaration on Environment (1991) Reprinted in *Beijing Review*, July, pp. 8–14.

Beijing Rundschau (1992) "Umwelt und Entwicklung im Blickpunkt der Welt". Rede von Li Peng auf dem Umweltgipfel in Rio, 29 (23.6.1992), pp. 7–10.

Benedick, Richard E. (1989) "Ozone Diplomacy", *Issues in Science and Technology*, Vol. 6, No. 3, pp. 43–50.

Benedick, Richard E. (1991) *Ozone Diplomacy – New Directions in Safeguarding the Planet*. Cambridge, MA: Harvard University Press.

Bergen Ministerial Declaration (1990) Bergen Ministerial Declaration on Sustainable Development in the ECE Region: Statement of Ministerial Conference Sponsored by the Government of Norway in Bergen,16 May 1990.

Bergesen H. O. et al. (1995) *Norge i det globale drivhuset* Oslo: Universitetsforlaget.

Bergesen, Helge Ole (1995) "A Global Climate Regime – Mission Impossible?", *Green Globe Yearbook*. pp. 51–58.

Bergesen, Helge Ole and Haigh, Nigel (1994) "EC Climate Policy – From Global Warming to Political Cooling". Unpublished paper.

Bergesen, Helge Ole and Sydnes, Anne-Kristin (1992) "Protection of the Global Climate: Ecological Utopia or Just a Long Way to Go?", *Green Globe Yearbook*. pp. 35–48.

Bernson, Vibeke (1993) *The Framework Convention on Climate Change: Analyzing the Role of Epistemic Communities and of Problem Uncertainty in the Outcome of the Negotiations*. Stockholm: Swedish National Chemicals Inspectorate.

Beuermann, Christiane. and Jäger, Jill. (1993) *Climate Change Policy in Germany – Development and Limitations*. Contribution to the EC project "Designing European Governing Institutions for Climate Futures", working draft.

Beuermann, Christiane and Jäger, Jill (1996) "Climate Change Politics in Germany: How Long Will Any Double Dividend Last", in Tim O'Riordan and Jill Jäger (eds) *Politics of Climate Change – A European Perspective*. London: Routledge, pp. 186–227.

Beukel, Erik (1993) *Global miljøbeskyttelse som kollektivt gode i international politik*. Odense: Odense Universitet, Institut for Erhvervsret og Politilogi – Politiske Skrifter No. 2.

Bleischwitz, Raimund et al. (1993) *US and EC Climate Change Policies Compared.*

Wuppertal: Wuppertal Institut – Wuppertal Papers No. 3.

BMU (1993) *National Report of the Federal Republic of Germany in Anticipation of Article 12 of the UN Framework Convention on Climate Change.* Bonn.

Bodansky, Daniel (1993) "The United Nations Framework Convention on Climate Change: A Commentary", *Yale Journal of International Law*, Vol. 18, No. 4, pp. 451–558.

Bodansky, Daniel (1994) "Prologue to the Climate Change Convention", in I. M. Mintzer and J. A. Leonard (eds) *Negotiating Climate Change: The Inside Story of the Rio Convention.* Cambridge: Cambridge University Press, pp. 45–76.

Boehmer-Christiansen, S. A. (1993) "Science Policy, the IPCC and the Climate Convention: The Codification of a Global Research Agenda", *Energy and Environment*, Vol. 4, No. 4, pp. 362–407.

Boehmer-Christiansen, S. A. (1995) "Britain and the Intergovernmental Panel on Climate Change: The Impacts of Scientific Advice on Global Warming" (2 parts), *Environmental Politics*, Vol. 4, No. 1, pp. 1–18, and No. 2, pp. 175–196.

Boehmer-Christiansen, S. A. and Skea J. F. (1994) *The Operation and Impact of the IPCC: Results of a Survey of Participants and Users.* Brighton: Centre for Science, Technology, Energy and Environment Policy, Discussion Paper no. 16.

Boehmer-Christiansen, S. A. (1994a) "Global Climate Protection Policy: The Limits of Scientific Advice, Part 1", *Global Environmental Change*, Vol. 4, No. 2, pp. 140–159.

Boehmer-Christiansen, S. A. (1994b) "Global Climate Protection Policy: The Limits of Scientific Advice, Part 2", *Global Environmental Change*, Vol. 4, No. 3, pp. 185–200.

Bohm, Peter (1993) "Incomplete International Cooperation to Reduce CO_2 Emissions: Alternative Policies", *Journal of Environmental Economics and Management*, Vol. 24, No. 3, pp. 258–271.

Bohm, Peter and Larsen, Bjørn (1994) "Fairness in a Tradable-Permit Treaty for Carbon Emissions Reductions in Europe and the former Soviet Union", *Environmental and Resource Economics*, Vol. 4, No. 3, pp. 219–239.

Bohm, Peter and Russell, Clifford S. (1985) "Comparative Analysis of Alternative Policy Instruments", in Allen V. Kneese and James L. Sweeney (eds) *Handbook of Natural Resource and Energy Economics, Vol. I.* Amsterdam and New York: North-Holland, pp. 395–460.

Bolin B. (1980) "Climatic Changes and Their Effects on the Biosphere". Geneva: Fourth WMO lecture, WMO-No. 542.

Bolin, B. (1993) "Energy and Climate Change", *WEC Journal*, July, pp. 42–43.

Bolin, B. (1993) *Global Climate Change: an Issue of Risk Assessment and Risk Management.* Paper presented at the NWO/Huygens lecture, 16 November.

Bolin, B. (1994) "Science and Policy Making", *Ambio*, Vol. 23, No. 1, pp. 25–29.

Bolin, B. (1995) "Politics of Climate Change", *Nature*, No. 374, 16 March, p. 208.

Bolin, B. et al. (1986) *The Greenhouse Effect, Climatic Change and Ecosystems.* Wiley, Chichester: SCOPE Report 29.

Bolin, B. and Hoghton J. (1995) "Berlin and Global Warming Policy", *Nature*, No. 375, 18 May, p. 176.

Bolin, B., Döös, B. R. Jäger, J. and Warrick, R. A. (1986) *The Greenhouse Effect, Climate Change and Ecosystems.* Chichester: John Wiley and Sons.

Bondi, Herman (1985) "Risk in Perspective", in M. G. Cooper (ed.) *Risk: Man-made Hazards to Man.* Oxford: Clarendon Press.

Børsting, Georg (1993) *Evalueringsmekanismer under ozon-regimet – en vurdering av institusjonell utforming* Oslo: Department of Political Science, University of Oslo – mimeo.

Børsting, Georg (1996) *The North–South Politics of the International Ozone Regime.* Oslo: University of Oslo, Department of Political Science – Cand.Polit. dissertation (forthcoming).

Böttcher, C. J. F. (1992) *Science and Fiction of the Greenhouse Effect and Carbon Dioxide.* The Hague: Global Institute for the Study of Natural Resources.

Bray, J. R. (1959) "An Analysis of the Possible Recent Change in Atmospheric Carbon Dioxide Concentration", *Tellus*, Vol. 11, No. 2, pp. 220–230.

Brewing, Christopher (1994) "The European Community: A Union of States without Union of Government", in Freidrich Kratochwil and Edward D. Mansfield (eds) *International Organization – A Reader.* New York: Harper Collins College Publishers, pp. 301–324.

Burtraw, Dallas and Michael A. Toman (1992) "Equity and International Agreements for CO_2 Containment", *Journal of Energy Engineering*, Vol. 118, No. 2, pp. 122–135.

Cain, Melinda L. (1983) "Carbon Dioxide and the Climate: Monitoring and a Search for Understanding", in David Kay and Harold J. Jacobson (eds) *Environmental Protection: The International Dimension.* Totowa, NJ: Allanheld, Osmun and Co, pp. 75–98.

Cairo Compact (1989) Statement from the World Conference for Preparing for Climate Change, Cairo, Egypt, December 1989.

Caldwell, Lynton Keith (1990) *International Environmental Policy: Emergence and Dimensions.* Durham: Duke University Press.

Callendar, G. S. (1958) "On the Amount of Carbon Dioxide in the Atmosphere", *Tellus*, Vol. 10, No. 2, pp. 243–248.

Calvert, Randall (1992). "Leadership and Its Basis in Problems of Social Coordination", *International Political Science Review*, Vol. 13, No. 1, pp. 7–24.

Carroll, John E. (ed.) (1988) *International Environmental Diplomacy: The Management and Resolution of Transfrontier Environmental Problems.* Cambridge: Cambridge University Press.

Cavender, Jeannine and Jäger, Jill (1993) "The History of Germany's Response to Climate Change", *International Environmental Affairs*, Vol. 5, No. 1, pp. 3–18.

CCAP (1993) "Climate Change Action Program". Washington, DC: US State Department.

Chayes, A. and Chayes, A. H. (1991) "Adjustment and Compliance Processes in International Regulatory Regimes", in T. Mathews (ed.) *Preserving the Global Environment: The Challenge of Shared Leadership.* New York: W. W. Norton & Co.

Chayes, Abraham and Chayes, Antonia Handler (1993) "On Compliance", *International Organization*, Vol. 47, No. 2, pp. 175–205.

Chen Yaobang (1995) *Statement at the Ministerial Segment of the First Session of Conference of the Parties to the United Nations Framework Convention of Climate Change.* 6 April 1995, Berlin.

China's Agenda 21 (1994) *White Paper on China's Population, Environment and Development in the 21st Century.* Beijing: Environmental Science Press.

Churchill, Robin and Freestone, David (eds) (1991) *International Law and Global Climate Change.* London: Graham and Trotman.

CILSS/UNSO/IGADD (1991) *Report of the First Consultative Meeting of Sudano-Sahelian Countries in the Context of UNCED 1992*, Ouagadougou, 13–15 February, 1991. Report submitted to the African Regional Preparatory

Conference on UNCED, Cairo, July 1991.

Clayton, A. (1994) *Governance, Democracy and Conditionality: What Role for NGOs.* London: Intrac.

Cline, William R. (1992) *The Economics of Global Warming.* Washington D.C.: Institute of International Economics.

Cline, William R. (1994) "Socially Efficient Abatement of Carbon Emissions", in Ted Hanisch (ed.) *Climate Change and the Agenda for Research.* Boulder, CO and San Francisco, CA: Westview Press, pp. 91–111.

Clover, C. (1993) *Daily Telegraph,* 3 April.

Cole, S. (1992) *Making Science.* Cambridge, MA: Harvard University Press.

COM(94) 67 final, 10 March.

Conca, Ken et al. (eds.) (1995) *Green Planet Blues: Environmental Politics from Stockholm to Rio.* Boulder, CO: Westview Press.

Craig, Harmon (1957) "The Natural Distribution of Radiocarbon and the Exchange Time of Carbon Dioxide Between Atmosphere and Sea", *Tellus,* Vol. 9, No. 1, pp. 1–17.

Dahl, Agnethe (1993) *Miljøhensyn i Vest-Europeisk energipolitikk: Reserver av fossile brensler, miljøbevegelsen og statlig styring av energisektoren som forklarende faktorer.* Lysaker: The Fridtjof Nansen Institute – Report No. 4.

Davidson, O. (1993) "Energy and Carbon Emissions – Sub-Sahara African Perspective", *Energy Policy,* January 1993.

Davidson, O. and Karakezi, S. (1992), *A New Environmentally-Sound Energy Strategy for the Development of Sub-Saharan Africa.* Nairobi: AFREPREN.

Den Elzen, M., Janssen, M., Rotmans, J., Swart, R. and De Vries, B. (1992) "Allocating Constrained Global Carbon Budgets", *International Journal of Global Energy Issues,* Special Issue on Energy and Sustainable Development, Vol. 4, No. 4, pp. 287–301.

DiPrimio, Juan C., Stein, G. and Wagner, H. F. (1992) "Climate – Verifying Compliance with an International Convention on Greenhouse Gases", *Environment,* Vol. 34, No. 2, pp. 4–5, 45.

Dornbush, R. and Poterba, J. M. (1991) *Global Warming: Economic Policy Responses.* Cambridge, MA: MIT Press.

EA (1992) *Climate Change Research in Japan.* Tokyo: Environment Agency.

Earth Negotiations Bulletin (1995a) "A Summary Report on the Eleventh Session of the INC for a Framework Convention on Climate Change", Vol. 12, No. 11, 18 February 1995.

Earth Negotiations Bulletin (1995b) "A Summary Report on the First Conference of the Parties to the Framework Convention on Climate Change", Vol. 12, No. 21, 10 April 1995.

EC (1990) "A Community Action Programme to Limit Carbon Dioxide Emissions and to Improve the Security of Energy Supply", *Communication to the Council,* 28 November.

EC (1992) "Court of Auditors: Special Report No. 3/92 concerning the Environment together with the Commission's Replies", *Official Journal,* No. C245, 23 September.

EC (1993) "A European Community Programme of Policy and Action in Relation to the Environment and Sustainable Development", *Official Journal,* No C138, 17 May.

EC (1994) "First Evaluation of Existing Programmes Under the Monitoring Mechanism of Community CO_2 and Other Greenhouse-Gas Emissions", *Report from the Commission,*

EC DG XI (1991) *Development of a Framework for the Evaluation of Policy Options to Deal with the Greenhouse Effect – Impacts, Adaptive Options and Cost Implications.* Draft Report from the Climate Research Unit (CRU) – University of East Anglia, and Environmental Resources Limited (ERL).

EC DG XI (1992) *Assessment of Possible Approaches to Sharing the Effort of Limiting Greenhouse-Gas Emissions Among Countries and Groups of Countries.* Final Report, Cambridge Decision Analysts Limited, March.

EC DG XI (1993) *Compliance of CO_2 Emission Objectives in National Programmes with the EC CO_2 Stabilization Target,* 4 March.

EC DG XI and DG XVII (1991) "A Community Strategy to Limit Carbon Dioxide Emissions and to Improve Energy Efficiency", *Communication from the Commission to the Council,* SEC(91) 1744 final, Brussels, 14 October.

ECA (1991a) *African Common Position on Environment and Development. Adopted at the Second African Regional Ministerial Preparatory Conference for UNCED,* Abidjan, November 1991.

ECA (1991b) "Transition to Sustainable Energy Development under Climate Change in Africa: Problems and Prospects", *Global Collaboration on Sustainable Energy Development.* Lyngby: Technical University of Denmark.

Eikeland, Per Ove (1993a) *US Energy Policy in the Greenhouse: From the North Slope Forests to the Gulf Stream Waters – This Land was Made for Fossil Fuels?* Lysaker: The Fridtjof Nansen Institute – EED Report No. 1.

Eikeland, Per Ove (1993b) "US Energy Policy at a Crossroads?", *Energy Policy,* Vol. 21, No. 10, pp. 987–999.

Eleri, E. O. (1994) "Africa's Decline and Greenhouse Politics", *International Environmental Affairs,* Vol. 6, No. 2.

Eleri, E. O. (1996) "The Energy Sector in Southern Africa – A Preliminary Survey of Post-Apartheid Challenges", *Energy Policy,* Vol. 24, No. 1.

Eleri, E.O. (1993) *Africa's Response to Climate Change: The Role of Governments, Societies and External Actor.* Lysaker: The Fridtjof Nansen Institute – EED Report No. 6.

Emsley, John (ed.) (1995) *The Global Warming Debate* London: European Science and Environment Forum.

Erikson, E. and Wilander, P. (1956) "On a Mathematical Model of the Carbon Cycle in Nature", *Tellus,* Vol. 8, No. 2, pp. 155–163.

Eskom (1994) *Eskom Environmental Report 1994.* Megawatt Park: Eskom.

ESSO (1994) *Energieprognose 1994: Mobil bleiben, Umwelt schonen.* Hamburg.

European Commission (1995) "For a European Union Energy Policy – Green Paper", COM(94) 659 final, 11 January 1995, *Energy in Europe.* Luxembourg: Office for Official Publications of the European Communities.

European Commission (1995) *An Energy Policy for the European Union,* COM(95) 682 final.

European Commission (1996) *Second Evaluation of Existing Programmes under the Monitoring Mechanism of Community CO_2 and Other Greenhouse-Gas Emissions.* Report from the Commission, COM (96).

Fairman, David (1996) "The Global Environment Facility: Haunted by the Shadow of the Future", in Robert Keohane and Marc Levy (eds) *Institutions for Environmental Aid: Pitfalls and Promise.* Cambridge, MA: MIT Press.

Fearnside, P. M. (1990a) "Deforestation in Brazilian Amazônia", reviewing D. Mahar: "Government Policies and Deforestation on (sic!) Brazil's Amazon

Region", *Conservation Biology*, Vol. 4, No. 4, pp. 449–460.

Fearnside, P. M. (1990b) "Environmental Destruction in the Brazilian Amazon", in D. Goodman and A. Hall (eds) *The Future of Amazonia: Destruction or Sustainable Development?* New York: St. Martin's Press, pp. 179–225.

Fearnside, P. M. (1986) *Human Carrying Capacity of Brazilian the Brazilian Rain Forest*. New York: Columbia University Press.

Fearnside, P. M. (1992) *Carbon Emissions and Sequestration in Forests: Case Studies From Seven Developing Countries. Volume 2: Brazil*. Washington DC/Berkeley CA: EPA/Lawrence Berkeley Laboratory.

Ferguson, H. and Jäger, J. (eds) (1991) *Climate Change: Science, Impacts, Policy: Proceedings of the Second World Climate Conference*. Cambridge: Cambridge University Press.

Fermann, Gunnar (1992) *Japan in the Greenhouse: Responsibilities, Policies and Prospects for Combating Global Warming*. Lysaker: The Fridtjof Nansen Institute – EED Report No. 13.

Fermann, Gunnar (1993) "Japan's 1990 Climate Policy Under Pressure", *Security Dialogue*, Vol. 24, No. 3, pp. 287–300.

Fermann, Gunnar (1994a) "Retorikeren, analytikeren og den prinsipielle observatør: Maktbruk og berettigelse i Gulf-konflikten", *Norsk Statsvitenskapelig Tidsskrift*, Vol. 10, No. 3, pp. 289–306.

Fermann, Gunnar (1994b) *Political Leadership and the Development of Problem-solving Capacity in the Global Greenhouse: Prospects of Germany, Japan and the United States Towards the 21st Century*. Lysaker: The Fridtjof Nansen Institute – EED Report No. 3.

Fermann, Gunnar (1995a) "Climate Change Policy-making in Japan: Vulnerabilities and Opportunities", *Current Politics and Economics of Japan*, Vol. 4, No. 2/3, pp. 103–128.

Fermann, Gunnar (1995b) *FN mellom gamle visjoner og nye realiteter: FN i norsk utenrikspolitikk etter den kalde krigen*. Trondheim: Senter for miljø og utvikling – rapport nr. 7.

Fischer, W. (1992) *Climate Protection and International Policy. The Rio-Conference Between Global Responsibility and National Interests*. Internationale Treibhausgasverifikation Nr. 4, Jül-2695, Jülich: Forschungszentrum Jülich.

Fish, Arthur L. and South, David W. (1994) "Industrialized Countries and Greenhouse-Gas Emissions", *International Environmental Affairs*, Vol. 6, No. 1, pp. 14–44.

Fisher, Anthony C. (1981) *Resource and Environmental Economics*. London and New York: Cambridge University Press.

Flynn (1993) "Collor, Corruption and Crisis: Time for Reflection", *Journal of Latin American Studies*, Vol. 25, part 2, May, pp. 351–373.

Foresta, R. A. (1991) *Amazon Conservation in the Age of Development: The Limits of Providence*. Gainesville: University of Florida Press.

Franck, Thomas M. and Sughrue, Dennis M. (1993) "The International Role of Equity-as-Fairness", *Georgetown Law Journal*, Vol. 81, No. 353, pp. 563–595.

Frankhauser, S. (1993) *Global Warming Economics: Issues and State of the Art*. London: University College London and University of East Anglia, CSERGE Working Paper GEC1993-28.

Freestone, David (1991) "European Community Environmental Law and Policy", in R. Churchill, L. Warren and J. Gibson (eds) *Law, Policy and the Environment*. Oxford: Basil Blackwell, pp. 135–154.

Freestone, David (1994) "The Road from Rio: International Environmental Law After the Earth Summit", *Journal of Environmental Law*, Vol. 6, No. 2, pp. 193–218.

Frohlich, Norman, Oppenheimer, Joe A. and Young, Oran R. (1971) *Political Leadership and Collective Goods*. Princeton, N.J.: Princeton University Press.

FUNATURA (1990) *Alternativas ao desmatamento na Amazônia: Conservação dos Recursos Naturais*. Brasilia: Fundação Pro-Natureza.

G-7 Economic Declaration (1989) Statement of Group of Seven Major Industrialized Countries in Paris, 16 July 1989.

Gan, Lin (1993) "The Making of the Global Environmental Facility. An Actor's Perspective", *Global Environmental Change*, Vol. 3, No. 3, pp. 256–275.

Gan, Lin (1995) *Energy Conservation and GHG Emission Reduction in China: The World Bank and UNDP Operation*. Paper presented at the 36th Annual Convention of the International Studies Association (ISA), Chicago, 21–25 February.

Garbo, Gunnar (1995) *Kampen om FN: Skal de store og sterke styre verden som de vil?* Oslo: Universitetsforlaget.

GEF (1992) *China: Issues and Options in Greenhouse Gas Emission Control*. Project document.

GER (1993) Vol. 5, No. 16, p. 4; (1993) Vol. 5, No. 24, p. 4.

Glaeser, Bernhard (1994) *Umwelt und Entwicklung in China: Zwischen Tradition und Moderne*. Berlin: Wissenschaftszentrum Berlin.

Glantz, M. H. (1992) "Global Warming and Environmental Change in Sub-Sahara Africa", *Global Environmental Change*, September 1992, pp. 183–204.

Gower, Barry S. (1995) "The Environment and Justice for Future Generations", in David E. Cooper and Joy A. Palmer (eds) *Just Environments*. London and New York: Routledge, pp. 49–58.

Green Globe Yearbook (1992, 1993, 1995) Oxford: Oxford University Press.

Grubb, Michael (1989) *The Greenhouse Effect: Negotiating Targets*. Dartmouth: The Royal Institute of International Affairs.

Grubb, Michael (1992a) "The Climate Change Convention: An Assessment", *International Environmental Reporter*, Vol. 15, No. 16, pp. 540–543.

Grubb, Michael (1995a) "European Climate-Change Policy in a Global Context", in Helge Ole Bergesen (ed.) *Green Globe Yearbook 1995*, pp. 41–50.

Grubb, Michael (1995b) "Seeking Fair Weather: Ethics and the International Debate on Climate Change", *International Affairs*, Vol. 71, No. 3, pp. 463–496.

Grubb, Michael et al. (1991) *Energy Policies and the Greenhouse Effect: Volume Two – Country Studies and Technical Options*. Dartmouth: The Royal Institute of International Affairs.

Grubb, Michael et al. (1992a) *The Earth Summit Agreements: A Guide and Assessment*, Dartmouth: The Royal Institute of International Affairs/Earthscan Publications Ltd.

Grubb, Michael et al. (1994) *Implementing the European CO_2 Commitment: A Joint Policy Proposal* London: Royal Institute of International Affairs.

Grubb, Michael, Sebenius, James, Magalhaes, Antonio and Subak, Susan (1992) "Sharing the Burden", in Irving M. Mintzer, (ed.) *Confronting Climate Change – Risks, Implications and Responses*. Cambridge: Cambridge University Press.

Guimarães, R. P. (1991) *The Ecopolitics of Development in the Third World: Politics and Environment in Brazil*. Boulder CO and London: Lynne Rienner Publishers.

Haas, Peter M. (1990) *Saving the Mediterranean: The Politics of International*

Environmental Cooperation. New York: Columbia University Press.

Haas, Peter M., Keohane, Robert O. and Levy, Marc A. (eds) (1993) *Institutions for the Earth: Sources of Effective International Environmental Protection.* Cambridge, MA: MIT Press.

Haas, Peter M., Levy, Marc A. and Parson, E. A. (1992) "How Should We Judge UNCED's Success?", *Environment*, Vol. 34, No. 8, pp. 6–11 and 26–33.

Hague Declaration (1989) Hague Declaration on the Atmosphere: Statement from the International Meeting Sponsored by the Government of the Netherlands in the Hague, 11 March 1989.

Haigh, Nigel (1991) "The European Community and International Environmental Policy", *International Environmental Affairs*, Vol. 3, No. 3, pp. 163–180.

Haigh, Nigel (1994) "The Development of EC Climate Policy from its Origins to EC Ratification of the FCCC", in *Study of the Possible Options for the Further Development of Commitments within the Framework Convention on Climate Change.* London: Institute for European Environmental Policy.

Hall, A. (1989) *Developing Amazonia. Deforestation and Social Conflict in Brazil's Carajás Programme.* Manchester: Manchester University Press.

Hall, D. O. and Rosillo-Calle, F. (1991) "African Forests and Grasslands: Sources or Sinks of Greenhouse Gases?", in S. H. Ominde and C. Juma (ed.) *A Change in the Weather – African Perspectives on Climatic Change.* Nairobi: Acts Press.

Handel, M. D. and Risbey, J. S. (1992) "An Annotated Bibliography on the Greenhouse Effect and Climate Change", *Climatic Change* (special issue), Vol. 21, No. 2, pp. 97–255.

Händl, Günther (1990) "International Efforts to Protect the Global Atmosphere: A Case of Too Little, Too Late?", *European Journal of International Law*, No. 3, pp. 250–257.

Hanisch, Ted et al. (1993a) "A Review of Country Studies on Climate Change". Oslo: Center for International Climate and Environmental Research, Oslo – CICERO Report No. 1.

Hanisch, Ted et al. (1993b) "Study to Develop Practical Guidelines for 'Joint Implementation' under the UN Framework Convention on Climate Change – A CICERO Study to the OECD Environment Directorate". Oslo and Paris: Center for International Climate and Environmental Research, and Mimeo from OECD, Group on Economic and Environment Policy Integration – CICERO Report No. 2.

Hansson, Sven Ove (1991) *The Burden of Proof in Toxicology.* Stockholm: Swedish National Chemicals Inspectorate – Report No. 9.

Hansson, Sven Ove (1993) "The False Promises of Risk Analysis", *Ratio*, Vol. 6, No. 1, pp. 16–26.

Hansson, Sven Ove (1996) "Decision-Making Under Great Uncertainty", *Philosophy of the Social Sciences*, Vol. 26, No. 4, pp. 369–386.

Hart, D. and Victor, D. (1993) "Scientific Elites and the Making of US Policy for Climate Change Research 1957-1974", *Social Studies of Science*, Vol. 23, No. 4, pp. 665–687.

Hatch, Michael T. (1993) "Domestic Politics and International Negotiations: The Politics of Global Warming in the United States", *Journal of Environment and Environment*, Vol. 2, No. 2, pp. 1–39.

Haugland, Torleif and Roland, Kjell. (1990) *Energy, Environment and Development in China.* Oslo: Fridtjof Nansen Institute, Report No. 17.

Haugland, Torleif (1993) *A Comparison of Carbon Taxes in selected OECD Countries.*

Paris: OECD Environment Monographs No. 78.

Hayes, P. and Smith, K. (eds) (1993) *The Global Greenhouse Regime – Who Pays?* London: Earthscan Publication Ltd.

Hayes, Peter (1993) "North–South Transfers", in Peter Hayes and Kirk Smith (eds) *The Global Greenhouse Regime: Who Pays?* London: Earthscan Publications, pp. 144–168.

Hecht, S. and Cockburn, A. (1989) *The Fate of the Forest: Developers, Defenders and Destroyers of the Amazon.* London: Verso.

Hegerl, G. C., Storch, H. von and Jones, P. D. (1996) "Detecting Greenhouse Gas-induced Climate Change with an Optimal Fingerprint Method", *J. Climate*, Vol. 9, No. 10, pp. 2281–2423.

Heinloth, K. (1985) *Zur Warnung des Arbeitskreises Energie der DPG vor einer drohenden, weltweiten Klima-katastrophe.* Bad Honnef: Deutsche Physikalische Gesellschaft.

Hernes, Helga et al. (1995) "Climate Strategy for Africa". Oslo: Center for International Climate and Environmental Research, Oslo – CICERO Report No. 3.

Hession, Martin (1995) "External Competence and the European Community", *Global Environmental Change*, Vol. 5, No. 2, pp. 155–156.

Hoel, Michael (1991). "Efficient International Agreements for Reducing Emissions of CO_2", *Energy Journal*, Vol. 12, No. 2, pp. 93–107

Hoel, Michael (1992) "Carbon Taxes – An International Tax or Harmonized Domestic Taxes?", *European Economic Review*, Vol 36, No.2/3, pp. 400–406.

Hoel, Michael and Isaksen, Ivar (1994) "Efficient Abatement of Different Greenhouse Gases", in Ted Hanisch (ed.) *Climate Change and the Agenda for Research.* Boulder, CO and San Francisco, CA: Westview Press, pp. 147–159.

Homer, J. B. (1991) *Natural Gas in Developing Countries – Evaluating the Benefits to the Environment.* Washington, DC: World Bank – Energy Sector Operations Division.

Houghton, J. T. (1993) "World Energy Council Commission 'Energy for Tomorrow's World' – the Realities, the Real Options and the Agenda for Achievement", *WEC Journal*, July, pp. 47–49.

Houghton, J. T., Callander, B. A. and Varney, S. K. (eds) (1992) *Climate Change 1992: The Supplementary Report to the IPCC Scientific Assessment.* Cambridge: Cambridge University Press.

Houghton, J. T., Henkins, G. J. and Ephraums J. J. (eds) (1990) *Climate Change: The IPCC Scientific Assessment.* New York: Cambridge University Press.

Houghton, John (1994) *Global Warming: The Complete Briefing.* Oxford: Lion Publishing.

Hovi, Jon (1991) "Overnasjonalitet", *Internasjonal Politikk*, Vol. 49, No. 1, pp. 5–17.

Hu Angang, Wang Yi (1991) "Current Status, Causes and Remedial Strategies of China's Ecology and Environment", *Chinese Geographical Sciences*, Vol. 1, No. 2, pp. 97–108.

Huang, J. P. (1991) "Fueling the Economy", *China Business Review*, Vol. 18, No. 2, pp. 22–29.

Hurrel, Andrew and Kingsbury, Benedict (eds) (1992) *The International Politics of the Environment: Actors, Interests and Institutions.* Oxford: Clarendon Press.

Ibe, A. C. and Awosika, L. F. (1991) "Sea Level Rise Impact on African Coastal Zones", in S. H. Ominde and C. Juma (eds) *A Change in the Weather. African*

Perspectives on Climate Change, Nairobi: Acts Press.

IEA (1992) *Climate-Change Policy Initiatives.* Paris: International Energy Agency.

IEA (1994) *World Energy Outlook.* Paris: International Energy Agency.

IEA (1995) *World Energy Outlook.* Paris: International Energy Agency.

IGBP (1994) *IGBP in Action: Work Plan 1994-1998.* Stockholm: Global Change Report No. 38.

Inoguchi, Takashi (1991) *Japan's International Relations.* London: Pinter.

Institute for European Environmental Policy (1995) *The 1996 Inter-Governmental Conference: Integrating the Environment into other EU Policies.* London: Institute for European Environmental Policy, April.

IPCC (1990) *Assessments – Climate Change*: Working Group I, *The IPCC Scientific-Assessment* (Cambridge Cambridge University Press); Working Group II, *The IPCC Impact Assessment* (Canberra: Australian Government Publishing Service); Working Group III, *The IPCC Response Strategies* (Geneva: WMO/UNEP).

IPCC (1990) *Climate Change – The IPCC Scientific Assessment.* Cambridge: Cambridge University Press.

IPCC (1990a) Potential Impacts of Climate Change, Report Prepared for IPCC by Working Group II. WMO and UNEP. IPCC, Geneva.

IPCC (1990b), Scientific Assessment of Climate Change. The Policymakers' Summary of the Report of Working Group I to the Intergovernmental Panel on Climate Change. WMO and UNEP. IPCC, Geneva.

IPCC (1992) *Climate Change – The IPCC Scientific Assessment – Supplement.* Cambridge: Cambridge University Press.

IPCC (1994) *IPCC Technical Guidelines for Assessing Climate Change Impacts and Adaptations.* London: Published for WMO/UNEP by the Department of Geography, University College London and the Centre for Global Environmental Research, Japan, CGER-1015-94.

IPCC (1994) *Radiative Forcing of Climate Change and an Evaluation of the IPCC IS92 Emission Scenarios.* Cambridge: Cambridge University Press.

IPCC (1994) Special Report, WG I, Chapters 1–5; WG III, Chapter 6. Cambridge: Cambridge University Press.

IPCC (1995) *Climate Change – The IPCC Scientific Assessment.* Cambridge: Cambridge University Press.

IPCC (1996) *Climate Change.* IPCC Second Assessment Report. *Synthesis Report. An Assessment of Scientific Information Relevant to Interpreting Article 2 of the UN Framework Convention on Climate Change* (Geneva: WMO). Working Group I, *The Science of Climate Change* (Cambridge: Cambridge University Press); Working Group II, *Scientific–Technical Analyses of Impacts, Adaptations and Mitigation of Climate Change* (Cambridge: Cambridge University Press); Working Group III, *Economics and Social Dimensions* (Cambridge: Cambride University Press).

Isaksen, Ivar S. (1995) "Drivhuseffekt og klimaendring", in Helge Ole Bergesen, Kjell Roland and Anne-Kristin Sydnes (eds) *Norge i det globale drivhuset.* Oslo: Universitetsforlaget, pp. 49–61.

Izumi, Kunihiko et al. (1994) *Climate Plan Evaluation, Japan.* Osaka: Citizens' Alliance for Saving the Atmosphere and the Earth – People's Forum.

Jacobsen, Harold K. (1984) *Networks of Interdependence: International Organizations and the Global Political System.* New York: Alfred A. Knopf.

Jacobson, H. K. and Price, M. F. (1990) *A Framework for Research on the Human Dimension of Global Environmental Change.* Paris: ISSC/UNESCO.

Jäger, J. and Ferguson, H. L. (eds) (1991) *Climate Change: Science, Impacts and*

Policy. Cambridge: Cambridge University Press.

Jäger, Jill, Cavender Bares, Jeannine and Ell, Renate (1994) "Vom Treibhauseffekt zur Klimakatastrophe: Eine Chronologie der Klimadebatte in Deutschland", in G. Altner, B. Mettler-Meibom, U. E. Simonis and E. U. von Weizsäcker (eds) *Jahrbüch Ökologie 1994.* München: Verlag C. H. Beck, pp. 252–262.

Jefferson, Michael (1995) "What is Happening to CO_2 Emissions?" *WEC Journal,* July, pp. 92–94.

Jellinek, Steven D. (1981) "On the Inevitability of Being Wrong", *Annals of the New York Academy of Sciences,* No. 363, pp. 43–47.

Jervis, Robert (1988) "Realism, Game Theory, and Cooperation", *World Politics,* Vol. 40, No. 3, pp. 317–349.

Jodha, N. S. and Maunder, W. John (1991) "Climate, Climate Change and the Economy", in Jäger, J. and Ferguson, H. L., *Climate Change: Science, Impacts and Policy.* Cambridge: Cambridge University Press, pp. 409–414.

Johnson, Stanley and Corcelle, Guy (1989) *The Environmental Policies of the European Communities* London: Graham and Trotman.

Jones, P. D. (1994) "Recent Warming in Global Temperature Series". *Geophys. Res. Lett.,* Vol. 21, pp. 1149–1152.

Jordan, A. (1994) "Financing the UNCED Agenda: The Controversy over Additionality", *Environment,* Vol. 36, No. 3, pp. 12–20, 31–36.

Jorgenson, Dale W. and Wilcoxen, Peter J. (1993) "Reducing US Carbon Emissions: An Econometric General Equilibrium Assessment", *Resource and Energy Economics,* Vol. 15, No.1, pp. 7–25.

Juma, C. (1993) "Promoting Transfer of Environmentally-sound Technology – The Case of National Incentive Schemes", *Green Globe Yearbook.*

Kasa, S. (1994) "Environmental Reforms in Brazilian Amazonia under Sarney and Collor: Explaining Some Contrasts", *IBERO-AMERICANA, Nordic Journal of Latin American Studies,* Vol. 24, No. 2, pp. 42–64.

Kasperson, Roger E. and Dow, Kirstin (1991) "Developmental and Geographical Equity in Global Environmental Change: A Framework for Analysis", *Evaluation Review,* Vol. 15, No. 1, pp. 149–171.

Katscher, W., Stein, G., Lanchberry, J. and Salt, S. (eds) (1994) *Greenhouse Gas Verification – Why, How and How Much?* Proceedings of a workshop in Bonn, 28–29 April 1994. Jülich: Forschungszentrum Jülich GmbH.

Kay, David A. and Jacobson, Harold K. (eds) (1993) *Environmental Protection: The International Dimension.* Totowa, NJ: Allenheld/Osmund.

Keeling, C. D. , Whorf, T. P., Wahlen, M. and Plicht, J. van der (1995) "Interannual Extremes in the Rate of Rise of Atmospheric Carbon Dioxide Since 1980", *Nature,* Vol. 375, pp. 666–670.

Kellogg, W. W. and Schware, S. (1981) *Climate Change and Society: Consequences of Increasing Carbon Dioxide.* Boulder, CO: Westview Press.

Kellogg, William W. (1987) "Mankind's Impact on Climate: The Evolution of An Awareness", *Climatic Change,* Vol. 10, pp. 113–136.

Kelly, P. M. and Hulme, M. (1992) *Climate Scenarios for the SADC Countries.* Climate Research Unit, School of Environmental Sciences, University of East Anglia.

Kennedy, Paul (1987) *The Rise and Fall of the Great Powers: Economic Change and Military Conflict From 1500 to 2000.* New York: Random House.

Kennedy, Paul (1993) *Preparing for the Twenty-First Century.* London: Harper Collins.

Keohane, R.O. and Nye, E. (1977) *Power and Interdependence: World Politics in Transition.* Boston: Little, Brown and Company.

Keohane, Robert O. (1989) *International Institutions and State Power.* Boulder, CO: Westview Press.

Kiefer, Thomas (1995) *Modernisierung in der VR China: Neue Konfliktpotentiale und immanente sowie globale Regelungsmechanismen.* Hamburg: Institut für Friedensforschung und Sicherheitspolitik, No. 90.

Klimatdelegationen (1994) *Rapport från Klimatdelegationen.* Stockholm: SOU No. 138 (in Swedish).

Kohler-Koch, B. (1994) *The Evolution of Organised Interests in the EC: Driving Forces, Co-evolution or New Type of Governance?* Paper presented at the XVth World Congress of the International Political Science Association, 21–25 August, Berlin.

Krämer, Ludwig (1990) *EEC Treaty and Environmental Protection.* London: Sweet and Maxwell.

Krasner, Stephen D. (1981) "Structural Causes and Regime Consequences: Regimes as Intervening Variables", *International Organization,* Vol. 36, No. 3, pp. 185–205.

Kverndokk, Snorre (1993) "Coalitions and Side Payments in International CO_2 Treaties". Oslo: Statistics Norway – Discussion Paper No. 97.

Lanchbery, John and Victor, David (1995) "The Role of Science in the Global Climate Negotiations", *Green Globe Yearbook 1995,* pp. 29–40.

Leggett, Jeremy (1992) "Global Warming: The Worst Case", *Bulletin of the Atomic Scientists,* Vol. 48, No. 1, pp. 29–33.

Lennon, S. J. (1993) "The Enhanced Greenhouse Effect and the South African Power Industry", *Journal of Energy in Southern Africa,* Vol. 4, No. 1.

Leonard, J. A. and Mintzer, Irving (eds) (1994) *Negotiating Climate Change: The Inside Story of the Rio Convention.* Cambridge: Cambridge University Press.

Li, X., Maring, H., Savoie, D., Voss, K. and Prospero M. (1996) "Dominance of Mineral Dust in Aerosol Light-scattering in the North Atlantic Trade Winds", *Nature,* Vol. 380, 4 April, pp. 416–419.

Lindzen, R. S. (1992) "The Origin and Nature of Alleged Scientific Consensus", *Energy and Environment,* special issue, No. 2, pp. 122–137.

List, Martin and Rittberger, Volker (1992) "Regime Theory and International Environmental Management", in Andrew Hurrell and Benedict Kingsbury (eds) *The International Politics of the Environment.* Oxford: Oxford University Press, pp. 85–109.

Loske, Reinhard (1993) "Chinas Marsch in die Industrialisierung", *Blätter für Deutsche und Internationale Politik,* Vol. 38, No. 12, pp. 1460–1472.

Lundli, Hans-Einar (1996) *The Politics of Ozone Depletion and Climate Change: Sources of Success and Failure.* Trondheim: Norwegian University of Science and Technology, Department of Sociology and Political Science – Cand.Polit. dissertation.

Lykke, Ivar (ed.) (1992) *Achieving Environmental Goals: The Concept and Practice of Environmental Performance Review.* London: Belhaven Press.

Mahar, D. J. (1989) *Government Policies and Deforestation in Brazil's Amazon Region.* Washington DC: The World Bank.

Malnes, Raino (1995a) "Leader" and "Entrepreneur" in International Negotiations: A Conceptual Analysis", *European Journal of International Relations,* Vol. 1, No. 1, pp. 87–112.

Malnes, Raino (1995b) *Valuing the Environment*. Manchester: Manchester University Press.

Manabe, S. and Welherald, R. T. (1975) "The Effects of Doubling the CO_2 Concentration on the Climate of a General Circulation Model". *J. Atmosph. Sci.*, Vol. 32, pp. 3–15.

Manne, Alan S. and Richels, Richard G. (1990) "Buying Greenhouse Insurance". Stanford University and Electric Power Research Institute – mimeo.

Manne, Alan S. and Richels, Richard G. (1991) "International Trade in Carbon Emission Rights: A Decomposition Procedure", *American Economic Review*, Vol. 81, No. 2, pp. 135–139.

Manne, Alan S. and Richels, Richard G. (1993) "The EC Proposal for Combining Carbon and Energy Taxes – The Implications for Future CO_2 Emissions", *Energy Policy*, January, pp. 5–12.

Margolis, M. (1992) *The Last New World: The Conquest of the Amazon Frontier*. New York and London: W. W. Norton.

Marland, G., Andres, R. J., Boden, T. A. (1994) "Global, Regional and National CO_2 Emissions", in T. A. Boden, D. P. Kaiser, R. J. Sepanski and F. W. Stoss (eds) *Trends '93. A Compendium of Data on Global Change* Oak Ridge: Carbon Dioxide Information Analysis Center, pp. 505–581.

Marshall Institute (1992) *Global Warming Update: Recent Scientific Findings*. Washington D.C.: George C. Marshall Institute.

Martine, G. (1990) "Rondônia and the Fate of Small Producers", in D. Goodman and A. Hall (eds) *The Future of Amazonia: Destruction or Sustainable Development*. New York: St. Martin's Press, pp. 27–45.

May, P. H. and Reis, E. J. (1993) "The User Structure in Brazil's Tropical Rain Forest", Kiel Working Paper No. 565. Kiel: Kiel Institute of World Economics.

McCormick, J. (1989) *The Global Environmental Movement: Reclaiming Paradise*. London: Belhaven Press.

McNeill, Jim (1990) "The Greening of International Relations", *International Journal*, Vol. 45, No. 1, pp. 1–35.

McTegart, W. J., Sheldon, G. W. and Griffiths, D. C. (eds) (1990) *Climate Change: The IPCC Impacts Assessments*. Canberra: Australian Government Publishing Service.

Miller, Jack et al. (1989). "Soviet Climatologist Predicts Greenhouse Paradise", *New Scientist*, Vol. 123, No. 1679, p. 24.

Ministerial Declaration (1990) *Second World Climate Conference* in Geneva, 29 October to 7 November 1990.

Mintzer, I. M. and Leonard, J. A. (eds) (1994) *Negotiating Climate Change: The Inside Story of the Rio Convention*. Cambridge: Cambridge University Press.

Mintzer, Irving M. (ed.) (1992) *Confronting Climate Change – Risks, Implications and Responses*. Cambridge: Cambridge University Press.

Mitchell, J. F. B., Johns. T. D., Gregory, J. M. and Tett, S. B. T. (1995) "Transient Climate Response to Increasing Aerosols and Greenhouse Gases". *Nature*, Vol. 376, pp. 501–504.

Mitchell, Ronald B. (1994) *International Oil Pollution at Sea: Environmental Policy and Treaty Compliance*. Cambridge, MA: MIT Press.

Miyamoto, S. (1989) "Diplomacia e militarismo: o Projeto Calha Norte e a ocupação do espaço amazônico", *Revista Brasileira de Ciência Política*, Vol. 1, No. 1, pp. 145–163.

Molion, L. C. B. (1976) *A Climatonomic Study of the Energy and Moisture Fluxes of the Amazon Basin With Considerations of Deforestation Effects*. São José dos Campos: Instituto de Pesquisas Espaciais.

Moltke, Konrad von (1991) "Three Reports of German Environmental Policy", *Environment*, Vol. 33, No. 7, pp. 13–19.

Monbiot, G. (1991) *Amazon Watershed: The New Environmental Investigation*. London: Michael Joseph.

Moore, B. and Braswell, B. H. (1994) "Planetary Metabolism: Understanding the Carbon Cycle", *Ambio*, Vol. 23, No. 1, pp. 4–12.

Morisette, P. M., Darmstadter, J., Plantinga, A. J. and Toman, M. A (1991) "Prospects for Global Greenhouse Gas Accord: Lessons From Other Agreements", *Global Environmental Change*, Vol. 1, No. 3, 209–223.

Moum, Knut et al. (1991) "Klimapolitikk og norsk økonomi", *Økonomiske analyser*, No. 3, pp. 12–24.

Mukherjee, Neela (1992) "Greenhouse Gas Emissions and the Allocation of Responsibility", *Environment and Urbanization*, Vol. 4, No. 1, pp. 89–98.

Myers, N. (1989) *Deforestation Rates in Tropical Forests and Their Climatic Implications*. London: Friends of the Earth.

Nazer, H. M. (1993) "Development of the Environment and the Environment for Development", *OPEC Bulletin*, February.

NEPA (1992) *Environmental Management Regulations and System in China*. Beijing: China Environmental Science Press.

Nitze, W. (1989) "The Intergovernmental Panel on Climate Change", *Environment*, Vol. 31, No. 1, pp. 44–45.

Nordhaus, William D. (1991) "To Slow or Not to Slow: The Economics of the Greenhouse Effect", *Economic Journal*, Vol. 101, No. 6, pp. 920–937.

Nordhaus, William D. (1993) "Optimal Greenhouse-Gas Reductions and Tax Policy in the DICE Model", *American Economic Review*, Vol. 83, No. 2, pp. 313–317.

Nordic Council of Ministers (1995) *Joint Implementation as a Measure to Curb Climate Change – Nordic Perspective and Priorities*. Copenhagen: Nordic Council of Ministers.

Nordwijk Declaration (1989) Nordwijk Declaration on Atmospheric Pollution and Climatic Change: Statement of Ministerial Conference Sponsored by the Government of the Netherlands in Nordwijk, 7 November 1989.

NRC (1992) *China and Global Change. Opportunities for Collaboration*. Washington: National Academy Press.

O'Riordan, Tim and Jäger, Jill (1996) "The History of Climate Change Science and Politics", in Tim O'Riordan and Jill Jäger, (eds) *Politics of Climate Change: A European Perspective*. London and New York: Routledge, pp. 1–31.

O'Riordan, Tim and Jäger, Jill (eds) (1996) *Politics of Climate Change: A European Perspective*. London: Routledge.

OAU (1989) *Kampala Declaration*. First African Regional Conference on Environment and Sustainable Development, Kampala, 12–16 June.

OAU/UNEP (1991) *Regaining the Lost Decade – a Guide to Sustainable Development in Africa*. Nairobi: UNEP.

Obasi, G. O. P. (1991) "Global Climate Change: African Perspectives", in S. H. Ominde and C. Juma (eds) *A Change in the Weather – African Perspectives on Climatic Change*. Nairobi: Acts Press.

Oberthür, Sebastian (1993) *Politik im Treibhaus. Die Entstehung des internationalen Klimaschutzregimes*. Berlin: Edition Sigma.

Oberthür, Sebastian and Ott, Hermann (1995) "Framework Convention on Climate Change: The First Conference of the Parties", *Environmental Policy & Law*, Vol. 25, No. 4, pp. 349–364.

OECD (1974) Council Communication No. 223, 14 November. Paris: OECD.

OECD (1991) The State of the Environment. Paris: OECD.

OECD (1992a) *Climate Change Policy Initiatives*. Paris: OECD/IEA.

OECD (1992b) *Energy Policies of IEA Countries*. Paris: OECD/IEA.

OECD (1993a) *Taxing Energy: Why and How*. Paris: OECD/IEA.

OECD (1993b) *Environmental Performance Review –Germany*. Paris: OECD.

OECD/IEA (1995) *World Energy Outlook*. Paris: OECD/IEA.

Official Journal of the European Communities (1967) No. L196, 16 August.

Official Journal of the European Communities (1992) No. C245, 23 September.

Official Journal of the European Communities (1993) Information and Notices, C 125, Vol. 36, 6 May; Legislation, L 119, Vol. 36, 14 May; Information and Notices, C 138, Vol. 36, 17 May.

Official Journal of the European Communities (1993) No. C138, 17 May.

Ojo, O. (1990) "Sociocultural implications of climate change and sea level rise in the West and Central African regions", in J. G. Titus (ed.) *Changing Climate and the Coast, Vol. 2: Western Africa, the Americas, the Mediterranean Basin, and the Rest of Europe*. Washington, DC: US-EPA.

Ojwang, J. B. and Karani, P. (1995) "Joint Implementation of GHG Abatement Commitments", in *A Climate for Development – Climate Change Policy Options for Africa*. Nairobi: Acts Press.

Orr, David W. and Soroos, Marvin S. (eds) (1979) *The Global Predicament: Ecological Perspectives on World Order*. The University of North Carolina Press.

Ottawa Meeting Statement (1989) Meeting Statement of Protection of the Atmosphere: International Meeting of Legal and Policy Experts, Ottawa, Canada, 20–22 February 1989.

Parson, Edward A. (1993) "Protecting the Ozone Layer", in Peter M Haas, et al. (eds) *Institutions for the Earth: Sources of Effective International Environmental Protection*. Cambridge, MA: MIT Press, pp. 27–74.

Parson, Edward A., Haas, Peter M. and Levy, Marc A. (1992) "A Summary of the Major Documents Signed at the Earth Summit and the Global Forum", *Environment*, Vol. 84, No. 8, October, pp. 12–15.

Paterson, Matthew (1992) "Global Warming", in Caroline Thomas (ed.) *The Environment in International Relations*. London: Royal Institute of International Affairs, pp. 155–198.

Paterson, Matthew (1996) "International Relations Theory: Neorealism, Neoinstitutionalism and the Climate Change Convention", in John Vogler and Mark F. Imber (eds) *The Environment and International Relations*. London: Routledge, pp. 59–76.

Pearce, David (1991) "A Sustainable World: Who Cares, Who Pays?", *RSA Journal*, July, pp. 493–505.

Pearce, F. (1995) "Price of Life Sends Temperatures Soaring", *New Scientist*, Vol. 147, No. 1972, p. 5.

Peck, Stephen C. and Teisberg, Thomas J. (1992) "CETA: A Model for Carbon Emissions Trajectory Assessment", *Energy Journal*, Vol. 13, No. 1, pp. 55–77.

Peck, Stephen C. and Teisberg, Thomas J. (1993) "Global Warming Uncertainties and the Value of Information: An Analysis Using CETA", *Resource and Energy Economics*, Vol. 15, No. 1, pp. 71–97.

Perlack, R. D., Russell, M. and Sheng, Z. (1993) "Reducing Greenhouse Gas Emissions in China. Institutional, Legal and Cultural Constraints and Opportunities", *Global Environmental Change*, Vol. 3, No. 3, pp. 78–100.

Perrin, Francis (1995) "Oil Taxation at the Heart of Discussions between the European Union and the GCC", *Arab Oil and Gas*, No. 543, pp. 28–30.

Plan Econ Report (1993) 19 December, p. 6.

Plan Econ Report (1995) 7 April, p. 9.

Plank, D. N. (1993) "Aid, Debt, and the End of Sovereignty: Mozambique and its Donors", *Journal of Modern African Studies*, Vol. 31, No. 3, pp. 407–430.

Plano, Jack O. and Olton, Roy (1988) *The International Relations Dictionary*. Harlow, Essex: Longman.

Polunin, N. (1972) *The Environmental Future*. London: MacMillan.

Pompermayer, J. M. (1984) "Strategies of Private Capital in the Brazilian Amazon", in M. Schmink and C. Wood (eds) *Frontier Expansion in Amazonia*. Gainesville: University of Florida Press.

Porter, Gareth and Brown, Janet Welsh (1991) *Global Environmental Politics*. Boulder, CO: Westview Press.

Potter, G. L., Ellsaesser, H. W., MacCracken, M. C. and Luther F. M. (1975) "Possible Climatic Impact of Tropical Deforestation", *Nature*, No. 258, pp. 697–698.

Powell, Robert (1991) "Absolute and Relative Gains in International Relations Theory", *American Political Science Review*, Vol. 85, No. 4, pp. 1303–1320.

Prado, A. C. (1986) "Uma avaliação dos incentivos fiscais do FISET – florestamento/reflorestamento", *Brasil Florestal*, February, pp. 3–25.

Presidential decree (1994) *O gosudarstvennoy strategii Rossiyskoy Federatsii po okhranye okruzhayushchey sredy i obyespecheniyu ustoychivogo razvitiya* (The National Strategy of the Russian Federation for the Protection of the Environment and the Securing of a Sustainable Development), No. 236, 4 February, Moscow.

Proctor, R. (1991) *Value Free Science? Purity and Power in Modern Knowledge*. Cambridge, MA: Harvard University Press.

Qu Geping (1990) "China's Environmental Policy and World Environmental Problems", *International Environmental Affairs*, Vol. 2, No. 2, pp. 103–108.

Qu Geping (1992) "China's Dual Thrust Energy Strategy. Economic Development and Environment Protection", *Energy Policy*, Vol. 20, No. 6, pp. 500–506.

Rahman, A. Atiq (1991) "Possible Criteria for Burden-sharing Under a Global Ceiling: Experience and Suggestions", in Ted Hanisch (ed.) *A Comprehensive Approach to Climate Change: Additional Elements from an Interdisciplinary Perspective*. Oslo: Center for International Climate and Energy Research, Oslo, pp. 140–151.

Ramakrishna, R. (1990) "North–South Issues, Common Heritage of Mankind and Global Climate Change", *Millennium*, Vol. 19, No. 3, Winter, pp. 429–446.

Raper, S. C. B., Wigley, T. M. L. and Warrick, R. A. (1995) "Global Sea Level Rise: Past and Future", in John Millman (ed.) *Proceedings of the SCOPE Workshop on Rising Sea Level and Subsiding Coastal Areas*. Dordrecht: Klüwer Academic Publishers.

Rapoport, Anatol (1989). *Decision Theory and Decision Behaviour: Normative and Descriptive Approaches*. Dordrecht: Kluwer.

Rawls, John (1971) *A Theory of Justice*. Cambridge, MA: Harvard University Press.

Rayner, S. (1994) *Governance and the Global Commons*. London: London School of

Economics, Centre for the Study of the Global Governance, Discussion Paper 8.

Rayner, Steve (1991) "The Greenhouse Effect in the US: The Legacy of Energy Abundance", in Michael Grubb et al.: *Energy Policies and the Greenhouse Effect: Country Studies and Technical Options*. London: Royal Institute of International Affairs, pp. 335–378.

Rayner, Steve (1993) "Prospects for CO_2 Emissions Reduction Policy in the USA", *Global Environmental Change*, Vol. 3, No. 1, pp. 12–31.

Redgwell, Catherine (1994) "Energy, Environment and Trade in the European Community", *Journal of Energy and Natural Resources Law*, Vol. 12, No. 1, pp. 128–150.

Revelle, Roger (1985) "Introduction: The Scientific History of Carbon Dioxide", in E. T. Sundquist and W. S. Broecker (eds) *The Carbon Cycle and Atmospheric CO_2*. Washington DC: American Geophysical Union – Geophysical Monograph Series Vol. 32.

Revelle, Roger and Suess, Hans E. (1957) "Carbon Dioxide Exchange Between Atmosphere and Ocean and the Question of an Increase of Atmospheric CO_2 During the Past Decades", *Tellus*, Vol. 9, No. 1, pp. 18–27.

Rich, B. M. (1988) *The Politics of Tropical Deforestation in Latin America: The Role of the Public International Financial Institutions*, mimeo.

Rich, B. M. (1994) *Mortaging the Earth*. Boston: Beacon Press.

Rose, Adam (1990) "Reducing Conflict in Global Warming Policy: The Potential of Equity as a Unifying Principle", *Energy Policy*, Vol. 18, No. 10, pp. 927–935.

Rose, Adam and Stevens, Brandt (1993) "The Efficiency and Equity of Marketable Permits for CO2 Emissions", *Resource and Energy Economics*, Vol. 15, No. 1, pp. 117–146.

Rosendal, G. Kristin (1992) *Biologisk mangfold og klimaendring: En studie av tilnærminger til skogproblematikken*. Lysaker: The Fridtjof Nansen Institute – Report No. 1.

Rosendal, G. Kristin (1995) "The Convention on Biological Diversity: A Viable Instrument for Conservation and Sustainable Use?" *Green Globe Yearbook*. pp. 69–82.

Ross, L. and Silk, M. A. (1986) "Post-Mao China and Environmental Protection: The Effects of Legal and Politico-economic Reform", *UCLA Pacific Basin Law Journal*, Vol. 4, No. 1, pp. 63–89.

Rotmans, Jan et al. (1994) "Climate Change Implications for Europe: Application of the ESCAPE Model", *Global Environmental Change*, Vol. 4, No. 2, pp. 97–124.

Rowlands, I. (1995) *The Climate Change Negotiations: Berlin and Beyond*. London: The Centre for the Study of Global Governance – Discussion Paper 17.

Rowlands, Ian (1995) *The Politics of Global Atmospheric Change*. Manchester: Manchester University Press.

Rowlands, Ian and Greene, Malory (eds) (1992) *Global Environmental Change and International Relations*. Basingstoke: Macmillan.

Rowlands, Ian H. (1992) "International Politics of Global Environmental Change", in Ian H. Rowlands and Malory Greene (eds) *Global Environmental Change and International Relations*. Hong Kong: Macmillan, pp. 19–37.

Rudner, Richard (1953) "The Scientist qua Scientist Makes Value Judgments", *Philosophy of Science*, Vol. 20, No. 1, pp. 1–6.

Rueschemeyer, D., Stephens, E. H. and Stephens, J. D. (1992) *Capitalist Development and Democracy*. Cambridge: Polity Press.

SADEN (1989) *Programa Nossa Natureza*, Brasilia: SADEN/PR.

Salati, E. and Ribeiro, M. N. G. (1979) "Floresta e Clima", *Acta Amazonica*, Vol. 9, No. 4, pp. 15–22.

Salati, E., Marquez, J. and Molion, L. C. B. (1978) "Origem e distribuição das chuvas na Amazônia", *Interciencia*, Vol. 3, No. 4, pp. 200–205.

Sand, P. H. (1992) *Effectiveness of International Environmental Agreements: Survey of Existing Legal Instruments*. Cambridge: Burlington Press.

Sandberg, K. and Smith, D. (1994) *Conflicts in Africa, – Between Conflicts and Development* (in Norwegian). Oslo: North/South Coalition – Information Booklet No. 2.

Santilli, M. (1989) "Tratado de Cooperação Amazônica: um instrumento diplomático a serviço de retorica nacionalista", *Tempo e Presença*, Vol. 11, No. 244/245, pp. 40–41.

Schipper, L. and Martinet, E. (1993) "Decline and Rebirth – Energy Demand in the Former Soviet USSR", *Energy Policy*, Vol. 21, No. 9, pp. 969–977.

Schmandt, Jürgen and Clarkson, Judith (eds) (1992) *The Regions and Global Warming: Impacts and Response Strategies*. New York: Oxford University Press.

Schreurs, Miranda A. (1993) *International Cooperation for the Environment in a State-Centric World: Comparing the Responses of Japan and Germany to Global Environmental Threats*. Paper prepared for the SSRC–MacArthur Foundation sponsored workshop on "Sovereignty and Security in Contemporary International Affairs", Harvard University, April 1–4.

Sebenius, J. K. (1991) "Negotiating a Regime to Control Global Warming", in R. E. Benedick et al. (eds) *Greenhouse Warming: Negotiating a Global Regime*. Washington DC: World Resources Institute.

Sebenius, James K. (1994) "Towards a Winning Climate Coalition", in Mintzer and Leonard (eds) *Negotiating Climate Change: The Inside Story of the Rio Convention*. Cambridge: Cambridge University Press, pp. 277–320.

Selrod, R. et al. (1995) *Joint Implementation Under the Convention on Climate Change – Opportunities for Development in Africa*. Nairobi: Acts Press.

Selrod, Rolf et al. (1995) "Joint Implementation – A Promising Mechanism for All Countries?" Oslo: Center for International Climate and Environmental Research, Oslo – CICERO Policy Note No. 1.

Seragelding (1990) *Saving Africa's Rainforests*. Washington, DC: World Bank.

Serrão, E. A. and Toledo, J. M. (1990) "The Search for Sustainability in Amazonian Pastures", in A. Anderson (ed.) *Alternatives to Deforestation. Steps Towards Sustainable Use of the Amazon Rainforest*. New York: Columbia University Press, pp. 195–215.

Setzer, A. W., Pereira, M. C., Pereira, A. C. and Almeida, S. A. O. (1988) *Relatorio de Atividades do Projeto IBDF-INPE 'SEQE' – Ano 1987*. Sao José dos Campos: Instituto Nacional de Pesquisas Espaciais.

Shorey, Helen (1992) "Maastricht and EC Environment Policy", *Land Management and Environmental Law Report*, Vol. 4, No. 3, pp. 85–87.

Shue, Henry (1992) "The Unavoidability of Justice", in Andrew Hurrell and Benedict Kingsbury (eds) *The International Politics of the Environment*. Oxford: Clarendon Press, pp. 373–397.

Siddiqi, T.A., Streets, D. G., Zongxin, W. and Jiankun, H. (1994) *National Response Strategy for Global Climate Change: Peoples Republic of China*. Tsinghua: East West Center, Argonne Laboratory, Tsinghua University.

Sjöberg, Helen (1995) "The Dialectics of the Global Environment Facility". Draft prepared for *The Greening of International Institutions* of the FIELD/Earthscan,

Series on International Law and Sustainable Development, Natural Resource Management Institute/Stockholm University, January 1995.

Sjöstedt, Gunnar (ed.) (1993) *International Environmental Negotiation*. London.

Skjærseth, Jon Birger (1992) *From Regime Formation to Regime Functioning "Effectiveness" — Coping with the Problem of Ozone Depletion*. Lysaker: The Fridtjof Nansen Institute – Report No. 7.

Skjærseth, Jon Birger (1993) *The Climate Policy of the EC – Too Hot to Handle: A Study of Interests and Preferences versus Problem-solving Capacity*. Lysaker: The Fridtjof Nansen Institute – EED Report No. 3.

Skodvin, T. and Underdal, A. (1994) *The Science-Politics Interface*. Paper presented at the International Studies Association Meeting, Washington D.C. 29 March to 1 April.

Smil, Vaclav (1993) *China's Environmental Crisis*. London: M. E. Sharpe.

Smil, Vaclav (1994) "China's Greenhouse Gas Emissions", *Global Environmental Change*, Vol. 4, No. 4, pp. 325–332.

Smith, K. R., Swisher, J. and Ahuja, D. R. (1993) "Who Pays to Solve the Problem and How Much?, in Peter Hayes & Kirk Smith (eds) *The Global Greenhouse Regime – Who Pays?* Tokyo: United Nations University Press, pp. 70–98.

Social Learning Group (1996) *Learning to Manage Global Environmental Risks: A Comparative History of Social Responses to Climate Change, Ozone Depletion, and Acid Rain* (in preparation).

Solomon, Barry D. and Ahuja, Dilip R.(1991) "International Reductions of Greenhouse-gas Emissions: An Equitable and Efficient Approach", *Global Environmental Change*, December, pp. 343–350.

Spector, Bertram I., Sjöstedt, Gunnar, and Zartman, William (eds) (1994) *Negotiating International Regimes: Lessons Learned from the United Nations Conference on Environment and Development*. London: Graham and Trotman.

Sprinz, Detlef and Vaahtoranta, Tapani (1994) "The Interest-based Explanation of International Environmental Policy", *International Organization*, Vol. 49, No. 1, pp. 77–105.

Statement of the Scientific and Technical Sessions (1990) *Second World Climate Conference* in Geneva, 29 October to 7 November, 1990.

Statistisches Bundesamt (1993) *Länderbericht Volksrepublik China 1993*. Stuttgart: Metzler-Poeschel.

Stepan, A. (1988) *Rethinking Military Politics: Brazil and the Southern Cone*. Princeton: Princeton University Press.

Strange, Susan (1986) *Casino Capitalism*. Oxford: Blackwell.

Strange, Susan (1988) *States and Markets*. London: Pinter Publishers.

Subak, S., Raskin, P. and Hippel, D. (1992) *National Greenhouse Gas Accounts: Current Anthropogenic Sources and Sinks*. Stockholm: Stockholm Environment Institute.

Subak, Susan (1993) "Assessing Emissions: Five Approaches Compared", in Peter Hayes and Kirk Smith (eds) *The Global Greenhouse Regime: Who Pays?* London: Earthscan, pp. 51–69.

Sun, De-Zheng and Lindzen, Richard, S. (1993) "Distribution of Tropical Tropospheric Water Vapor", *Journal of the Atmospheric Sciences*, Vol. 50, No. 12, pp. 1643–1660.

Susskind, Lawrence (1992) *Environmental Diplomacy: Negotiating Effective International Agreements*. New York: Oxford University Press.

Tarlock, Thomas (1992) "Environmental Protection: The Potential Misfit Between

Equity and Efficiency", *University of Colorado Law Review*, Vol. 63, pp. 871–900.

Tata Conference Statement (1989) Tata International Conference on Global Warming and Climate Change: Perspectives from Developing Countries, New Delhi, 21–23 February 1989.

Tegen, Ina, Lacis, Andrew A., and Fung, Inez (1996) "The Influence on Climate Forcing of Mineral Aerosols From Disturbed Soils", *Nature*, Vol. 380, 4 April, pp. 419–422.

Thacher, Peter S. (1992) "Evaluating the 1992 Earth Summit – An Institutional Perspective", *Security Dialogue*, Vol. 23, No. 3, pp. 117–126.

Thomas, Caroline (ed.) (1992) *The Environment in International Relations*. Dartmouth: The Royal Institute of International Affairs.

Thorsrud, E. (1972) "Policy-making as a Learning Process", in A. B. Cherns et al. (eds) *Social Science and Government*. London: Tavistock, pp. 121–133.

Todd, Daniel and Zhang Lei (1994) "Ports and Coal Transfer. Hub of China's Energy Supply Policy", *Energy Policy*, Vol. 22, No. 7, pp. 609–622.

Toichi, Tsutomu (1993) "World Energy Futures Groping for a New World Order", *Energy in Japan*, No. 120, pp. 25–32.

Tomitate, Takao (1992) "Japanese Energy Policy and the International Arena", *Oxford Energy Forum*, No. 10, pp. 8–9.

Tomitate, Takao (1993) "Japan's Climate Policy: A Rejoinder", *Security Dialogue*, Vol. 24, No. 3, pp. 301–304.

Toronto Conference Statement (1988) The Changing Atmosphere: Implications for Global Security – Statement from International Meeting Sponsored by the Government of Canada in Toronto, 27–30 June 1988.

Torvanger, Asbjørn et al. (1994) "Joint Implementation under the Climate Convention: Phases, Options and Incentives". Oslo: Center for International Climate and Environmental Research, Oslo – CICERO Report No. 6.

Tripp, James T. B. (1988) "The UNEP Montreal Protocol: Industrialized and Developing Countries Sharing the Responsibility for Protecting the Stratospheric Ozone Layer", *International Law and Politics*, Vol. 20, pp. 733–752.

Tsukuda, Koji (1992) "Long-term Trends of Rand on CO_2 Emission Control Technologies in Major Countries", *Energy in Japan*, No. 117, pp. 27–44.

Turner, B. L. et al. (eds) (1990) *The Earth as Transformed by Human Action*. Cambridge: Cambridge University Press.

UN (1991) *Proceedings of the Fourth Inter-Governmental Negotiating Committee on Climate Change*. New York: United Nations.

Underdal, A. (1992) "Designing Politically Feasible Solutions", in Raino Malnes and Arild Underdal (eds) *Rationality and Institutions*. Oslo: Scandinavian University Press.

Underdal, Arild (1987) "Transforming 'Needs' into 'Deeds'", *Journal of Peace Research*, Vol. 24, No. 2, pp. 167–183.

Underdal, Arild (1990) *Negotiating Effective Solutions: The Art and Science of Political Engineering*. Department of Political Science – University of Oslo (draft).

Underdal, Arild (1991a) "Solving Collective Problems: Note on Three Models of Leadership", in *Challenges of a Changing World – Festschrift to Willy Østreng*. Lysaker, Norway: The Fridtjof Nansen Institute, pp. 139–153

Underdal, Arild (1991b) "International Cooperation and Political Engineering", in Stuart S. Nagel (eds) *Global Policy Studies: International Interaction Toward Improving Public Policy*. London: MacMillan.

Underdal, Arild (1992a), "De juridiske og politiske forpliktelser: Viktige fremskritt er gjort", *Cicerone*, Vol. 1, No. 2, p. 2.

Underdal, Arild (1992b) "The Concept of Regime Effectiveness", *Cooperation and Conflict*, Vol. 27, No. 3, pp. 227–240.

Underdal, Arild (ed.) (1996) *The International Politics of Environmental Challenge*. Dordrecht: Kluwer.

UNDP (1995) *Human Development Report 1995*. Oslo: Universitetsforlaget.

UN-DPI (1990). "African LDCs", *Africa Recovery*. New York: United Nations Briefing Paper, No. 2.

UNEP (1993) *Environmental Data Report 1993–94*. Oxford: Blackwell.

UNEP Dec. SSII/3.C (1990), UNEP Governing Council Decision SSII/3.C, 3 August 1990.

UNFCCC (1995) "Conclusion of Outstanding Issues and Adoption of Decisions. Proposal on Agenda Item 5 (a) (iii) submitted by the President of the Conference. Review of the adequacy of Article 4, paragraph 2 (a) and (b) of the Convention, including proposals related to a protocol and decisions on follow-up", Berlin: Conference of the Parties, First session, 28 March to 7 April.

UNFPA (1992) *The State of World Population*. Oxford: New International Publications Cooperative.

UNGA (1988) Resolution 43/53 on "Protection of Global Climate for Present and Future Generations of Mankind", 6 December.

UNGA (1989), Resolution 44/207 on "Protection of Global Climate for Present and Future Generations of Mankind", 22. December.

UNGA (1990) Resolution 45/212 on "Protection of Global Climate for Present and Future Generations of Mankind", 21 December.

United Nations (1993) Secretary-General SM/5186, ENV/DEV/236, 21 December.

United States National Research Council (1983) *Towards an International Geosphere–Biosphere Program: A Study of Global Change*. Washington D.C.: National Science Foundation.

Untawale, Mukund G. (1990) "Global Environmental Degradation and International Organization", *International Political Science Review*, Vol. 11, No. 3, pp. 371–383.

US Climate Action Network/Climate Network Europe (1995) *Independent NGO Evaluations of National Plans for Climate Change Mitigation*. Third Review, February 1995.

USCSP (1995). *Interim Report on Climate Change Country Studies*. Washington, DC: USCSP.

Utkin, E. F. (1994) Anketa analiza nacionalnoy politiki v oblasti izmeneniya klimata (Analysis of the National Policy in the Sphere of Climate Change). Moscow: Russian Ministry of Ecology.

van der Esch, Bastian (1995) "Integrating the Environment into the Energy Policy of the European Union", *Journal of Energy and Natural Resources Law*, Vol. 13, No. 1, pp. 14–20.

Veggeberg, S. (1992) "Global Warming Researchers Say They Need Breathing Room", *The Scientist*, 11 May.

Vellinga, Pier and Grubb, Michaele (eds) (1993) *Climate-Change Policy in the European Community*. London: Royal Institute of International Affairs.

Vermeer, E. B. (1990) "Management of Environmental Pollution in China", *China Information*, Vol. 5, No. 1, pp. 34–65.

Vernon, Raymond (1993) "Behind the Scenes – How Policymaking in the

European Community, Japan, and the United States Affects Global Negotiations", *Environment*, Vol. 35, No. 5, pp. 13–43.

Victor, David and Salt, Julian (1994) "From Rio to Berlin: Managing Climate Change", *Environment*, Vol. 36, No. 10, December, pp. 6–15, 25–32.

Vieira, R. S. (1993) "Brazilian Environmental Law Relating to Amazonia", in M. Bothe, T. Kurzidem and C. Schmidt (eds) *Amazonia and Siberia: Legal Aspects of the Preservation of the Environment and Development in the Last Open Spaces.* London: Graham and Trotman/ Martinus Nijhoff Publishers, pp. 107–129.

Viola, E. J. (1988) "The Ecologist Movement in Brazil (1974–86): From Environmentalism to Ecopolitics", *International Journal of Urban and Regional Research*, Vol. 12, June, pp. 211–227.

Viola, E. J. (1992) "Social Responses to Contemporary Environmental Change in Brazil: The Development of a Multisectoral Environmental Movement", paper for the conference on *The New Europe and the New World. Latin America and Europe 1992: The Environmental Dimensions.* Latin American Centre, Oxford University, 9–11 September.

Viola, E. J. (1993) *A Expansão do Ambientalismo Multissetorial e a Globalização da Ordem Mundial, 1985–1992,* Brasilia: ISPN, Documento de Trabalho No. 16.

Visek, Acharo (1995) "Implementation and Enforcement of EC Environmental Law", *Georgetown International Environmental Law Review*, Vol. 7, No. 2, pp. 377–420.

Warren, Andrew (1993) "The Rise and Fall of SAVE", *Environment Risk*, March, pp. 29–30.

Watson, Robert T., Zinyowera, Marufu C. and Moss, Richard H. (eds) (1996) *Climate Change 1995. Impacts, Adaptation and Mitigation of Climate Change: Scientific–Technical Analysis. Contribution of Working Group II to the Second Assessment Report of the Intergovernmental Panel on Climate Change.* Cambridge: Cambridge University Press.

WB (1989) *Sub-Sahara Africa – From Crisis to Sustainable Growth. A Long-term Perspective Study.* Washington, DC: The World Bank.

WB (1990) *World Development Report 1990: Poverty.* Washington, DC: World Bank.

WB (1992) *World Development Report 1992: Development and the Environment.* Oxford: Oxford University Press.

WB (1994) *Brazil: The Management of Agriculture, Rural Development and Natural Resources.* Vol. II: Background papers. Washington DC: World Bank.

WCED (1987) *Our Common Future.* Oxford: Oxford University Press.

Weart, S. (1992) "From the Nuclear Frying Pan into the Global Fire", *The Bulletin of Atomic Scientists*, June, pp. 19–27.

WEC (1993) *Energy for Tomorrow's World.* Düsseldorf: World Energy Council.

WEC (1995) *Global Energy Perspectives to 2050 and Beyond.* London: World Energy Council.

Weiss, Edith Brown (1993) "International Environmental Law: Contemporary Issues and the Emergence of a New World Order", *International Environmental Law*, Vol. 81, pp. 675–710.

Wettestad, Jørgen (1990) *Effektiv verifikasjon av internasjonale drivhusavtaler: Teknisk oppnåelig, men politisk komplisert?* EED publication No. 19-1990. Lysaker: Fridtjof Nansen Institute.

Wilkinson, David (1992) "Maastricht and the Environment: The Implications for the EC's Environment Policy of the Treaty on European Union", *Journal of*

Environmental Law, No. 2, pp. 221–239.

Wiman, Bo (1995) "Metamophors, Analogies, and Models in Communicating Climate-Change Uncertainties and Economics to Policy: A Note on a pre-UNCED US Case", *Ecological Economics*, Vol. 15, pp. 21–28.

WMO (1986) *Report of the International Conference on the Assessment of the Role of CO_2 and Other Greenhouse Gases in Climate Change and Associated Impacts*, Geneva: WMO.

WMO (1986) *Report of the International Conference on the Assessment of CO_2 and Other Greenhouse Gases in Climate Variations and Associated Impacts*, Geneva: WMO.

WMO (1990) *Report on Numerical Experimentation*. Geneva: Sixth Session of CAS/JSC, WCRP-53.

WMO (1991a) *Eleventh World Meteorological Congress*. Geneva: Abridged Report, WMO No. 756.

WMO (1991b) *Report on the Global Climate Observing System*. Winchester, UK: Session of CAS/JSC, WCRP-56.

WMO (1992) *Climate Change, Environment and Development. World Leaders' Viewpoints: His Excellency Li Peng, Premier of the State Council of China*. Geneva: WMO.

WMO and UNEP (1988) *Developing Policies for Responding to Climate Change*, Geneva: WMO.

WMO Res. 8 1990/WMO Executive Council Resolution 8 (EC-XLII), June 1990.

WMO/ICSU (1990) *Global Climate Change: A Scientific Review*. Geneva: World Climate Research Programme.

Wong, C. S. (1978) "Atmospheric Input of Carbon Dioxide From Burning Wood", *Science*, No. 200, pp. 197–200.

World Commission on Environment and Development (1987) *Our Common Future*. Oxford: Oxford University Press.

WRI (1990) *World Resources 1990–91*. New York: Oxford University Press.

WRI (1992) *World Resources 1992–93*. New York: Oxford University Press.

WRI (1994) *World Resources 1994–95*. New York: Oxford University Press.

WRI (1996) *World Resources 1996-1997*. New York: Oxford University Press.

Wu, Zongzing; and Wei, Zhihong (1991) "Policies to Promote Energy Conservation in China", *Energy Policy*, Vol. 19, No. 12, pp. 934–939.

Wynne, Brian (1993) "Implementation of Greenhouse Gas Reductions in the European Community: Institutional and Cultural Factors, *Global Environmental Change*, March, pp. 101–128.

Yearbook of International Organizations (1993/94) Vol. I, 1993.

Young, H. P. (1990) "Sharing the Burden of Global Warming", *Philosophy and Public Policy*, Vol. 10, No. 3/4, pp. 6–9.

Young, H. P. and Wolf, Amanda (1992) "Global Warming Negotiations: Does Fairness Matter?', *The Brookings Review*, spring, pp. 46–51.

Young, Oran (1994) *International Governance: Protecting the Environment in a Stateless Society*. Ithaca: Cornell University Press.

Young, Oran and Osherenko, G. (eds) (1993) Polar Politics: Creating International Environmental Regimes. Ithaca, NY: Cornell University Press.

Young, Oran R. (1989) "The Politics of International Regime Formation: Managing Natural Resources and the Environment", *International Organization*, Vol. 43, No. 3, pp. 349–375.

Young, Oran R. (1991) "Political Leadership and Regime Formation: On the

Development of Institutions in International Society", *International Organization*, Vol. 45, No. 3, pp. 281–308.

Zha Keming (1995) "Energie-Entwicklungspolitik in China unter besonderer Berücksichtigung der Elektrizitätswirtschaft", *Elektrizitätswirtschaft*, Vol. 94, No. 19, pp. 1170–1179.

Zheng-Kang, Cheng (1990) "Equity, Special Considerations, and the Third World", *Colorado Journal of International Environmental Law and Policy*, Vol. 1, No. 1, pp. 57–68.

Zhong Xiang Zhang (1994) "Analysis of the Chinese Energy System: Implications for Future CO_2 Emissions", *International Journal of Environment and Pollution*, Vol. 4, No. 3/4, pp. 181–198.

Zirker D. and Henberg, M. (1994) "Amazônia: Democracy, Ecology and Brazilian Military Prerogatives in the 1990s", *Armed Forces and Society*, Vol. 21, No. 2, pp. 259–281.

Zou Jiahua (1992) *Energy Horizons in a World of Nine Billion Inhabitants*. Madrid: World Energy Council, 15th Congress.

Index